T0335854

METHODS IN MOLECULAR BIOLOGY

Series Editor
John M. Walker
School of Life and Medical Sciences
University of Hertfordshire
Hatfield, Hertfordshire, AL10 9AB, UK

For further volumes:
http://www.springer.com/series/7651

Poliovirus

Methods and Protocols

Edited by

Javier Martín

Division of Virology, NIBSC, Hertfordshire, UK

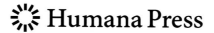

Editor
Javier Martín
Division of Virology
NIBSC
Hertfordshire, UK

ISSN 1064-3745   ISSN 1940-6029   (electronic)
Methods in Molecular Biology
ISBN 978-1-4939-3291-7   ISBN 978-1-4939-3292-4   (eBook)
DOI 10.1007/978-1-4939-3292-4

Library of Congress Control Number: 2015959077

Springer New York Heidelberg Dordrecht London

© Springer Science+Business Media New York 2016
This work is subject to copyright. All rights are reserved by the Publisher, whether the whole or part of the material is concerned, specifically the rights of translation, reprinting, reuse of illustrations, recitation, broadcasting, reproduction on microfilms or in any other physical way, and transmission or information storage and retrieval, electronic adaptation, computer software, or by similar or dissimilar methodology now known or hereafter developed.
The use of general descriptive names, registered names, trademarks, service marks, etc. in this publication does not imply, even in the absence of a specific statement, that such names are exempt from the relevant protective laws and regulations and therefore free for general use.
The publisher, the authors and the editors are safe to assume that the advice and information in this book are believed to be true and accurate at the date of publication. Neither the publisher nor the authors or the editors give a warranty, express or implied, with respect to the material contained herein or for any errors or omissions that may have been made.

Printed on acid-free paper

Humana Press is a brand of Springer
Springer Science+Business Media LLC New York is part of Springer Science+Business Media (www.springer.com)

Preface

Polioviruses belong to the genus Enterovirus in the family Picornaviridae. Enteroviruses have been traditionally distinguished within the Picornaviridae family on the basis of their physical properties such as buoyant density in caesium chloride and stability in weak acid. However, recent advances in molecular and cell biology have changed the focus to the analysis of the genomic structure and nucleotide sequence, details of the viral replication cycle, and antigenic/immunogenic properties. The viral genome of poliovirus consists of a single RNA strand of positive polarity of about 7500 nucleotides in length containing the coding sequences for structural and nonstructural viral proteins flanked by 5′ and 3′ noncoding sequences that modulate RNA replication and translation. The virus particle consists of 60 protomers each containing a single copy of each of the four capsid proteins (VP1 to VP4) arranged in icosahedral symmetry. Poliovirus exists in three serotypes based on specific neutralization reactions with immune sera. Each serotype is defined by the inability of antisera raised against the other two serotypes to completely neutralize infectivity.

Polioviruses are also distinguished from other enteroviruses by neutralization with serotype-specific sera. However, the main distinctive properties of poliovirus are its ability to bind CD155, a member of the immunoglobulin superfamily, as a receptor for cell entry and the propensity to cause paralysis in humans. The most important route of transmission is fecal-oral, and the virus replicates efficiently in the intestinal tract with shedding in feces typically lasting for 2–4 weeks. The examination of stool samples from AFP patients for the presence of poliovirus has been the driving force of the GPEI allowing to link poliovirus isolates to specific individuals and to focus the investigations and public health interventions to particular communities. Large numbers of excreted poliovirus particles remain infectious in the environment for varying lengths of time depending on the immediate conditions. The presence of virus in samples from the wastewater system may be detected by a variety of laboratory methods for concentration, separation, and identification. Environmental surveillance has indeed proven to be very successful in detecting poliovirus circulation in specific populations even in the absence of associated poliomyelitis cases and is becoming a very useful supplementary tool for the surveillance for poliovirus.

The fact that only a fraction of poliovirus infections leads to paralytic disease reinforced the need to devise very sensitive and reliable laboratory techniques to isolate and identify poliovirus from samples of acute flaccid paralytic (AFP) cases, which is also associated with several other syndromes and diseases. This led to the establishment of strict quality criteria for AFP surveillance that included the detection of a minimal number of paralytic cases in children less than 15 years of age due to other causes but polio, the timely sampling of at least 80 % of AFP cases, and the analysis of AFP samples in a fully accredited laboratory using standardized protocols. Poliovirus has also been and continues to be one of the most widely used viruses in research, and work in many laboratories worldwide has helped understanding many viral and biological processes such as virus cell entry, RNA replication and translation, and viral antigenicity.

The present book describes the most common laboratory procedures for isolation, identification, and characterization of polioviruses used in clinical and research laboratories.

Hertfordshire, UK *Javier Martin*

Contents

Contributors

MUHAMMAD MASROOR ALAM • *WHO Regional Reference Laboratory for Poliomyelitis, National Institute of Health, Islamabad, Pakistan*

HUMAYUN ASGHAR • *WHO Eastern Mediterranean Regional Office, Cairo, Egypt*

MAËL BESSAUD • *INSERM U994, Institut National de Santé et de La Recherche Médicale, Paris, France; Institut Pasteur, Biologie des Virus Entériques, Paris, France*

SOILE BLOMQVIST • *Virology Unit, National Institute for Health and Welfare, Helsinki, Finland*

BRUNO BLONDEL • *INSERM U994, Institut National de la Santé et de la Recherche Médicale, Paris, France; Institut Pasteur, Biologie des Virus Entériques, Paris, France*

ERIKA BUJAKI • *National Institute for Biological Standards and Control, Medicines and Healthcare Products Regulatory Agency, Potters Bar, Hertfordshire, UK*

CARA C. BURNS • *Polio and Picornavirus Laboratory Branch, Division of Viral Diseases, National Center for Immunization and Respiratory Disease, Centers for Disease Control and Prevention, Atlanta, GA, USA*

QI CHEN • *Polio and Picornavirus Laboratory Branch, Division of Viral Diseases, National Center for Immunization and Respiratory Disease, Centers for Disease Control and Prevention, Atlanta, GA, USA*

KONSTANTIN M. CHUMAKOV • *Center for Biologics Evaluation and Research, US Food and Drug Administration, Silver Spring, MD, USA*

FRANCIS DELPEYROUX • *INSERM U994, Institut National de la Santé et de la Recherche Médicale, Paris, France; Institut Pasteur, Biologie des Virus Entériques, Paris, France*

GLYNIS DUNN • *National Institute for Biological Standards and Control, Medicines and Healthcare Products Regulatory Agency, Potters Bar, Hertfordshire, UK*

DAVID FRANCO • *KU Leuven – University of Leuven, Department of Microbiology and Immunology, Rega Institute for Medical Research, Laboratory of Virology and Chemotherapy, Leuven, Belgium*

RIMKO TEN HAVE • *Institute for Translational Vaccinology, Bilthoven, The Netherlands*

MUSA HINDIYEH • *Department of Epidemiology and Preventive Medicine, School of Public Health, Sackler Faculty of Medicine, Tel Aviv University, Tel Aviv, Israel; Molecular Assay Section, Central Virology Laboratory, Public Health Services, Chaim Sheba Medical Center, Tel Hashomer, Israel*

JANE C. IBER • *Polio and Picornavirus Laboratory Branch, Division of Viral Diseases, National Center for Immunization and Respiratory Disease, Centers for Disease Control and Prevention, Atlanta, GA, USA*

JAUME JORBA • *Polio and Picornavirus Laboratory Branch, Division of Viral Diseases, National Center for Immunization and Respiratory Disease, Centers for Disease Control and Prevention, Atlanta, GA, USA*

GIDEON KERSTEN • *Institute for Translational Vaccinology, Bilthoven, The Netherlands*

OLEN M. KEW • *Polio and Picornavirus Laboratory Branch, Division of Viral Diseases, National Center for Immunization and Respiratory Disease, Centers for Disease Control and Prevention, Atlanta, GA, USA*

DAVID R. KILPATRICK • *Polio and Picornavirus Laboratory Branch, Division of Viral Diseases, National Center for Immunization and Respiratory Disease, Centers for Disease Control and Prevention, Atlanta, GA, USA*

SATOSHI KOIKE • *Neurovirology Project, Tokyo Metropolitan Institute of Medical Science, Tokyo, Japan*

CÉLINE LACROIX • *KU Leuven – University of Leuven, Department of Microbiology and Immunology, Rega Institute for Medical Research, Laboratory of Virology and Chemotherapy, Leuven, Belgium*

PIETER LEYSSEN • *KU Leuven – University of Leuven, Department of Microbiology and Immunology, Rega Institute for Medical Research, Laboratory of Virology and Chemotherapy, Leuven, Belgium*

LARISSA VAN DER MAAS • *Institute for Translational Vaccinology, Bilthoven, The Netherlands*

YOSSI MANOR • *National Center for Viruses in the Environment, Central Virology Laboratory, Public Health Services, Chaim Sheba Medical Center, Tel Hashomer, Israel*

ELLA MENDELSON • *Department of Epidemiology and Preventive Medicine, School of Public Health, Sackler Faculty of Medicine, Tel Aviv University, Tel Aviv, Israel; Central Virology Laboratory, Public Health Services, Chaim Sheba Medical Center, Tel Hashomer, Israel*

PHILIP D. MINOR • *National Institute for Biological Standards and Control, Medicines and Healthcare Products Regulatory Agency, Potters Bar, Hertfordshire, UK*

NORIYO NAGATA • *National Institute of Infectious Diseases, Musashimurayama, Tokyo, Japan*

JOHAN NEYTS • *KU Leuven – University of Leuven, Department of Microbiology and Immunology, Rega Institute for Medical Research, Laboratory of Virology and Chemotherapy, Leuven, Belgium*

M. STEVEN OBERSTE • *Polio and Picornavirus Laboratory Branch, Division of Viral Diseases, National Center for Immunization and Respiratory Diseases, Centers for Disease Control and Prevention, Atlanta, GA, USA*

MARK A. PALLANSCH • *Polio and Picornavirus Laboratory Branch, Division of Viral Diseases, National Center for Immunization and Respiratory Diseases, Centers for Disease Control and Prevention, Atlanta, GA, USA*

ARMANDO DE PALMA • *KU Leuven – University of Leuven, Department of Microbiology and Immunology, Rega Institute for Medical Research, Laboratory of Virology and Chemotherapy, Leuven, Belgium*

ISABELLE PELLETIER • *INSERM U994, Institut National de la Santé et de la Recherche Médicale, Paris, France; Institut Pasteur, Biologie des Virus Entériques, Paris, France*

JASON A. ROBERTS • *National Enterovirus Reference Laboratory, Victorian Infectious Diseases Reference Laboratory, Doherty Institute, Melbourne, VIC, Australia*

MERJA ROIVAINEN • *Virology Unit, National Institute for Health and Welfare, Helsinki, Finland*

SALMAAN SHARIF • *WHO Regional Reference Laboratory for Poliomyelitis, National Institute of Health, Islamabad, Pakistan*

LESTER M. SHULMAN • *Environmental Virology Laboratory, Central Virology Laboratory, Public Health Services, Chaim Sheba Medical Center, Tel Hashomer, Israel; Department of Epidemiology and Preventive Medicine, School of Public Health, Sackler Faculty of Medicine, Tel Aviv University, Tel Aviv, Israel*

DANIT SOFER • *Central Virology Laboratory, Public Health Services, Chaim Sheba Medical Center, Tel Hashomer, Israel*

BRUCE R. THORLEY • *National Enterovirus Reference Laboratory, Victorian Infectious Diseases Reference Laboratory, Doherty Institute, Melbourne, VIC, Australia*

ALOYS TIJSMA • *KU Leuven – University of Leuven, Department of Microbiology and Immunology, Rega Institute for Medical Research, Laboratory of Virology and Chemotherapy, Leuven, Belgium*

WILLIAM C. WELDON • *Polio and Picornavirus Laboratory Branch, Division of Viral Diseases, National Center for Immunization and Respiratory Diseases, Centers for Disease Control and Prevention, Atlanta, GA, USA*

JANNY WESTDIJK • *Institute for Translational Vaccinology, Bilthoven, The Netherlands*

THOMAS WILTON • *Crucell Holland B.V., Leiden, The Netherland*

WENBO XU • *China Center for Disease Control and Prevention, Beijing, People's Republic of China*

SYED SOHAIL ZAHOOR ZAIDI • *WHO Regional Reference Laboratory for Poliomyelitis, National Institute of Health, Islamabad, Pakistan*

YONG ZHANG • *China Center for Disease Control and Prevention, Beijing, People's Republic of China*

Chapter 1

An Introduction to Poliovirus: Pathogenesis, Vaccination, and the Endgame for Global Eradication

Philip D. Minor

Abstract

Poliomyelitis is caused by poliovirus, which is a positive strand non-enveloped virus that occurs in three distinct serotypes (1, 2, and 3). Infection is mainly by the fecal–oral route and can be confined to the gut by antibodies induced either by vaccine, previous infection or maternally acquired. Vaccines include the live attenuated strains developed by Sabin and the inactivated vaccines developed by Salk; the live attenuated vaccine (Oral Polio Vaccine or OPV) has been the main tool in the Global Program of Polio eradication of the World Health Organisation. Wild type 2 virus has not caused a case since 1999 and type 3 since 2012 and eradication seems near. However most infections are entirely silent so that sophisticated environmental surveillance may be needed to ensure that the virus has been eradicated, and the live vaccine can sometimes revert to virulent circulating forms under conditions that are not wholly understood. Cessation of vaccination is therefore an increasingly important issue and inactivated polio vaccine (IPV) is playing a larger part in the end game.

Key words Poliovirus, Picornaviruses, Human enterovirus C species pathogenesis, Live and killed vaccines, The Global Poliomyelitis Eradication Initiative, Vaccine derived polioviruses, Stopping vaccination

---

1 Introduction

Poliovirus is a positive strand RNA virus classified with certain Coxsackie A viruses within the Human Enterovirus C species of the picornaviridae. The genome is about 7500 bases in length enclosed in a non-enveloped capsid consisting of 60 copies each of four proteins designated VP1, 2, 3, and 4 and there are three serotypes (1, 2, and 3) as shown first in the 1950s by the demonstration that infection of monkeys with one serotype did not protect against the other two. The serotypes are most generally identified by their reaction with specific sera. Polio virus was shown to be a transmissible agent in 1909 by Landsteiner and Popper by infection and paralysis of old world monkeys [1] and this remained the only

Javier Martín (ed.), *Poliovirus: Methods and Protocols*, Methods in Molecular Biology, vol. 1387,
DOI 10.1007/978-1-4939-3292-4_1, © Springer Science+Business Media New York 2016

method of detection until the 1950s. Subsequently cell culture methods were developed, and currently molecular methods play a central role in detecting and accurately identifying strains.

2   Pathogenesis and History

There are very few recognizable cases of paralytic poliomyelitis reported in Western medical literature until the end of the nineteenth century when large seasonal epidemics in young children began to be recognized. However, the first case generally considered to be poliomyelitis is depicted on the funerary stele of the priest Rom from about 1300 BCE (Fig. 1) which shows the typical withered limb and down flexed foot that follows destruction of motor innervation and muscle wasting. Poliovirus derives its name from the fact that it specifically attacks the grey matter (Greek polios = grey, myelos = matter) of the anterior horn of the spinal cord. The clinical presentation is so striking that it seems reasonable to conclude that cases were very rare in the western world [2] until the epidemic pattern became established at the end of the nineteenth century and beginning of the twentieth.

Fig. 1 Funerary stele of the priest Rom ca. 1300 BCE showing the withered limb and down flexed foot typical of poliomyelitis. Printed by permission of the Ny Carlsberg Glyptotek, Copenhagen

Early epidemiological studies in Sweden showed that as well as frank paralysis (the major disease) there was often an earlier milder syndrome of sore throat and malaise typical of a general viral infection (the minor disease). Most infections were entirely silent, however. In fact the accepted view is that there are about 200 silent infections for every paralytic case depending on the virus type and strain, with type 1 being the most dangerous and type 2 the least. Initially because of the use of particular animal models of infection and the disease presentation it was thought that polio grew mostly in the central nervous system. However in the 1940s it became clear that this was not the case and in fact there was good evidence from very early on that the virus grows chiefly in the gut. From there it can break out into the blood and spread to unknown systemic sites [3] where it replicates before spreading to other areas including the Central Nervous system. There are therefore two viremic steps, one immediately following infection of the gut and the second about a week later which spreads the infection to other sites including typically the throat and tonsils. Viremic spread is greatly inhibited if not prevented altogether by the presence of humoral antibody. A study in the 1950s showed that passive antibody was sufficient to give protection from disease [4]. This in turn provides an explanation of the emergence of polio in major epidemics. Where a baby is exposed to the virus while protected by maternal antibody, infection is confined to the gut and thus harmless. As hygiene standards improve as at the end of the nineteenth century exposure of the baby to the virus happens later in life when maternal protection has been lost, allowing disease to occur. This is almost certainly an oversimplification as in areas where poliovirus circulates intensively, such as in Northern India before eradication, there were a very large number of cases despite the high level of serum antibodies in the mothers and the early exposure of the babies. It could be that maternal protection can be overwhelmed by high intensity exposure to the virus.

The model requires that a child's gut becomes infected with poliovirus in the presence of protective levels of maternal antibody until the child's immune system develops. Maternal antibody cannot provide full protection against infection of the infant gut or the child would become infected only when maternal antibody had declined and would therefore be susceptible to disease at the same time. Improving hygiene standards would then only delay disease not cause an increase in cases or an epidemic pattern to emerge. It is not known whether passive antibodies will protect against gut infection. More important to the eradication program, the effectiveness of immunity induced by inactivated vaccine (IPV) at preventing gut infection remains at least a matter for debate as IPV may be chiefly associated with the induction of humoral antibodies.

In summary poliovirus is shed in the feces and the main route of spread is fecal–oral, although respiratory routes are possible; most infections are silent and if the level of immunity is high, the proportion of symptomatic infections may be even lower unless immunity prevents gut infection. Counting cases of poliomyelitis is an indication of virus circulation but not necessarily the most secure, sensitive, or reliable method in the end stages of eradication.

3   Antigenic Forms of Polio Virus

In the 1950s it was shown that when a culture of poliovirus was fractionated on a sucrose gradient of five steps (A to E) there were two peaks of antigenic activity, one associated with the C step, the other with the D. The C antigen reacted more strongly with acute phase human sera and the D antigen with convalescent sera [5]. The two antigens were therefore termed C and D antigen and correspond to a first approximation to empty capsids and full infectious virus particles respectively. Because a convalescent patient is assumed to be protected, it was concluded that antibodies against the D antigen were protective while those against the C antigen were not; antibodies reacting with D antigen neutralize infectivity. It was also concluded that only D antigen could stimulate protective antibody; it is now known that neutralizing antibodies exist that react with both C and D antigen which can be induced by either particle type, so that this is not the case although it sometimes assumed that antigenicity and immunogenicity are interchangeable. As most IPV is now purified to exclude empty capsids, potency may be based on measurement of D antigen content. The D antigen is converted into a C or C like form by relatively mild treatments such as heating, adsorption to surfaces, or mild UV treatment [6].

4   Vaccines Against Poliomyelitis-Inactivated Polio Vaccine (IPV)

Vaccines were developed in the 1950s and are the key to the control of poliomyelitis and ultimately the Global Polio Eradication Initiative. The first to be licensed was the inactivated vaccine developed by Jonas Salk, which consisted of largely unpurified preparations of tissue culture grown virus treated with diluted formalin under careful conditions so that while infectivity was lost, the ability to induce a protective immune response was retained. There were many issues associated with the vaccine initially. The first batches included preparations that were not fully inactivated and caused poliomyelitis in recipients [7]. This was a failure of the production process; specifically there was a need for adequate filtration of the vaccine before and after inactivation to make sure that aggregates of virus not penetrated by the formaldehyde were removed. Introduction of filtration steps ensured that this did not happen again, but also reduced the amount of antigen present leading to the need to measure and standardize the potency by some means. Initially this was done by assessing immunogenicity in animal models such as guinea pigs or young chickens, assays that are variable and not always quantitative. The virus was originally grown on primary monkey kidney cells from wild caught old world monkeys, mostly rhesus macaques. They are susceptible to many simian viruses and one polyoma virus, SV40, was present in many batches and only poorly inactivated by the formalin treatment. It was also hard to detect as it did not cause a cytopathic effect in rhesus monkey cells, although it did so in cells from cynomolgus monkeys. SV40 is able to cause tumors in certain animal models and it is clear that many batches of IPV contained live virus. The data suggest that the contamination had no long term consequences in practice but it was a major concern which has still not been resolved to everyone's satisfaction [8]. Supply was also a practical issue, because of the scale of production required. Finally there was a long running debate between those who favored the killed vaccines and those who preferred live vaccines that would imitate natural infection.

Current production technology for IPV was developed in the 1980s; it depends on large scale cell cultures usually in continuous cell lines, giving high titre preparations. Assay of potency is most commonly by measurement of D antigen content or immunogenicity in rats, in which the response is quantitative. The issues with the original IPV of safety, supply, contamination and assay have therefore been dealt with.

The effect of IPV on the incidence of poliomyelitis in the USA is shown in Fig. 2 where cases were reduced by over 95 % and the downward trend continued. Nonetheless in the early 1960s the

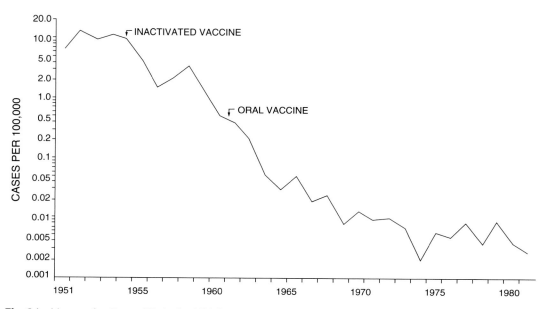

Fig. 2 Incidence of poliomyelitis in the USA from 1950 to 1970s showing the effect of killed and live vaccines

live attenuated vaccines developed by Albert Sabin replaced IPV, and they have formed the main tool in the Global Polio Eradication program of WHO. However IPV is playing an increasingly significant role in the end of polio.

5 Live Attenuated Polio Vaccines (OPV)

There was a long running argument between those who favored the killed vaccines of the Salk type and those who favored a live attenuated vaccine that would mimic natural infection. The view was that live vaccines would break transmission by inducing the full range of immunity including gut immunity where IPV would not. It was also argued that a live vaccine that infected the recipient and was shed for some time could also infect and immunize contacts and therefore have a bigger effect than IPV. Spread was therefore thought to be a positive argument for OPV.

By the late 1970s polio was largely eliminated from developed countries, most of whom used OPV as developed by Sabin. There was a residual number of cases, about 1000-fold less than in the pre-vaccine era and it was shown that apart from a few imported cases they were caused by the vaccine strains, either in recipients or in their immediate contacts. The vaccine could therefore revert and the spread to others was a mixed blessing. There is thus a need to test the vaccine before it is used to ensure that it is safe, and for many years this involved primates, following modified versions of the tests used by Sabin in developing the strains in the first place. Currently there are alternatives in the form of transgenic mice bearing the polio receptor so that unlike normal mice they are susceptible to infection, or molecular tests for consistency.

The vaccine replicates for an average of about 4 weeks in recipients [9] and over that time the properties of the virus excreted change. The type 2 and type 3 components become more virulent in animal models and many changes can be followed at the molecular level. Nonetheless OPV was remarkably safe and very effective in temperate climates.

On the other hand the effect of OPV in developing countries and tropical climates was poor in so far as it could be tracked. Part of this may have been due to vaccine quality and delivery systems, but part was probably due to the epidemiology of polio virus infections, which are seasonal, particularly in temperate climates. Thus immunizing in a routine program in the low season in temperate climates can reduce the number of susceptibles below a sustainable level so that transmission is reduced in the high season. Where transmission is year round and high it is a matter of chance whether the vaccinee gets the wild type or the vaccine first, so transmission continues. The solution was to run a campaign of immunization where a large proportion of susceptibles were immunized at the same time so reducing the pool rapidly. This method had been used in the Southern States of America as "Sabin Sundays," but it was introduced into Latin America in the form of National Immunisation Days. The effect was dramatic and by 1988 some Latin American countries were polio free and others were making clear inroads. In 1988, the 41st World Health Assembly recognized the progress made and adopted resolution 41–28 that committed WHO to the eradication of polio by the year 2000. While this target was obviously not met, by 2003 endemic transmission of polio was confined to four countries: India, Pakistan, Afghanistan, and Nigeria. There were repeated instances of reintroduction of polio into countries that had eliminated; in one instance immunization in Northern Nigeria was stopped for a period for local reasons and polio spread across the whole of Central Africa and was exported to Yemen and Indonesia. It is clear that if one country has polio, the entire world is at risk.

At the time of writing however there has not been a case of type 2 polio caused by a wild type strain since 1999 other than some cases in India as a result of contamination of OPV with a laboratory strain. It is very likely that type 2 polio in the wild has been eradicated. Similarly there has not been a case of type 3 polio in the world for the last 8 months, and it is conceivable that it has also been eradicated. The last case of type 1 polio in India was in 2011 which given the epidemiological circumstances was a colossal achievement. Polio eradication is therefore moving into the endgame. If it is eventually achieved it will have been the most ambitious vaccination activity of all time, with a cost so far of about $9 billion.

6　The Endgame

It is hard to justify continuing a vaccination program against a disease that has been eradicated, but the cessation of vaccination against polio will be a very complex process. Firstly it is necessary to be sure that the disease has truly been eradicated, and this means eradication of the virus. Effective surveillance for poliovirus will therefore be essential after eradication is declared and may focus increasingly on environmental searches rather than recording cases of disease. Until it is certain that the virus has gone, vaccine production must continue and as this currently requires growth of poliovirus this risks reintroduction. The apparent eradication of wild type 2 virus suggests that vaccination could stop one serotype at a time; in fact the type 2 strain of OPV dominates the response to vaccination and most campaigns now use bivalent vaccine containing only type 1 and 3 and serious consideration is being given to stopping all usage of type 2 OPV in both routine and campaign based programs.

In addition a major feature of OPV is that the strains which are already capable of some transmission from recipients to contacts can evolve to become as transmissible as the wild type. Circulating vaccine derived polioviruses (cVDPVs) have caused numerous outbreaks in areas where vaccine coverage is imperfect although the conditions required for their emergence are still not really clear. In addition individuals deficient in humoral immunity (hypogammaglobulinemics) can become excrete the vaccine strains for periods of many years during which the virus typically becomes virulent. Surveillance of sewage in many countries has also revealed the presence of strains of derived from the vaccine but of unknown specific origin; they are probably from unidentified long term excreters. The use of OPV therefore poses a hazard to the eradication of polio and the need for surveillance covers the vaccine derived strains as well as the wild type.

Consequently countries are moving to the use of IPV. Western Europe now uses IPV exclusively as does Northern America; countries of South America and Russia are also adopting it. Current IPV production requires the growth of strains that are known from accidental past exposures to be highly paralytic; production must therefore be strictly contained in a polio free world, and given the scale of virus growth required this is extremely difficult. There are currently three main producers of IPV in the world of which two provide the majority. It is hard to believe that other countries and producers will fail to become involved.

The World Health Organisation has developed a number of relevant documents to address some of the issues of what needs to be contained, when containment is required and how it should be done.

The eradication plan (Draft of February 2013) lists four phased objectives based on the hypothesis that wild type polio will be eradicated by the end of 2014.

1. Poliovirus detection and interruption; Plan for the last wild type poliovirus for the end of 2014, followed by outbreak responses, particularly for cVDPVs.

2. Strengthen routine immunization, address the prerequisites for withdrawal of type 2 OPV. From end 2015 to end 2016 complete IPV introduction and type 2 OPV withdrawal.
   From end 2016 IPV and bOPV are in routine use.

3. Finalize containment plans by the end of 2015, complete containment by end 2018. It is implicit that this applies only to the wild type as OPV will still be in use.

4. Legacy planning.

Global certification would be at the end of 2018, and bOPV use would cease during 2019.

The global action plan describes risk elimination and risk management procedures once polio is eradicated. It relates specifically to containment and safe laboratory use of poliovirus and the production of IPV.

1. Primary safeguards include containment, management structures in the facilities, protection of staff through immunization and contingency plans in the event of an incident. Containment covers design of the plant including airflows, showering, autoclaves where the demands of GMP and containment are not all coincident (e.g., for GMP airflows may need to be positive to prevent contamination of the product, while containment would require negative pressure to ensure that nothing escaped).

2. Secondary safeguards relate to the epidemiology in the country which should have high (>90 %) vaccine coverage and an effective immunization program for children. This should reduce spread in the event of an escape.

3. Tertiary safeguards relate to public hygiene including sewage systems and treatment.

There are four phases to the implementation of polio containment:

1. National surveys and an inventory of wild type polio virus holdings and destruction of unnecessary stocks.

2. Establishment of long term policy relating to the need to retain polio capability; WHO envisage no more than 20 laboratories worldwide still holding and using polio.

3. Global destruction and containment of wild poliovirus to start 1 year after the last wild type poliovirus is isolated. Implementation of primary secondary and tertiary safeguards for laboratories still using the wild type strains.

4. Recall and destroy OPV stocks. Implement primary and secondary safeguards for use of OPV/Sabin strains at the time of cessation of OPV use.

Primary, secondary and tertiary safe guards are required for wild type virus after eradication. Primary and secondary safeguards are required for OPV/Sabin strains after cessation of routine OPV use.

Guidelines for the safe production and quality control of inactivated poliomyelitis vaccine manufactured from wild poliovirus (Addendum, 2003, to the Recommendations for the Production and Quality Control of Poliomyelitis Vaccine (Inactivated)) provides detailed guidance for the development of BSL3 (polio) facilities. They are very stringent.

The issues of containment have raised the possibility that the Sabin strains might be a safer way to make IPV. Apart from differences in immunogenicity that have become apparent the safety added is by no means clear given the issues with cVDPVs and other vaccine derived strains. The guideline on vaccine production refers explicitly to production using the Sabin strains, which do not require BSL3/IPV containment provided they are produced under conditions that would make them suitable for oral vaccine use; this implies testing which may introduce unacceptable extra burdens on production as the wild type currently requires no control of this kind.

Consequently there is interest in developing alternative strains and procedures for vaccine production, including strains that are genetically stable, that are attenuated by so many small mutations that there reversion is essentially impossible or eventually stable empty capsids. All are at a relatively early stage of development and the choice of approach and the containment required remain to be determined.

7   Summary and Conclusions

Poliomyelitis was a major public health issue for developed countries in the middle of the twentieth century. Vaccines make its eradication possible and the progress to this end is outstanding. The last steps in the process involve containment among other things and are complex and challenging. In view of the stage of the eradication program and the changes in vaccination they need to be addressed.

References

1. Landsteiner K, Popper E (1909) Ubertragung der poliomyelitis scuta auf affen. Z Immunitatdforsch Exp Therap Orig 2:377–390
2. Paul JR (1971) A history of poliomyelitis. Yale University Press, New Haven
3. Minor PD (1997) Poliovirus. In: Nathanson N, Ahmed R, Ganzalez-Scarano F, Griffin DE, Holmes KV, Murphy FA, Robinson HL (eds) Viral pathogenesis. Lipincott Raven, Philadelphia, pp 555–574
4. Hammon WD, Coriell LL, Wehrle PF, Stokes J (1953) Evaluation of Red Cross gammaglobulin as a prophylactic agent for poliomyelitis. 4. Final report of results based on clinical diagnosis. JAMA 151:1272–1285
5. Mayer MM, Rapp HJ, Roizman B, Klein SW, Cowan KM, Lukens D, Schwerdt CE, Schaffer FL, Charney J (1957) The purification of poliomyelitis virus as studied by complement fixation. J Immunol 78:435–455
6. Le Bouvier GL (1959) The modification of poliovirus antigens by heat and ultraviolet light. Lancet 269(6898):1013–1016
7. Offit P (2005) The cutter incident: how America's first polio vaccine led to the growing vaccine crisis. Yale University Press, New Haven. ISBN 0-300-10864-8
8. Shah K, Nathanson N (1976) Human exposure to SV40: review and comment. Am J Epidemiol 103:1–12
9. Minor PD (2012) The polio eradication programme and issues of the endgame. J Gen Virol 93:457–474

Chapter 2

Poliovirus Laboratory Based Surveillance: An Overview

Syed Sohail Zahoor Zaidi, Humayun Asghar, Salmaan Sharif, and Muhammad Masroor Alam

Abstract

World Health Assembly (WHA) in 1988 encouraged the member states to launch Global Polio Eradication Initiative (GPEI) (resolution WHA41.28) against "the Crippler" called poliovirus, through strong routine immunization program and intensified surveillance systems. Since its launch, global incidence of poliomyelitis has been reduced by more than 99 % and the disease squeezed to only three endemic countries (Afghanistan, Pakistan, and Nigeria) out of 125. Today, poliomyelitis is on the verge of eradication, and their etiological agents, the three poliovirus serotypes, are on the brink of extinction from the natural environment. The last case of poliomyelitis due to wild type 2 strain occurred in 1999 in Uttar Pradesh, India whereas the last paralytic case due to wild poliovirus type 3 (WPV3) was seen in November, 2012 in Yobe, Nigeria. Despite this progress, undetected circulation cannot fully rule out the eradication as most of the poliovirus infections are entirely subclinical; hence sophisticated environmental surveillance is needed to ensure the complete eradication of virus. Moreover, the vaccine virus in under-immunized communities can sometimes revert and attain wild type characteristics posing a big challenge to the program.

Key words Poliovirus, Poliomyelitis, Vaccine virus, Surveillance systems

---

1 Immunization

The large epidemics of poliomyelitis during nineteenth and early twentieth century throughout Europe and the Americas gave rise to an extraordinary public reaction and mobilization; revolutionizing medical philanthropy and encouraging the development of new methods to prevent and treat the disease. During 1955, Jonas Salk and his colleagues introduced first inactivated poliovirus vaccine (IPV) and demonstrated its safety and efficacy [1]. IPV was the only licensed vaccine in the USA, Canada, and Western Europe till 1961–1962. Albert Sabin in 1959 conducted large clinical trails of first ever oral poliovirus vaccine (OPV) in the Soviet Union, Poland, and Czechoslovakia and got licensure in 1961 of monovalent vaccine against poliovirus type 1, 2 and 3 in 1962 [2]. However, later on trivalent oral poliovirus vaccine (tOPV) was introduced and

Javier Martín (ed.), *Poliovirus: Methods and Protocols*, Methods in Molecular Biology, vol. 1387,
DOI 10.1007/978-1-4939-3292-4_2, © Springer Science+Business Media New York 2016

used widely in many countries for their routine immunization activities. Both vaccines (OPV and IPV) effectively prevent poliomyelitis.

IPV triggers excellent immune response and provide only humoral immunity whereas OPV confers high level of both humoral and secretory immunity. Considering in view the benefits of OPV over IPV, it is used as a powerful tool in immunization activities against poliovirus in the Global Polio Eradication Initiative. To achieve protective level of immunity against the poliovirus, GPEI uses supplementary immunization activities in addition to routine immunization, with the support of national government, WHO, and other partner organizations. These supplementary activities are mass immunization campaigns also known as National Immunization Days (NIDs). The purpose of NIDs is to interrupt circulation of poliovirus by immunizing every child less than 5 years of age with at least two doses of oral polio vaccine, regardless of previous immunization status. NIDs are conducted through house-to-house activity to administer the vaccine directly to every child. The idea is to cover children either with no immunization, or partially immunized and to boost immunity in those who have been immunized. In this way the virus will be deprived of susceptible host as enough people in a community are immunized, and will eventually die out.

2   Global Polio Laboratory Network (GPLN)

The laboratory plays a crucial role to ensure that the public health program meet its objectives. World Health Organization established GPLN in parallel with the development of acute flaccid paralysis surveillance to support Global Polio Eradication Initiative (GPEI) [3]. Currently the GPLN consists of 145 laboratories that describe the structure and responsibilities into three levels of institutions: National Laboratories (NL), Regional Reference Laboratories (RRL), and Global Specialized Laboratories (GSL). Currently the GPLN comprises 122 National and Subnational Laboratories, 16 Regional Reference Laboratories, and 7 Global Specialized Laboratories (Fig. 1). The key role for GPLN is to provide timely and accurate virological information that can be used to target and cease the wild poliovirus spread and importation. This information may also be used to guide and focus immunization activities to achieve poliovirus eradication goal. The GPLN is guided by expert virologists who serve as Global and Regional laboratory coordinators. The GPLN follows "WHO Polio Laboratory Manual" for isolation, identification, and characterization of polioviruses [4] which is regularly updated as new methods are developed and tested for suitability of its use by GPLN laboratories.

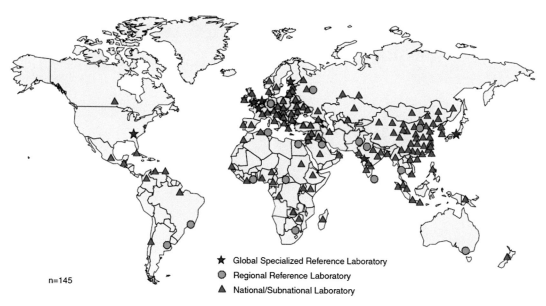

n=145

★ Global Specialized Reference Laboratory
◯ Regional Reference Laboratory
▲ National/Subnational Laboratory

Fig. 1 Geographical distribution of Global Poliovirus Laboratories Network. *Stars, circles,* and *triangles* indicate Global Specialized Reference laboratories, Regional Reference Laboratories, and National and Subnational Laboratories working for Polio eradication program. *Source*: http://www.cdc.gov/mmwr/preview/mmwrhtml/mm4908a3.htm

3 AFP Surveillance System

The objective of AFP surveillance relies on detailed investigation and reporting of any cases of floppy weakness in a child less than 15 years of age. This system is key to the poliovirus eradication initiative and forms the basis of the documentation needed for certification of polio-free status. The AFP system has both field and laboratory components; the field investigators including clinicians and epidemiologists examine the reported paralytic cases and send their appropriate specimen to the laboratory for viral investigations.

To obtain accurate and timely results, correct clinical specimen's collection, timing, and transportation under optimal conditions plays a vital role. This requires close coordination and cooperation between clinicians, epidemiologists, and virologists. Trainings, designation of responsibilities, and detailed planning are needed to achieve the polio eradication initiative. The GPEI has described clear guidelines and training procedures for proper specimen collection and transportation to the laboratory via "reverse cold chain" [4]. Stool samples and sewage samples are the specimens of choice for AFP and environmental poliovirus surveillance, respectively. Two stool specimens of sufficient quantity (~8 g) are collected from each AFP case under 5 years at least 24 h

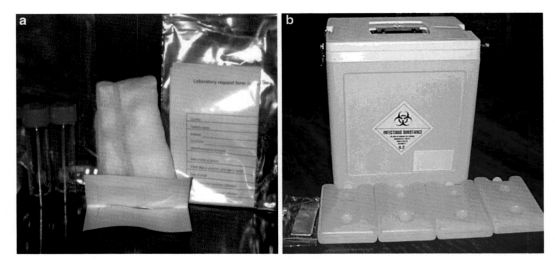

Fig. 2 Standard Stool collection kit as recommended by WHO guidelines. (**a**) Indicates the two vials, absorbing material, labels, and laboratory investigation form in a ziplock bag. (**b**) Shows ice packs and stool carrier box used for transportation of samples to the laboratory

apart, within 14 days after the onset of paralysis for laboratory analysis. A standard sized kit including sample collection vials, absorbing material, labels, laboratory investigation form, and the stool carrier box is used, as shown in Fig. 2. The specimens must arrive at the laboratory within 48 h of second stool collection, maintaining the reverse cold chain with proper documentation.

4   Virus Isolation in Cell Culture

Cell culture is still considered to be the gold standard in virology. It is considered as the most critical and basic procedure in virus isolation. Although polioviruses can be grown in a wide range of human cells (RD, HeLa, HEp-2, WI-38, MRC-5, HEK293) and simian cells (from rhesus macaques and African green monkeys; Vero, LLC-MK2, Buffalo Green Monkey) (http://www.atcc.org/) [5], GPLN recommends two cell lines for poliovirus isolation in routine testing: (1) RD cells (a continuous cell line from human rhabdomyosarcoma) [6] and (2) L20B cells (a derivative of mouse L cell line with genetically engineered human poliovirus receptor, CD 155)—a highly selective cell line for growth of poliovirus [7]. Usually all polioviruses grow on L20B cells; however, some Coxsackie A viruses and other enteroviruses can also grow in L20B [7]. All viruses with growth on L20B cells are further characterized through molecular identification method.

5 Molecular Identification Method

Real-time reverse transcriptase polymerase chain reaction (qRT-PCR) is currently used in GPLN for Intratypic differentiation (ITD) of polioviruses. A series of primer pairs and specific fluorescent probes have been developed that identify isolates groups: (1) enteroviruses (panEV), (2) polioviruses (panPV), (3) poliovirus serotype (Serotype-1, Serotype-2, Serotype-3), and (4) vaccine related (Sabin1, Sabin2, Sabin3). The qRT-PCR kits are supplied to all GPLN laboratories by the Centers for Disease Control and Prevention (CDC), Atlanta, USA and are usually supplemented with additional qRT-PCR reagents that can facilitate screening of genetically divergent vaccine derived polioviruses (VDPV1, VDPV2, VDPV3) [8–10]. Isolates with Sabin negative reaction and panEV, panPV, and serotype positive reactions are screened as wild polioviruses. However, isolates having Sabin-like results are further subjected to VDPV screening against their respective serotype.

6 Nucleotide Sequencing of Poliovirus Isolates

ITD screens out OPV-like polioviruses, wild polioviruses, and VDPVs. All the isolates with wild poliovirus and suspected VDPVs are sequenced for Viral Protein 1 region (VP1) (~900-nucleotide stretch) of the poliovirus genome following standardized procedures using standardized sequencing primer sets. VP1 region encodes several serotype specific antigenic sites [11] and evolve primarily by successive fixation of nucleotide substitutions rather than by recombination [12]. Heterotypic and homotypic poliovirus mixtures are specifically amplified by using serotype- and genotype-specific primer sets [13].

Genetic sequence obtained is analyzed by using a software package that is capable of reading sequence files, construct contigs and allows performs base pair editing. Examples of such packages are Sequencher analysis package distributed by Genes Codes Corporation Ann Arbor, Michigan; Lasergene package distributed by DNASTAR Inc. Madison, Wisconsin; SeqScape by Applied Biosystems, Foster City, CA; and Bionumerics by Applied Maths Inc., Austin, Texas. Clean sequences are then compared with respective prototype Sabin strain to distinguish between OPV-like polioviruses, Vaccine-Derived polioviruses and Wild polioviruses (WPV). VDPVs and WPVs sequences are blasted against their respective databank to identify the sequence relationships among poliovirus isolates and are then summarized in the form of phylogenetic trees using bioinformatic tools like MEGA 6, Geneious, etc.

7 Environmental Surveillance

The Global Polio Eradication Initiative (GPEI) relies mainly on sensitive population surveillance to rapidly detect all circulating polioviruses and to guide eradication activities. Although the case based clinical surveillance will remain the primary mechanism for detection of poliovirus, the pivotal role of environmental surveillance has been immensely recognized especially in the remaining three endemic countries, i.e., Pakistan, Afghanistan, and Nigeria. The virologic data generated through AFP and environmental surveillance guides to monitor the impact of immunization activities and underscores areas that require improvements. The outcomes of such supplemental surveillance tool have been proven quite useful thus facilitating the more rapid identification of outbreaks and providing useful information to validate the interruption of virus transmission [14].

Environmental surveillance is mainly conducted in areas with suboptimal AFP surveillance, inadequate immunization activities or areas with a high-risk of wild-poliovirus importations [15]. The establishment of environmental surveillance for poliovirus is usually planned in consultation with the World Health Organization country and regional offices, experts from national Expanded Program on Immunization, national polio laboratory, local sanitary engineering staff, and the country's local health departments. These bodies assess and validate the WHO recommended guidelines ensuring the proper selection of sewage sample collection sites, availability of laboratory resources, advance training of the personnel and validation of adopted laboratory procedures. The important components of planning thus include site, length and time of sampling, target population, adequate laboratory space, verified testing procedures, data management, reporting, and quality assurance [16].

The purpose of environmental surveillance may define the length and time schedule of sampling in different settings. Areas with a need to confirm the elimination of wild poliovirus requires a regular sampling program with at least one sample collected per site per month and continued for 3 years after the last wild poliovirus isolation. If the surveillance is initiated to detect the WPV1 importation or detection of VDPVs, a more selective approach in high-risk areas should be preferred to detect any silent circulation. Depending upon the population size, single or composite samples may be collected from the inlets to sewage treatment plants or other collector sewer points serving target population.

The two principle methods for collection of environmental surveillance includes trap and grab sampling. In grab sampling, at least 1 l of sample volume is collected at a point-of-time or at different scheduled timings to make a composite sample considering the peak-hours of household sewage flow. Trap method utilizes a bag made of nonselective absorbing material that is hanged into the sewage stream for a period of 24–48 h and transported to the laboratory for virus isolation. Based on practical evaluation, WHO recommends "grab method" to be used for environmental sample collection, due to its increased sensitivity and feasibility. Upon receipt in the laboratory, samples are processed and concentrated using any of the methods such as two-phase separation method or ultrafiltration. The concentrated samples are then tested for the presence of poliovirus using a similar algorithm as for AFP samples including isolation of poliovirus using L20B and RD cells, PCR for ITD, and genomic sequencing of wild and vaccine-derived polioviruses.

Environmental surveillance system has been successfully established in five out of six WHO regions, viz., Eastern Mediterranean Region (EMR), South East Asian Region (SEAR), African Region (AFR), American Region (AMR) and European Region (EUR) (Fig. 3). In these settings, this supplementary activity has provided vital information to strengthen the routine AFP surveillance, for instance, detection of Northeast African genotype of WPV1 in Egypt [17], Southeast Asian genotype in Israel during 2013 [18], and rapid detection of WPV1 importations into Finland and the Netherlands [19]. Thus environmental surveillance plays a pivotal role in monitoring the imperceptible circulation of poliovirus in both polio-endemic and polio-free communities.

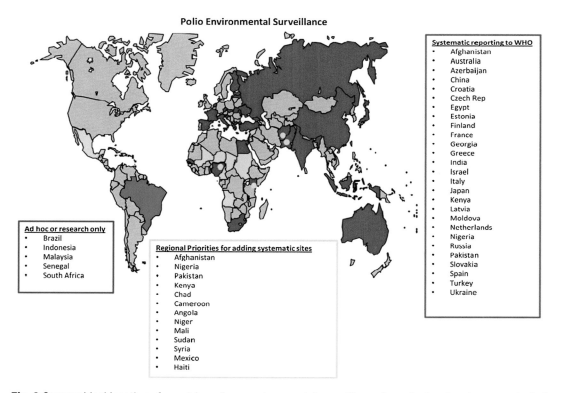

Fig. 3 Geographical location of countries where environmental surveillance for poliovirus has been conducted as a research or routine activity under Global Polio Eradication Initiative (adapted from Ref. [20])

References

1. Salk JE et al (1954) Formaldehyde treatment and safety testing of experimental poliomyelitis vaccines. Am J Public Health Nations Health 44(5):563–570

2. Sabin AB (1985) Oral poliovirus vaccine: history of its development and use and current challenge to eliminate poliomyelitis from the world. J Infect Dis 151(3):420–436

3. Hull BP, Dowdle WR (1997) Poliovirus surveillance: building the global Polio Laboratory Network. J Infect Dis 175(Suppl 1): S113–S116

4. World Health Organization (1997) Field guide for supplementary activities aimed at achieving polio eradication, 1996 revision. Geneva, Switzerland

5. Melnick JL (1997) Poliovirus and other enteroviruses. In: Evans AS, Kaslow RA (eds) Viral infections of humans, epidemiology and control, 4th edn. Plenum, New York, pp 583–663

6. McAllister RM et al (1969) Cultivation in vitro of cells derived from a human rhabdomyosarcoma. Cancer 24(3):520–526

7. Polio laboratory manual—World Health Organization (2004) WHO/IVB/04.10, 2004. 4th ed.

8. Kilpatrick DR, Nottay B, Yang CF, Yang SJ, Da Silva E, Peñaranda S, Pallansch M, Kew O (1998) Serotype-specific identification of polioviruses by PCR using primers containing mixed-base or deoxyinosine residues at positions of codon degeneracy. J Clin Microbiol 36(2):352–357

9. Kilpatrick DR, Nottay B, Yang CF, Yang SJ, Mulders MN, Holloway BP, Pallansch MA, Kew OM (1996) Group-specific identification of polioviruses by PCR using primers containing mixed-base or deoxyinosine residue at positions of codon degeneracy. J Clin Microbiol 34(12):2990–2996

10. Kilpatrick DR, Ching K, Iber J, Campagnoli R, Freeman CJ, Mishrik N, Liu HM, Pallansch MA, Kew OM (2004) Multiplex PCR method for identifying recombinant vaccine-related polioviruses. J Clin Microbiol 42(9):4313–4315

11. Minor PD (1990) Antigenic structure of picornaviruses. Curr Top Microbiol Immunol 161:121–154

12. Jorba J et al (2008) Calibration of multiple poliovirus molecular clocks covering an extended evolutionary range. J Virol 82(9):4429–4440

13. Manor Y et al (1999) A double-selective tissue culture system for isolation of wild-type poliovirus from sewage applied in a long-term environmental surveillance. Appl Environ Microbiol 65(4):1794–1797

14. Alam MM et al (2014) Detection of multiple cocirculating wild poliovirus type 1 lineages through environmental surveillance: impact and progress during 2011-2013 in Pakistan. J Infect Dis 210(Suppl 1):S324–S332

15. Blomqvist S et al (2012) Detection of imported wild polioviruses and of vaccine-derived polioviruses by environmental surveillance in Egypt. Appl Environ Microbiol 78(15):5406–5409

16. World Health Organization (2003) Guidelines for environmental surveillance of poliovirus circulation. Department of Vaccines and Biologicals, World Health Organization, Geneva, Switzerland. http://www.who.int/vaccines-documents/DoxGen/H5-Surv.htm

17. Hovi T et al (2005) Environmental surveillance of wild poliovirus circulation in Egypt—balancing between detection sensitivity and workload. J Virol Methods 126(1-2):127–134

18. Anis E et al (2013) Insidious reintroduction of wild poliovirus into Israel, 2013. Euro Surveill 18(38):pii: 20586

19. van der Avoort HG et al (1995) Isolation of epidemic poliovirus from sewage during the 1992-3 type 3 outbreak in The Netherlands. Epidemiol Infect 114(3):481–491

20. Asghar H et al (2014) Environmental surveillance for polioviruses in the global polio eradication initiative. J Infect Dis 210(Suppl 1): S294–S303

Chapter 3

Isolation and Characterization of Enteroviruses from Clinical Samples

Soile Blomqvist and Merja Roivainen

Abstract

Enterovirus infections are common in humans worldwide. Enteroviruses are excreted in feces during infection and can be detected from stool specimens by isolation in continuous laboratory cell lines. Characterization of enteroviruses is based on their antigenic and/or genetic properties.

Key words Enterovirus, Virus isolation, Cell culture, Neutralization, Antiserum, Sequencing, Genetic typing

1 Introduction

Four species of enteroviruses (*Human enterovirus A* to *D*, genus *Enterovirus*, family *Picornaviridae*) cause infections in humans. They include more than 100 antigenically or genetically defined serotypes or types [1]. Clinical manifestations differ from asymptomatic or mild respiratory infections to severe diseases of central nervous system. Typical enterovirus diseases include hand-foot-and-mouth disease and viral meningitis.

The golden standard for enterovirus diagnostics is the isolation of viruses in cell culture. Most of the enteroviruses multiply rapidly in several continuous cell lines routinely used in diagnostic laboratories. The obtained virus isolates have high titers (cell culture infectious doses, $CCID_{50}$) and they can be further typed by either antigenic (neutralization with serotype-specific antisera) or genetic properties (capsid protein VP1 sequence identities determine the type of the virus). The results of these two typing assays correlate well [2, 3] and the choice of the method relies on the facilities available in the laboratory.

After the molecular typing method became available some years ago, it has been used increasingly because it overcomes some restrictions of the traditional neutralization assay. Type-specific

Javier Martín (ed.), *Poliovirus: Methods and Protocols*, Methods in Molecular Biology, vol. 1387,
DOI 10.1007/978-1-4939-3292-4_3, © Springer Science+Business Media New York 2016

antisera are not available for all of the recently described "new" enterovirus types. Besides, enteroviruses undergo continuous antigenic evolution and the currently circulating viruses may escape the neutralization by the antisera produced about 50 years ago. In addition, the genetic typing is based on the sequences obtained from the VP1 capsid protein coding region and these sequences can be used for molecular epidemiological analysis besides to the identification of the type of the virus.

Enteroviruses multiply in the human intestine and are excreted into feces in high concentrations. The stool specimens are considered the best choice for virus isolation. The number of viruses excreted in the feces may vary from day to day during the infection and so it is recommended that two samples taken at 1–2 days interval should be analyzed for each patient to increase the sensitivity of the assay [4]. Before inoculation into cell cultures, the stool samples are treated with chloroform in order to inactivate the bacteria always present in feces. In addition to the intestine, enteroviruses may infect many other organs and the virus isolation can be attempted also from other clinical specimens, like spinal fluid, respiratory samples, and swap samples from skin lesions. These specimens are collected according to the symptoms of the disease.

We describe here the traditional virus isolation from stool specimens [4], the antigenic typing assay [4] and the VP1 sequence-based type identification [5]. We assume that the readers have already good knowledge on the maintenance of laboratory cell lines, on working on the biosafety level 2 laboratories, and on the principles of the basic molecular biology methods, such as handling of RNA. In addition, the facilities for DNA sequencing and the software for sequence analysis are necessities.

2   Materials

2.1   Enterovirus Isolation

1. Phosphate buffered saline (PBS) with calcium and magnesium ions, pH 7.2–7.4. Store at +4 °C. Supplement the PBS solution with antibiotics to a final concentration of 100 U/ml of penicillin G, 100 µg/ml of streptomycin sulfate, and 50 µg/ml of gentamicin. Store the PBS-antibiotics—solution at +4 °C for up to 1 week.

2. Ethanol-stabilized chloroform (Merck). Store at room temperature, protected from light, in a fume hood.

3. Continuous cell lines that support the growth of enteroviruses (*see* **Note 1**) seeded and grown in monolayer tube cultures (*see* **Note 2**).

4. The maintenance media for cells. The media is cell-dependent; for RD(A) mix 100 ml Eagle's MEM (GIBCO), 2 ml 1 M Hepes pH 7.4, 2 ml 1 M $MgCl_2$, 2 ml heat-inactivated fetal

bovine serum (Integro), 25 U/ml mycostatin, and the same antibiotics in the same final concentrations as in the PBS above.

2.2 Enterovirus Typing by Pools of Antisera

1. Antisera pools for enterovirus typing (*see* **Note 3**). Dilute and store the pools according to the supplier's instructions.

2. The same cell lines and maintenance media used in the enterovirus isolation (see above). The cells should be seeded and grown in cell culture flasks.

3. PBS without magnesium and calcium ions. Supplement with antibiotics to a final concentration of 100 U/ml of penicillin G, 100 µg/ml of streptomycin sulfate, and 50 µg/ml of gentamicin. Store the PBS-antibiotics—solution at +4 °C for up to 1 week.

4. 0.05 % trypsin solution. 0.5 g trypsin (DIFCO), 8.0 g NaCl, 0.4 g KCl, 0.35 g $NaHCO_3$, 1.0 g D(+)-glucose, 0.005 g phenol red sodium salt, 0.2 g EDTA in 1000 ml purified water. Store at –20 °C.

2.3 Enterovirus Typing by Partial VP1 Sequencing

Store all reagents at room temperature unless indicated otherwise.

1. RNeasy™ Mini Kit (Qiagen) for extraction of RNA from the virus isolates (*see* **Note 4**). Make the working solutions according to the manufacturer's instructions.

2. Oligonucleotide primers for the RT-PCR and sequencing reactions: 292-sense (5′-MIG CIG YIG ARA CNG G-3′) and 222-antisense (5′-CIC CIG GIG GIA YRW ACA T-3′) [5]. Store the stock solutions at –20 °C. Dilute to the 25 µM concentration for RT-PCR and to the 5 µM concentration for sequencing with RNase-free H_2O (Sigma) and store at –20 °C.

3. 10× RT-PCR buffer. 670 ml 1 M Tris–HCl pH 8.8, 170 ml 1 M $(NH_4)_2SO_4$, 0.120 ml 0.5 M EDTA pH 8.0, and 159.88 ml H_2O. Store at –20 °C.

4. Mixture of four dinucleotides (dATP, dCTP, dGTP, and dTTP; Roche). Make a solution, which has a 2 mM final concentration of each of the nucleotides in RNase-free H_2O (Sigma). Store at –20 °C.

5. 25 mM $MgCl_2$ (supplied with the Taq-polymerase, Integro BV). Store at –20 °C.

6. Enzymes for RT-PCR reactions: AMV reverse transcriptase, 20 U/µl (Finnzymes), RNasin Inhibitor-enzyme, 40 U/µl (Promega), and Taq polymerase-enzyme, 5 U/µl (Integro BV). Store at –20 °C.

7. Agarose powder (Lonza).

8. Ethidium bromide (EtBr; BDH Prolabo). One drop from the EtBr bottle into 50 ml of the gel solution gives an EtBr concentration of 0.5 µg/ml.

9. 10×TBE-buffer. 61.83 g Tris (Merck), 121.1 g boric acid (ICN), and 7.44 g disodium salt (Titriplex III, Merck), adjust to 1000 ml with distilled H_2O. Make the 1:10 (=1×TBE) working solution with distilled water.

10. 100 bp DNA size marker ladder (Generuler, BDH Prolabo). The product is delivered with 6× loading buffer. Make the working dilution of the DNA marker according to the manufacturer's instructions before use. Store at +4 °C.

11. QIAquick Gel Extraction Kit (QIAGEN) (*see* **Note 5**). Add 200 ml 99.5 % ethanol into 50 ml of Buffer PE before use.

12. PCR thermocycler, for example GeneAmp 9700 (Applied Biosystems), and tubes, strips, or 96-well plates for RT-PCR reactions, compatible with the PCR thermocycler (*see* **Note 6**).

13. Equipment needed for preparing the agarose gels and for their electrophoresis. The ultraviolet light supply connected to a camera for viewing and taking pictures of the gel.

3   Methods

3.1   Enterovirus Isolation

1. Add 10 ml PBS (supplemented with antibiotics), about 1 g glass beads (approx. 3 mm in diameter) and 1 ml chloroform to a 15 ml chloroform-resistant centrifuge tube. Transfer approximately 2 g of fecal specimen into the tube with a wooden spatula. Close the tube carefully and shake vigorously for 20 min either manually or using a mechanical shaker. Spin for 20 min at 1500×*g* in a refrigerated (+4 °C) centrifuge. Transfer the supernatant (the stool extract) into a storage tube (*see* **Note 7**). Store at −20 °C.

2. Use at least RD(A) cells for enterovirus isolation (*see* **Note 8**). Examine the cell cultures microscopically to be sure that the cells are healthy and the monolayer is confluent. Use two parallel cell monolayer tubes for one specimen. Use one tube of each cell type as a negative control. Remove the growth medium from the cells and replace it with 1 ml of warm maintenance medium (+36 °C). Inoculate each tube (excluding the negative control) with 0.2 ml of stool extract, close the caps, and incubate in the stationary position at +36 °C.

3. Examine the cultures with a microscope for the appearance of cytopathic effect (CPE) at least at every 2–3 days. The characteristic enterovirus CPE (rounded cells, which detach from the surface of the tube) develops usually within 2–5 days. When 75–100 % of the cells are affected, freeze the tube at −20 °C. Thaw the tube and do the second passage in order to enrich the virus by inoculating 0.2 ml into a tube containing

the same cell line that was used in the first inoculation and 1 ml of maintenance medium. When 100 % CPE is ready, store the tube at –20 °C. Use this second passage virus material for typing.

4. If no CPE appears after 7 days of the first inoculation, perform a blind passage by inoculating 0.2 ml of the first passage into two cell culture tubes and continue examination for a further 7 days. If no CPE is seen, the specimen is recorded as negative in enterovirus isolation (*see* **Note 9**).

3.2 Enterovirus Typing by Pools of Antisera

Each enterovirus isolate is tested in two dilutions against the pooled enterovirus antisera. The titer of the virus should be approx. 100 $CCID_{50}$, which is usually achieved by using 1:100 and 1:1000 dilutions of the virus. The exact procedure of the typing assay depends on the pools used; in principle, equal volumes of diluted virus and diluted antisera pools are mixed and incubated for neutralization. The cell suspension is added and the cells are incubated and observed microscopically. The specific antiserum prevents the growth of a given enterovirus and the absence of CPE is recorded as a positive result. The interpretation of the typing results is done according to the supplier's instructions.

1. Plan the pipetting schedule (antisera, virus dilutions, virus controls, cell controls, and virus titration) and label the 96-well flat-bottomed cell culture microtitre plate accordingly (*see* **Note 10**).

2. Add 50 µl diluted antisera to the appropriate wells. Add 50 µl maintenance medium to virus control wells. Add 100 µl maintenance medium to cell control wells.

3. Prepare 1:100 and 1:1000 dilutions of the unknown virus. Add 50 µl of either virus dilution to all wells containing antisera and to the virus control wells. Perform a titration of the virus on the same plate (*see* **Note 11**). Cover the plate with the lid and incubate for 1 h at +36 °C.

4. Detach the cell monolayer from the cell culture flask by using the 0.05 % trypsin solution (*see* **Note 12**). Prepare a suspension of 100,000–200,000 cells per 1 ml maintenance medium. Distribute 100 µl cell suspension into all wells of the typing plate. Cover the plate with a sealer tape and incubate at +36 °C.

5. Examine daily (or if not possible at 2 or 3 day intervals) by a microscope and record the CPE. Continue recording until 24 h after CPE in the virus control wells reaches 100 %. The virus is identified by the pattern of inhibition of CPE (negative wells have healthy cell monolayers) by antiserum pools (*see* **Note 13**).

3.3 Enterovirus Typing by Partial VP1 Sequencing

See first **Note 14** for good laboratory techniques.

1. Perform RNA extraction according to instructions given by the manufacturer of the extraction kit (*see* **Note 15**). If RNEasy (Qiagen) is used, elute the RNA into a volume of 40 μl (*see* **Note 16**). Store at −20 °C until used and at −70 °C for longer time.

2. Prepare the RT-PCR reaction mix in a volume sufficient for all reactions (calculate the number of samples and positive and negative controls) by mixing all reagents in one tube on ice. For one reaction, use 5 μl 10× RT-PCR-buffer, 5 μl 2 mM mixture of dNTPs, 3 μl 25 mM $MgCl_2$, 2 μl 25 μM sense primer 292, 2 μl 25 μM antisense primer 222, 0.3 μl RNasin Inhibitor (40 U/μl), 0.2 μl AMV-reverse transcriptase (20 U/μl), 0.3 μl Thermoperfect DNA polymerase (5 U/μl), and 31 μl H_2O. Mix thoroughly by pipetting up and down. Transfer 49 μl of the RT-PCR reaction mix into tubes or into wells of the PCR strip or plate.

3. Add 1 μl sample RNA, 1 μl positive control RNA, or 1 μl water (negative control) into the tubes or wells containing the RT-PCR mix. Mix by pipetting several times up and down. Seal the caps or cover the plate with a foil.

4. Start the RT-PCR program in the PCR thermocycler (+50 °C for 30 min, +94 °C for 3 min, 35 cycles of (+94 °C for 30 s, +42 °C for 30 s, and +72 °C for 60 s), +72 °C for 5 min, and +4 °C to hold). When the program has reached the first temperature (+50 °C), put the tubes or the strip into the PCR machine and let the program run until the final +4 °C temperature is reached.

5. Visualize the RT-PCR amplicons by using an agarose gel electrophoresis. Assemble the gel casting tray and put the sample comb into its place. Prepare the agarose gel by mixing 4.0 g agarose powder and 200 ml 1× TBE-buffer. Heat in the microwave oven until the solution almost boils. Move it into the fume hood and let the solution cool by gently mixing the bottle until the solution is about +60 °C. Add 4 drops of ethidium bromide; the final concentration of ethidium bromide should be 0.5 μg/ml. Roll the bottle slowly to mix the red EtBr thoroughly into the solution. Pour the gel solution into the casting tray and let it solidify at room temperature for about 30 min. Remove the comb and transfer the gel with its container into the electrophoresis chamber, which contains 1× TBE buffer (*see* **Note 16**).

6. Prepare the samples by mixing 5 μl RT-PCR product and 2 μl 6× loading buffer. Mix by pipetting or vortexing. Pipet the

samples into the wells of the agarose gel carefully, avoiding the formation of the air bubbles. Add 7 μl DNA ladder into the outermost wells of the gel. Run the gel at 140 mA for 20–60 min until the fastest color front has moved 3–4 cm from the sample wells. Stop the power supply. Remove the gel from the chamber and take a photo of the gel under the ultraviolet light. The expected size of the RT-PCR amplicon with the primers 292 and 222 is about 340 base pairs (*see* **Note 17**).

7. Purify the RT-PCR amplicons before sequencing (*see* **Note 18**). If only one band with a correct size is visible in the gel, use QIAquick PCR purification procedure. If more bands are present, the band of a correct size must be sliced from the gel using the QIAquick Gel Extraction procedure. Perform the purification according to the protocol presented in the manufacturer's handbook. The final volume of water in the DNA elution step is 40 μl.

8. Sequencing of the purified DNA amplicons may be performed in several ways depending on the facilities available in the laboratory, but, importantly, two sequencing reactions, one with the sense primer and one with the antisense primer, should always be performed for each of the DNA amplicons. We use 3 μl purified DNA amplicon and 1.6 μl 5 μM sense primer for one reaction and 3 μl DNA and 1.6 μl 5 μM antisense primer for the second reaction and add 2.4 μl water into each reaction to reach the final volume of 7 μl. The following steps (the BigDye terminator cycle sequencing reactions, the purification of the amplicons and the analysis of the amplicons on an automated capillary sequencer) are purchased as a full-service from another laboratory.

9. A software package (e.g., Geneious, created by Biomatters and available from www.geneious.com, or Sequencher from Gene Codes Corporation, available from www.genecodes.com) is needed for analysis of the sequence files. Check carefully the quality of the raw sequence data: the chromatograms should consist of clearly distinguishable sharp calls for each of the four bases. Remove several of the first and the last base calls from the sequences; these are usually not of high quality enough. Assemble the sense and antisense sequences to a contig, check for and edit the ambiguous base calls sometimes seen between the two sequences, and, finally, save the sequence in a text format. Compare the sequence to the GenBank nucleotide data base with BLAST [6]. The enterovirus prototype sequence, which poses the highest nucleotide identity with the query sequence, determines the type of the enterovirus. This identity should be more than 75 % at the nucleotide level (*see* **Note 19**).

4 Notes

1. Primarily, we use the wide-spectrum rhabdomyosarcoma, RD(A), cells and, if needed, increase the sensitivity of the assay with other cell lines such as HeLa, green monkey kidney (GMK), Vero, and/or CaCo-2. The ICAM-expressing HeLa cells are especially good for those species C enteroviruses that use ICAM-1 as their cellular receptor.

2. The cells can be seeded, grown, and inoculated also in 6-, 12-, or 24-well plates, but be aware of the risk for cross-contamination when handling multiple specimens at the same time.

3. The laboratories in the WHO Polio Laboratory Network use pooled horse antisera prepared at the National Institute of Public Health and the Environment (RIVM), Bilthoven, The Netherlands. This set of enterovirus typing antisera contains seven pools for the 20 most common echoviruses and Coxsackie virus A9, one pool for six Coxsackie B viruses and one trivalent pool for three poliovirus serotypes. Enterovirus typing pools are also commercially available. The compositions of antisera and thus the capability of typing the different enterovirus serotypes may vary in different pools. In addition, antisera for single enterovirus serotypes are available for example from the Swedish Institute for Infectious Disease Control (SMI).

4. There are many other traditional and commercially available methods, which can be used for the extraction of viral RNA as well. Typically, the virus isolates obtained by propagation in cell cultures contain so large amount of viruses that the sensitivity of the RNA extraction method is not a limiting factor in the assay.

5. The QIAQuick (Qiagen)-kit is used for purification of the RT-PCR products before sequencing. The corresponding methods are available from many manufacturers. We have good experience also on the PCR CheckIT gels from Elchrom Scientific.

6. Any thermocycler with adjustable temperatures and incubation times can be used. Tubes, strips or 96-well plates should be compatible with the PCR machine. We use 96-well plates from Thermo-Fast and close the plates with foil seals from Abgene.

7. Be careful not to take with the chloroform when pipetting the supernatant. The chloroform is toxic for the cells in the virus isolation step. If the supernatant is not clear after the centrifugation, repeat the chloroform treatment and the centrifugation.

8. RD(A) cells support growth of most of the enteroviruses. The sensitivity of the virus isolation assay can be increased by using

simultaneously more cell lines. However, the cell preferences of different enterovirus species, types or strains cannot be guaranteed. According to our experience, Coxsackie B viruses multiply rapidly and efficiently in HeLa cells, some but not all of the Coxsackie A viruses grow better in HeLa than in RD(A) cells, echoviruses grow well in RD(A) cells, but also in GMK and Vero cells, CaCo-2 cells support growth of many echoviruses, and enterovirus 71 multiplies slowly but still induces visible CPE both in Vero and in RD(A) cells. The L20B cells are more specific for polioviruses and they are used in the clinical (acute flaccid paralysis, AFP) surveillance for polioviruses.

9. Some enterovirus strains may need longer incubation times than 7 days for the full CPE. For example, enterovirus 71 produces CPE slowly in Vero cells and the CPE should be monitored for up to 10 days. If the 75–100 % CPE is not achieved before the cells begin to degenerate due to their age, do the third passage in freshly seeded cells. Some enteroviruses do not infect/produce visible CPE in the laboratory cell lines. The clinical specimen may be positive in diagnostic enterovirus PCR while being negative in the virus isolation.

10. You can find an example for the working chart from the Polio Laboratory Manual [4].

11. The titer of the unknown virus should be about 100 $CCID_{50}$/ well. If the titer is lower, the negative result (no CPE) is not reliable. If the titer is higher, the amount of antisera may not be sufficient for total neutralization (CPE appears).

12. The trypsinization time depends on the cell line; for RD(A) the time is approximately 30 s.

13. Sometimes the typing assay fails to identify the analyzed virus. The virus may be newly discovered or new and thus the specific antiserum for a given virus is not included in the antisera pools. Alternatively, the virus may have drifted antigenically and the sera fail to identify it. The virus isolate may contain a mixture of two or more enterovirus types, which interferes with the interpretation of the results. The virus may be in an aggregated form which prevents the neutralization by specific antisera; in this case the isolate can be retested after dissolving the aggregates by the chloroform treatment.

14. We use four separate laboratory rooms for the different phases of the procedure to avoid contamination. The virus RNA extraction is done in the biosafety level 2 laboratory; the reagents and mixtures for the RT-PCR are prepared in a clean pre-PCR laboratory, the RNA template is added in the third laboratory and, finally, the RT-PCR products are handled in the post-PCR laboratory. All laboratory safety cabins are cleaned thoroughly before and after work. The filter tips are used in all

pipets. The most obvious contaminants are infectious viruses, which readily spread via aerosol, and RT-PCR products, which contain a lot of DNA and can be spread for example by aerosol or contaminated gloves.

15. We always include a negative control (nuclease-free water) in the extraction step. This control is analyzed in the RT-PCR and it should not give any visible amplicons on an agarose gel.

16. Be careful when extracting and handling RNA, because RNA is readily degraded by RNases. Be sure that all the reagents and plastic ware are RNase-free.

17. The electrophoresis buffer should rise about 0.5 cm above the gel. The buffer should be changed once a week. If the wells contain air bubbles, remove them for example with the pipet tip.

18. If no enterovirus amplicons (size of approx. 340 base pairs) are seen, reproduce the RT-PCR with a second set of pan-enterovirus primers, 040, 011, and 012 [7]. The enterovirus amplicon obtained with these primers has a size of approx. 450 base pairs.

19. Because of the extensive genetic evolution some currently circulating enterovirus strains of species C may have less than 75 % nucleotide identities with the prototype strains [8]. In these cases, the typing can be based on the most closely matching enterovirus isolate sequences in the GenBank. If the query sequence is not closely related to any of the GenBank sequences (nucleotide identities <75 % and amino acid identities <85 %), you may have succeeded in isolating a new enterovirus not characterized before. If you suspect a new enterovirus type, please contact the Picornavirus Study Group (www.picornastudygroup.com) for further instructions.

References

1. Smura T, Savolainen-Kopra C, Roivainen M (2011) Evolution of newly described enteroviruses. Future Virol 6:109–131
2. Oberste MS, Maher K, Kilpatrick DR et al (1999) Molecular evolution of the human enteroviruses: correlation of serotype with VP1 sequence and application to picornavirus classification. J Virol 73:1941–1948
3. Blomqvist S, Paananen A, Savolainen-Kopra C et al (2008) Eight years of experience with molecular identification of human enteroviruses. J Clin Microbiol 46:2410–2413
4. World Health Organization (2004) Polio laboratory manual, 4th edn. WHO, Geneva, Switzerland
5. Oberste MS, Nix WA, Maher K et al (2003) Improved molecular identification of enteroviruses by RT-PCR and amplicon sequencing. J Clin Virol 26:375–377
6. Altschul SF, Gish W, Miller W et al (1990) Basic local alignment search tool. J Mol Biol 2215:403–410
7. Oberste MS, Maher K, Kilpatrick DR et al (1999) Typing of human enteroviruses by partial sequencing of VP1. J Clin Microbiol 37:1288–1293
8. Brown BA, Maher K, Flemister MR et al (2009) Resolving ambiguities in genetic typing of human enterovirus species C clinical isolates and identification of enterovirus 96, 99 and 102. J Gen Virol 90:1713–1723

Chapter 4

Isolation and Characterization of Poliovirus in Cell Culture Systems

Bruce R. Thorley and Jason A. Roberts

Abstract

The isolation and characterization of enteroviruses by cell culture was accepted as the "gold standard" by clinical virology laboratories. Methods for the direct detection of all enteroviruses by reverse transcription polymerase chain reaction, targeting a conserved region of the genome, have largely supplanted cell culture as the principal diagnostic procedure. However, the World Health Organization's Global Polio Eradication Initiative continues to rely upon cell culture to isolate poliovirus due to the lack of a reliable sensitive genetic test for direct typing of enteroviruses from clinical specimens. Poliovirus is able to infect a wide range of mammalian cell lines, with CD155 identified as the primary human receptor for all three seroytpes, and virus replication leads to an observable cytopathic effect. Inoculation of cell lines with extracts of clinical specimens and subsequent passaging of the cells leads to an increased virus titre. Cultured isolates of poliovirus are suitable for testing by a variety of methods and remain viable for years when stored at low temperature.

This chapter describes general procedures for establishing a cell bank and routine passaging of cell lines. While the sections on specimen preparation and virus isolation focus on poliovirus, the protocols are suitable for other enteroviruses.

Key words Poliovirus, Enterovirus, Virus isolation, Cytopathic effect, Cell culture, Continuous cell lines, Cryopreservation of mammalian cells, Cell bank, Extraction of clinical specimens, Immunofluorescence

1 Introduction

The devastating impact of worldwide poliomyelitis outbreaks during the twentieth century led to intensive study of the causative agent, poliovirus. The description by Enders, Weller and Robbins in 1949 of the propagation of poliovirus type 2 Lansing strain in non-neural tissue was a breakthrough, not only for research into poliomyelitis but also for diagnostic virology [1]. The propagation of poliovirus was confirmed by the induction of paralysis in mice and monkeys after intracerebral inoculation of passaged cultures and neutralization of poliovirus activity by antiserum specific for the Lansing strain. The following year the trio published their findings that cytopathogenic effect, later termed cytopathic effect (CPE), of poliovirus

Javier Martín (ed.), *Poliovirus: Methods and Protocols*, Methods in Molecular Biology, vol. 1387,
DOI 10.1007/978-1-4939-3292-4_4, © Springer Science+Business Media New York 2016

propagation could be observed microscopically as degeneration of tissue cultures [2]. In a subsequent series of papers, the three authors refined the procedures to cultivate poliovirus in tissue culture from stool specimens using antibiotics and roller cultures that obviated confirmation of virus propagation by more complex techniques involving animals [3, 4]. In 1954, the three authors received The Nobel Prize in Physiology or Medicine for this seminal work.

Further milestones in the history of cell culture procedures occurred in the 1950s, including the introduction of trypsin to generate single cell suspensions for subculture, the establishment of the first human cell line, HeLa, precluding the reliance on primary cell culture, and the development of defined culture media [5]. All of these events were necessary for the large scale propagation of poliovirus in vitro and enabled the development of various polio vaccines, with the inactivated and live-attenuated products developed by Salk and Sabin, respectively, the most well-known.

The World Health Assembly established the Global Polio Eradication Initiative in 1988. The World Health Organization (WHO) polio eradication program is built upon maintaining high vaccine coverage with the Sabin oral polio vaccine to stop person-to-person transmission of poliovirus, surveillance for cases of acute flaccid paralysis (polio-like illness) in children and the testing of stool specimens from children presenting with acute flaccid paralysis by virus culture in a laboratory accredited by WHO. Virus culture of stool specimens presents challenges due to toxicity and an abundance of other microorganisms. Furthermore, non-polio enteroviruses can be isolated by culture from up to 20 % of stool specimens of acute flaccid paralysis cases, depending on the geographical location, and their cytopathic effect in virus culture cannot be distinguished from that caused by poliovirus [6, 7]. All cell culture isolations of enterovirus, including poliovirus, require confirmation by techniques such as immunofluorescence, antisera neutralization, ELISA, and reverse transcription polymerase chain reaction (RT-PCR).

Continuous mammalian cell lines should be ordered from a reliable source that can provide evidence of cell authentication [8]. With careful handling and use of appropriate equipment, cell lines can be stored for many years in liquid nitrogen and revived for routine passaging as required. The procedures in this chapter describe the receipt of cell lines, cell banking, specimen extraction for virus culture and virus isolation for poliovirus and other enteroviruses.

2 Materials

1. Cell lines. While poliovirus types 1, 2 and 3 can be cultured in a wide variety of cell lines, the WHO recommends use of the L20B and RD mammalian cell lines:

   (a) L20B. Mouse fibroblast L-cells genetically engineered for cell surface expression of the human poliovirus receptor,

CD155 [9]. The cell line facilitates the isolation of poliovirus in preference to non-polio enteroviruses but some strains of Coxsackie virus A (CV-A), in particular CV-A8 and CV-A10, are capable of growth in L20B with a characteristic enterovirus cytopathic effect (CPE) [10, 11]. Adenoviruses and reoviruses are also capable of infecting L20B with a CPE that is distinct from enteroviruses.

(b) RD. The human rhabdomyosarcoma (RD) cell line has been utilized for the growth of a wide range of enteroviruses. RD cells are particularly useful for the isolation of type A Coxsackie viruses and echoviruses but strains of type B Coxsackie virus grow poorly. Polioviruses can grow to a titre of greater than 10^8 $CCID_{50}/50$ μl in the RD cell line. The suffix "-A" in RD-A denotes the cell line was sourced from a member of the WHO Global Polio Reference Laboratory network.

2. Cell culture media (*see* **Note 1**). Poliovirus growth in continuous cell lines may be in an open system with an atmosphere that includes CO_2 or in a closed system with the lid of the flasks and tubes tightened. Complete medium consists of Eagle's minimum essential medium (MEM) as the base constituent and is supplemented with serum, balanced salts (Earle's or Hank's), buffering agents (sodium bicarbonate and HEPES), L-glutamine (an essential amino acid), and antibiotics (Tables 1 and 2). Growth medium includes 10 % fetal bovine serum as a source of nutrients for metabolism and cell division. The proportion of fetal bovine serum is reduced to 2 % in holding or maintenance media for sustenance while limiting cellular division. The specific media requirements of cell lines other than L20B and RD should be confirmed to optimize growth of both the cell line and virus.

Table 1
Eagle's complete MEM for poliovirus culture in the presence of CO_2 (open system)

Component	Growth medium	Maintenance medium
Eagle's minimum essential medium (Earle's salts with phenol red, no bicarbonate)	83.5 ml	90.5
L-glutamine (200 mM)	1.0 ml	1.0 ml
Fetal bovine serum (heat inactivated)	10.0 ml	2.0 ml
Sodium bicarbonate, 7.5 % solution	3.5 ml	4.5 ml
HEPES 1 M	1.0 ml	1.0 ml
Penicillin (1×10^4 U/ml)/streptomycin (10 mg/ml) solution	1.0 ml	1.0 ml
Total volume	100 ml	100 ml

Table 2
Eagle's complete MEM for poliovirus culture in the absence of CO$_2$ (closed system)

Component	Growth medium	Maintenance medium
Eagle's minimum essential medium (Earle's salts, no bicarbonate)	85.5 ml	92.5
L-glutamine (200 mM)	1.0 ml	1.0 ml
Fetal bovine serum (heat inactivated)	10.0 ml	2.0 ml
Sodium bicarbonate, 7.5 % solution	1.5 ml	2.5 ml
HEPES 1 M	1.0 ml	1.0 ml
Penicillin (1 × 10^4 U/ml)/streptomycin (10 mg/ml) solution	1.0 ml	1.0 ml
Total volume	100 ml	100 ml

3. Sterile cell culture flasks: 25 cm^2, 75 cm^2, 175 cm^2.

4. Sterile tubes with externally threaded screw capped lids:

    (a) Polypropylene 1.5 ml, 50 ml.

    (b) Polycarbonate 5 ml.

    (c) Cryogenic 1.8 ml ampoules.

    (d) Sterile cell culture tubes.

5. Sterile nontoxic polyethylene transfer pipettes.

6. Sterile serological pipettes: 5, 10, and 25 ml.

7. Hemocytometer (e.g., improved Neubauer) and glass coverslips.

8. Gloves. Examination gloves are made from latex or nitrile that can be of varying quality. Select a brand that will not easily tear and that does not cause skin irritation from repeated use.

9. Cell freezing container. Insulated containers designed to facilitate the freezing of cell lines at a controlled rate are available.

10. Sterile wooden tongue depressor.

11. Laboratory sealing film.

12. Glass beads of 2 and 5 mm diameter, used to process postmortem and stool specimens, respectively. The beads should be washed and dried before use.

13. Microscope slide staining jar.

14. Anti-poliovirus blend mouse antibodies and anti-mouse IgG-FITC labeled secondary antibody.

15. Equipment:

   (a) Variable volume pipettes and filter tips: 2–20 and 20–200 μl.

   (b) Inverted microscope.

   (c) Refrigerated bench centrifuge with buckets suitable for cell culture tubes and 50 ml tubes.

   (d) Biological safety cabinet class II.

   (e) Liquid nitrogen storage container and personal protective equipment for cryogenic procedures.

   (f) Refrigerator, –20 and –80 °C freezers.

   (g) Bench top vortex.

   (h) Mechanical shaker.

   (i) Warm air incubator or CO_2 incubator with shelving and a mechanical roller.

   (j) Plastic racks to hold tube cultures.

   (k) Hand-held pipette aid for serological pipettes.

16. Solutions:

   (a) Eagle's minimum essential medium (Earle's or Hank's salts with phenol red, no bicarbonate).

   (b) L-glutamine (200 mM). An essential amino acid that is unstable at 36 °C.

   (c) Fetal bovine serum (FBS) (*see* **Note 2**).

   (d) Sodium bicarbonate (7.5 %).

   (e) HEPES (1 M). Provides additional buffering capacity to sodium bicarbonate at physiological pH.

   (f) Penicillin (10^4 U/ml) and streptomycin (10 mg/ml).

   (g) Sterile phosphate buffered saline (PBS) solution. Available commercially with (complete PBS) and without (incomplete PBS) calcium and magnesium ions. Complete PBS is used for specimen extraction with the divalent cations stabilizing the enterovirus particles. Incomplete PBS is used when passaging cell lines to avoid the divalent cations interfering with the activity of the EDTA–trypsin solution.

   (h) Trypsin–EDTA solution.

   (i) Ethanol.

   (j) Chloroform. Stabilized with ethanol and use within 2 years of opening.

   (k) Trypan blue stain (0.4 %).

   (l) Dimethyl sulfoxide (DMSO). Use within 3 months of opening. Multiple ampoules of smaller volume can be purchased rather than a larger bottle.

(m) Acetone.

(n) Glycerol.

(o) Disinfectant: a number of commercial disinfectants are available that have proven virucidal activity. Alcohols do not inactivate poliovirus.

3   Methods

3.1   Establishing a Cell Bank for the Isolation of Poliovirus Using Continuous Mammalian Cell Lines

The proper handling and storage of continuous cell lines is important to maintain their sensitivity to poliovirus infection (*see* **Note 3**). Cell lines should be stored either submerged or within the vapor phase of liquid nitrogen to retain long-term viability. Storage in a −80 °C freezer is not recommended. All handling and manipulations with the cell lines described in this chapter should be performed within a class 2 biological safety cabinet (BSC). While a cell culture laboratory is considered a "clean" environment when compared to a virus isolation laboratory that processes clinical specimens, many cell culture reagents such as serum, trypsin and the cell lines are derived from animal sources and thus, all items should be considered as potentially infectious. Appropriate personnel protective equipment should be worn when handling cell culture material.

Cell lines should be received from a reliable commercial or laboratory source. The supplier should provide the history of the cell line indicating its original source, test results for the presence of adventitious agents such as mycoplasma, the culture media and supplements recommended for growth and maintenance of the cell line and the seeding level for routine subculture or passage. The supplier should also specify the absolute passage number and confluency of the cells at the time of shipment, the culture media in the flask and whether the cells were grown in an open or closed system.

Continual passage of cell lines will result in genotypic and phenotypic variation from the cell line originally received. The intention when establishing a new cell bank is to expand the number of cells exponentially for storage of multiple ampoules in liquid nitrogen but in as few passages as possible so as to retain the characteristics of the cell line. Once a cell line has been revived from liquid nitrogen it should only be passaged a set number of times to retain the cell phenotype and also minimize the risk of contamination with mycoplasma or other agents through repeated handling. The number of ampoules required for storage in liquid nitrogen will depend upon how routinely the cell line will be used. WHO recommend that L20B and RD cell lines should not be passaged more than 15 times or beyond 3 months after revival from liquid nitrogen, whichever occurs first [12]. For example, if a cell line is passaged once a week, a fresh ampoule should be revived every 3 months or 12 passages. This means four ampoules will be required every year and preparation of a cell bank of 50 ampoules will last for more than 7 years. If a cell bank needs to be replenished it will require two to three passages to expand the number of cells and the master cell bank passage number will be further from the original as a consequence.

1. Upon receipt of the cell line from the sending laboratory, inspect the flask for cracks or evidence of leakage and examine the condition of the cells microscopically (*see* **Note 4**). Record the morphology and confluency, which should also be reported to the sender.

2. Incubate the flask overnight at 36 °C to acclimatize the cells.

3. Replace the culture media the next day according to the confluency: <75 % use growth media, >75 % use maintenance media. Return the flask to the incubator until the confluency is >80 %, but do not allow to reach 100 % so as to maintain exponential cell growth, at which point the cells are passaged.

4. Passage the cells by preparing a single cell suspension after incubation with trypsin–EDTA, which detaches cells from the surface of the flask. Tip off the culture media into a waste discard container and wash the monolayer twice with incomplete PBS that does not contain the divalent cations, magnesium and calcium (*see* **Note 5**).

5. Add sufficient trypsin–EDTA to just cover the cell monolayer; approximately 0.5 ml for a 25 cm^2 flask and 1 ml for a 75 cm^2 flask. Gently roll the flask from side to side to ensure the trypsin–EDTA covers the cells and place in a 36 °C incubator to assist the enzymatic activity of trypsin. Remove the flask from the incubator every minute to observe the cell monolayer under a microscope. Each time return the flask to the incubator until the cells round up and start to detach from the surface (*see* **Note 6**).

6. Add 10 ml of growth media and detach any remaining cells by repeatedly flushing the surface gently using a transfer pipette. Transfer the cell suspension to a 50 ml tube and disperse cell aggregates by gently pipetting up and down with a transfer pipette.

7. Perform a cell count using a hemocytometer (*see* **Note 7**).

   (a) Clean the surface of the counting chamber and the glass coverslip with ethanol to remove any grease or dust and polish dry. Place the coverslip over the two sets of gridlines on the chamber. Dilute 0.2 ml of the single cell suspension with 0.2 ml of Trypan blue and mix well (*see* **Note 8**).

   (b) Using a micropipette with a narrow bore tip, gently expel 10 μl of the dye–cell suspension to one side of the counting chamber by touching the tip to the underside of the coverslip. Remix the dye–cell suspension to ensure independent sampling and load 10 μl to the other side of the counting chamber.

(c) Record the number of viable (unstained) cells in one half of the hemocytometer and repeat the counting procedure for the second half (*see* **Note 9**). Determine the mean of the two cell counts. Check that the individual counts are within 20 % of the mean, to indicate accurate sampling and prepare a fresh Trypan blue cell suspension if they are not.

(d) The viable cell concentration per ml is determined by the formula:

$$iCC = t \times df \times \frac{1}{4} \times 10^4$$

iCC = initial cell concentration per ml

t = total cell count of both sides of the counting chamber

df = dilution factor of cell suspension with Trypan blue

$\frac{1}{4}$ = correction factor for mean number of cells per corner grid

10^4 = conversion factor of counting chamber

Example: t = 480; df = 2; iCC = $480 \times 2 \times \frac{1}{4} \times 10^4 = 2.4 \times 10^6$ per ml.

8. Determine the dilution (d) required to seed the new flasks by dividing the working cell concentration (wCC) by the initial cell concentration (iCC) from **step 6d**.

wCC = 2×10^5 cells per ml

Example: $d = 2 \times 10^5 / 2.4 \times 10^6$ per ml = 1/12

The working cell concentration is prepared by adding one ml of the cell suspension to a new flask and diluting with 11 ml of growth media.

9. Seed as many flasks as required with the working cell concentration, add the appropriate volume of growth media and incubate at 36 °C. It is preferable to expand the number of cells by using large cell culture flasks (175 cm²) to minimize the number of passages required to establish the cell bank.

10. Assess the confluency of the cells by microscopic observation each day. When the confluency is 80–90 %, detach cells from the flasks with trypsin–EDTA. Perform a cell count and determine the total number of cells harvested from the flasks by multiplying the cell concentration by the volume of the cell suspension. Determine the number of ampoules that can be cryopreserved at a cell concentration of 5×10^6 cells per ml (*see* **Note 10**). If the number of ampoules is less than required seed a further set of cell culture flasks and repeat **step 8**.

11. Cryopreservation of cell lines. The general rule for the cryopreservation of continuous cell lines is to freeze slowly, thaw quickly and dilute slowly. A cryoprotectant such as dimethyl sulfoxide (DMSO) is required to prevent ice crystal formation that would rupture cells during freeze-thawing. DMSO is water mis-

cible and a powerful solvent that readily penetrates cell membranes and also human skin. Chemical-resistant gloves or two layers of examination gloves should be worn when handling DMSO and immediately changed and hands washed if the chemical is spilt on the gloves. DMSO is hygroscopic (absorbs moisture from the air) and should be used within 3 months of opening. It is recommended to purchase individual ampoules of a small volume (10 ml) rather than a single large bottle. The serum concentration in the growth medium used for cryopreservation of cells can be increased to 20 % to assist with cell recovery.

(a) Label the number of ampoules required for cryopreservation, determined at **step 9**, with the cell line, passage number, and date of cell banking. Ensure the ampoules are permanently labeled either with an ethanol-resistant marking pen by hand or with cryogenic labels made for the purpose. Uncap the ampoules and stand upright in a rack.

(b) Prepare sufficient freezing mix for the total volume of cells available consisting of growth media with 20 % serum and supplemented with 10 % DMSO. Chill the freezing mix prior to use.

(c) Centrifuge the cell suspension at $1000 \times g$ for 5 min. Immediately discard the supernatant and drain the residual media on tissue. Cap and hold the top of the tube with one hand while flicking the bottom of the tube hard with the other hand to dislodge the cell pellet.

(d) Add the freezing mix and gently swirl the cells checking that they are uniformly suspended and, if not, gently mix with a transfer pipette.

(e) Add 1 ml of cell suspension in freezing mix per labeled ampoule and securely cap the tube. The ideal cooling rate for the cells is 1 °C/min, which can be achieved in a number of ways. Specially designed apparatus that fit within the neck of a liquid nitrogen container can be used to lower the cells through the gaseous phase at a controlled rate. Insulated containers with either sealed coolant or requiring the addition of alcohol are available. Alternatively, the cell cooling rate can be slowed by insulating the cells within polystyrene foam and a towel. Do this by placing the tubes in a polystyrene foam rack and cover with a second rack. Tape the racks together and wrap in a large, thick bath towel and secure by running tape around the bundle ensuring the ampoules are kept upright during the process. Using either commercial containers or racks wrapped in towels to control the cooling rate, transfer the cells to a −20 °C freezer for 2 h, followed by overnight equilibration to −80 °C. The next day remove the ampoules from the insulated material and transfer to liquid nitrogen (*see* **Note 11**).

(f) Within a week, retrieve a single ampoule to assess the viability of the cells stored in liquid nitrogen. If the cell recovery is low a new cell bank should be prepared.

(g) Maintain an accurate and up to date inventory of the cell type, number of ampoules and location stored in liquid nitrogen.

12. Retrieval of cell lines from liquid nitrogen. Cells must be thawed quickly to prevent intracellular ice crystal formation leading to cell death. Have everything in readiness in the BSC before retrieving the cells from liquid nitrogen.

(a) Label the cell culture flasks. Have a plastic beaker with warm water and a spray bottle of 70 % ethanol on hand.

(b) Retrieve the ampoules from liquid nitrogen checking the label to confirm the details and later update the inventory.

(c) Transfer to the cell culture laboratory in a plastic container rather than a glass beaker in case liquid nitrogen entered the ampoule and ruptures on warming.

(d) In the BSC, hold the ampoule and rapidly move the frozen cells backwards and forwards in the beaker of warm water.

(e) Place the ampoule with the thawed cells on tissue and spray the outside of the ampoule with 70 % ethanol.

(f) Open the ampoule and transfer the thawed cells to a labeled flask.

(g) Add growth media to the cells dropwise while gently moving the flask backwards and forwards to facilitate dilution of the cells. The gradual dilution of DMSO avoids osmotic shock that will reduce cell recovery. The rate of addition can be slowly increased once more than 2–3 ml of media has been added to the flask. The cells should be diluted in a final volume 10–20 times that stored in liquid nitrogen to reduce the cytotoxic effects of DMSO.

(h) Cap the flask and incubate at 36 °C.

(i) The next day, observe the cell morphology under a microscope and record the confluency of the monolayer. Discard the media containing the diluted DMSO and replace with fresh growth media in readiness for subculturing and routine use.

13. Mycoplasma testing. Mycoplasma are ubiquitous intracellular microorganisms of human and animal origin that cannot be observed by microscopy but can modify a cell's phenotype. Continuous cell lines can become infected with mycoplasma

from the operator, other cell lines and reagents from animal sources such as serum and trypsin. Regular screening of routine cell cultures, at least once per revival from liquid nitrogen, should be performed by microbiological culture, fluorescent staining, PCR, or ELISA using either published protocols or commercial kits. Cell lines positive for mycoplasma should be discarded.

3.2 Routine Passaging of Continuous Mammalian Cell Lines and Preparation of tube Cultures for Virus Isolation

After retrieval from liquid nitrogen stocks, cell lines are maintained in flasks and used to prepare tube cultures for virus isolation (*see* **Note 12**). To reduce genotypic and phenotypic variation of the cell line from that originally supplied and the likelihood of contamination by mycoplasma or other agents, do not passage more than 15 times or beyond 3 months after retrieval from liquid nitrogen, whichever occurs first. Depending on the number of specimens received for processing, the tube cultures are prepared weekly or twice weekly.

1. Microscopically assess the confluency of the cell monolayer in flasks. The confluency of the monolayer should be less than 90 % so that the cells are within the exponential growth phase and should not overgrow before the next passage.

2. Remove the cells from the flask with trypsin–EDTA and perform a cell count following **steps 4–7** in Subheading 3.1.

3. Seed the required number of flasks with cell suspension in growth media according to Table 3. Record the passage number on the flask (*see* **Note 13**). Incubate the flasks at 36 °C until the cell monolayer is approximately 80 % confluent after 5–7 days when the growth media should be replaced with maintenance media unless the cells will be split on that day.

4. Seed the required number of culture tubes for poliovirus isolation according to Table 3. The seeding level should be adjusted based on experience to achieve a cell confluency of approximately 75 % after 48 h in growth media.

Table 3
Volumes and seeding levels for culture of the L20B and RD cell lines

Cell culture container	Volume (ml)	Seeding level (total cell number)
Tube: 125 × 16 mm	1	1×10^5
Small flask: 25 cm²	10	1×10^6
Medium flask: 75 cm²	30	2.5×10^6
Large flask: 175 cm²	75	5×10^6

Determine the dilution (d) required to seed the tubes by dividing the working cell concentration (wCC) by the initial cell concentration (iCC) determined by the cell count.

wCC = 1×10^5 cells per ml

Example: $d = 1 \times 10^5 / 2.4 \times 10^6$ per ml = $1/24$

The working cell concentration is prepared by adding one ml of the cell suspension to a new flask and diluting with 23 ml of growth media.

5. Uncap the cell culture tubes and aliquot 1 ml of the cell suspension into each tube using a sterile graduated transfer pipette.

6. Recap the culture tubes and place in a rack with the orientation marker turned to the top (*see* **Note 14**).

7. Place the tube rack in a 36 °C incubator.

8. Once the cells reach the required confluency change the culture media in the tubes with maintenance media and again before inoculation with specimen extract, unless this is on the same day. The tubes may be used for virus isolation for 5–7 days after the original seeding.

3.3 Preparation of Clinical Specimens for the Isolation of Poliovirus

The main mode of poliovirus transmission is via the fecal–oral route and the preferred type of specimen for virus isolation is stool. While poliovirus can be isolated from throat swabs and nasopharyngeal aspirates for up to two weeks early on in the infection, the virus is shed for up to 6 weeks in stools. To optimize the isolation of poliovirus, it is recommended to collect two stool specimens more than 24 h apart—due to intermittent virus shedding—and within 14 days of the onset of symptoms when the virus titre is greatest. While it is possible to isolate poliovirus from rectal swabs, whole stool specimens are preferred as the quantity of virus, and hence test sensitivity, is likely to be greater within a whole specimen.

A period of viraemia can lead to replication of poliovirus in the central nervous system and present as paralytic poliomyelitis, the clinical condition the general public most associate with the virus. Despite an association with neurological disease, poliovirus isolation from CSF is uncommon and considered not worth attempting if stool specimens are available. Poliovirus infection can be fatal, especially with bulbar paralysis involving the respiratory muscles, and can also present as encephalitis and meningitis. This section describes the preparation of stool specimens, swabs, and postmortem tissue for poliovirus culture.

1. Turn on the class II BSC and allow the airflow to equilibrate. Maintain the minimal number of items required for the procedure in the cabinet. Never place items across the front grill and do not make rapid movements within the cabinet or when

moving items into and out of the cabinet as this disrupts the airflow curtain that provides protection to the worker. Always manipulate fluids slowly and gently to avoid creating aerosols.

2. Prepare a container for disposal of potentially infectious waste (*see* **Note 15**).

3. Organize the work area so that sterile reagents and samples are to one side of the cabinet, waste container to the other and specimens to be processed located centrally.

4. Cabinets should be periodically tested for filter integrity (e.g., oil mist test) and operator protection (e.g., potassium iodide release test). Testing should conform to international standards such as AS 2252 (Australia), EN 12469 (European), JIS K 3800 (Japan), and NSF 49 (USA) and should be repeated when the cabinet is moved or the laboratory layout altered, particularly involving changes to the air flow within the room.

5. Clean and decontaminate the cabinet inner surfaces (both horizontal and vertical) after every working session and periodically (e.g., once per month) decontaminate and clean the tray under the BSC working surface.

6. Preparation of stool specimens. Ideally 5 g of stool specimen should be collected and 2 g (approximately the size of a small thumbnail joint) processed for virus culture (*see* **Note 16**).

   (a) Label the original specimen containers and individual 50 ml tubes with unique identifying codes. The use of 50 ml tubes is recommended to ensure adequate mixing of the extract. Also label two 5 ml tubes for long-term storage of the stool extract at **step (j)**.

   (b) Add 10 ml complete PBS, approximately 1 g of 5 mm glass beads and 1.0 ml chloroform per tube and reseal the caps (*see* **Note 17**). If 1 g or less of stool specimen is received, the quantities may be halved to avoid dilution of any viruses present; use 5 ml complete PBS, 0.5 g 5 mm glass beads, and 0.5 ml chloroform.

   (c) Working with one specimen at a time, transfer approximately 2 g of stool specimen to a labeled tube with a sterile wooden tongue depressor. The tongue depressor may be broken in half lengthways and used for two specimens. Ensure that the unique identifying code of the original sample matches the code on the 50 ml tube.

   (d) Retain the remaining original specimen for storage at −20 or −80 °C.

   (e) Close the tube securely and wrap laboratory sealing film in a clockwise direction around the lid to prevent leakage and loosening of the cap.

(f) Vortex the specimen extract for 1 min to assist with dispersal of the specimen.

(g) Shake tubes vigorously using a mechanical shaker for 20 min (*see* **Note 18**).

(h) Ensure the caps of the 50 ml tubes are secure before transferring to centrifuge buckets that can be sealed with an aerosol-resistant lid. Spin for 20 min at $1500 \times g$ in a refrigerated centrifuge at 4 °C.

(i) Transfer the centrifuge buckets to the BSC and open the aerosol-resistant lids. Check for leakage and decontaminate all surfaces if necessary.

(j) Working with one tube at a time, transfer the upper aqueous phase into two pre-labeled 5 ml storage vials from (**a**). Avoid disturbing the interface and the lower phase containing the chloroform. If the upper phase appears turbid or a well-defined interface was not observed after centrifugation, transfer the upper phase to a clean 50 ml tube, relabel and repeat the extraction by adding 1 ml chloroform and follow **steps** (**e**)–(**j**).

(k) Store one tube of stool extract at –20 °C as a backup and the other at 4 °C ready to inoculate onto cell monolayers.

(l) Discard the biological waste material, remove items from the BSC and decontaminate the work surface with disinfectant.

7. Preparation of swabs and postmortem specimens. Poliovirus can be isolated from throat, nose and rectal swabs, although they are considered not to be ideal specimens for virus isolation due to lower virus yield compared to whole stool and the shorter duration of virus replication in nasal and pharyngeal tissue. Postmortem specimens such as brain, bowel, and spinal cord may be referred from fatal cases of suspected poliovirus infection. A simple extraction method for clinical specimens is to pulverize the tissue or swab with 2 mm glass beads rather than using mortar and pestle.

(a) Working in a BSC, place postmortem tissue in a sterile petri dish and slice into small pieces with a sterile scalpel. Transfer the tissue slices to a small sterile bottle containing glass beads of 2 mm diameter and 2 ml complete PBS and secure the lid. For swabs, place the item in a bottle containing 2 mm glass beads and snap the shaft so that the swab fits within the bottle. Add 2 ml complete PBS and secure the lid.

(b) Vortex the bottle vigorously for 3 min.

(c) Transfer the liquid to a labeled 1.5 ml tube and centrifuge for 1 min at $15,000 \times g$.

(d) Transfer the supernatant to a second labeled 1.5 ml tube and store at 4 °C until ready to inoculate the cell monolayers and at −20 or −80 °C for long-term storage.

3.4 Virus Isolation

The three serotypes of poliovirus grow readily in a wide variety of continuous mammalian cell lines but the WHO recommends using L20B and RD, which are sensitive to poliovirus infection and enables the standardization of techniques and comparison of results between laboratories.

Poliovirus replication produces an observable cytopathic effect in sensitive cell lines that is characteristic of enteroviruses.

1. Microscopically examine recently prepared tube cultures to ensure the cell monolayer has the appearance of normal morphology and is approximately 75–90 % confluent. Refer to Subheading 3.2.

2. Discard the media and replace with 1 ml maintenance media.

3. Label two tubes each of RD and L20B cells with a unique code for each specimen extract to be inoculated and indicate the history of the virus culture (*see* **Note 19**). The use of two tubes per cell line will increase the sensitivity of poliovirus isolation. Label an uninoculated negative control tube of each cell line with the passage number and the date of virus culture (*see* **Note 13**). Ensure labeling will not obscure the monolayer when observed under a microscope.

4. Place the labeled cell culture tubes upright in a rack and in the same order as the specimen extract tubes in another rack.

5. Working with one specimen at a time, remove the cap of the specimen extract tube and the caps of the appropriately labeled culture tubes. Using a micropipette and plugged filter tips, inoculate each culture tube with 0.2 ml of specimen extract (*see* **Note 20**). Re-cap all tubes and repeat for each specimen.

6. Place the inoculated L20B cells and negative cell control tube in a stationery rack, the same as used for the seeding of the cell line. Place the RD cells and negative cell control tube in a tube-holder roller drum assembly that turns at approximately 0.2 rpm. Both the virus culture rack and drum should be inclined approximately 5° towards the back compared to a level horizontal surface so the media covers the cell monolayer at the bottom of the tube and does not flow into the cap.

7. Examine the tube cultures each day under a microscope. Start with the negative control tube to check that the appearance of the cell morphology is normal. Carefully examine the entire cell monolayer of the tubes inoculated with the specimen extract for the appearance of CPE (Figs. 1 and 2). Record the appearance of the cell monolayer as negative or grade the CPE according to the proportion of the cell monolayer that is

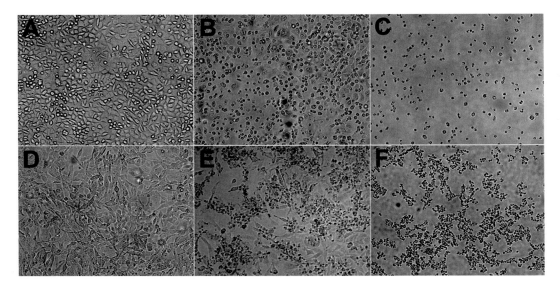

Fig. 1 L20B and RD cell lines. The L20B (panels **a**, **b**, and **c**) and RD (panels **d**, **e**, and **f**) cells are depicted for uninoculated (**a**, **d**), 2+ CPE (**b**, **e**) and 4+ CPE (**c**, **f**) due to growth of poliovirus

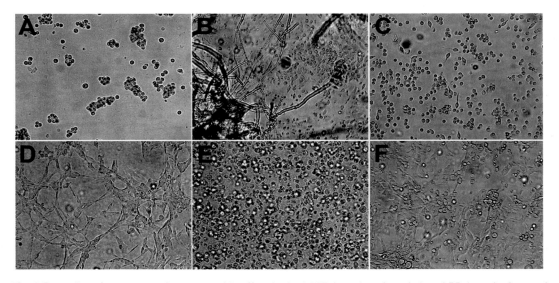

Fig. 2 Examples of non-enterovirus cytopathic effect in the L20B (panels **a**, **b** and **c**) and RD (panels **d**, **e**, and **f**) cell lines are depicted for toxicity (**a**, **d**), fungus (**b**), bacteria (**e**), and adenovirus (**c**, **f**)

affected: 1+ up to 25 %, 2+ up to 50 %, 3+ up to 75 %, and 4+ up to 100 %. Note the appearance of the CPE as to whether the cells are lysed, rounded, granular, non-adherent, etc. CPE may initially be observed as a focal point and should be left to develop until it covers more than 50 % of the cell monolayer (3+). A permanent record of the cell culture readings should be made either electronically or in a book.

8. If the tube cultures of both cell lines are lysed (4+) within 24 h of the primary inoculation, the specimen extract is toxic and, potentially, there was insufficient time for virus infection to be established. Discard the tubes and repeat the inoculation of the specimen extract diluted 1/10 with complete PBS to reduce the toxicity.

9. When CPE is observed over more than half of the cell monolayer (3+), passage the tube culture to a fresh monolayer of cells (*see* **Note 21**).

    (a) Lay the tubes horizontally in a plastic container with the orientation marker facing upwards so the media covers the remaining adherent cells. Place the container in a freezer at ≤−20 °C until the media is frozen so as to lyse any remaining intact cells and release virus particles.

    (b) Remove the container from the freezer and place the tubes upright in a rack leaving to thaw at ambient temperature. **Steps** (**a**) and (**b**) may be repeated to ensure complete lysis of the remaining intact cells, in particular if the CPE had not progressed beyond 75 % of the monolayer (4+).

    (c) Pellet the cellular debris by centrifuging at $1500 \times g$ for 5 min in a refrigerated bench centrifuge.

    (d) Transfer the supernatant to a labeled screw-capped storage vial.

    (e) Label new tube cultures with the unique code and the passage history: R_1L_1, R_2, etc.

    (f) Discard the media from the required number of new tube cultures, including a negative cell control tube, and replace with 2 ml maintenance media.

    (g) Inoculate the tube cultures with 0.2 ml of harvested supernatant from step (c).

    (h) Incubate tube cultures at 36 °C and observe daily under a microscope for CPE and record the reading.

10. Characteristic enterovirus CPE is observed as cells becoming rounded and refractile and may appear granular, leading to lysis and detachment as the CPE progresses (Fig. 1). Harvest the tube culture lysates displaying enterovirus-like CPE by following **steps 9** (**a**)–(**d**). Poliovirus does not have a characteristic CPE that can be distinguished from that caused by other enteroviruses. Further tests must be performed to confirm the presence of poliovirus including immunofluorescence (refer to Subheading 3.4), which can be used to quickly detect poliovirus in the cell lysate.

11. If no CPE is observed after 5–7 days incubation of the primary inoculation of specimen extract at **step 5**, perform a blind passage by harvesting the cell lysate and inoculating 0.2 ml on a

new monolayer of the same cell line as described at **step 9**. Observe the passaged cultures for a further 5–7 days (10–14 days total incubation) and if no CPE develops, discard the tubes and record the specimen as negative for poliovirus by cell culture.

12. Microorganisms other than enterovirus can cause CPE in tube cultures, in particular when stool specimen extracts are incubated (Fig. 2). Fungal contamination can be observed by the naked eye or microscopic observation of distinctive hyphae (Fig. 2b). Turbid culture media can be evidence of bacterial contamination that is observed microscopically as a haze and may create a barely perceptible movement in the media; the cells may display signs of degeneration (Fig. 2e). Discard the contaminated culture tubes and record the nature of the contamination. If primary inoculation tubes were contaminated after **step 5**, return to the original specimen extract, re-treat with chloroform and inoculate fresh culture tubes, recording this as a repeat primary inoculation; L_1, R_1, etc. Additional antibiotics may also be added to the culture media. If the contamination of a passaged culture occurred (L_2, R_1L_1, etc.), filter the primary inoculation through a 0.2 μm membrane using a syringe and collect the filtrate in a newly labeled tube. Inoculate 0.2 ml of the filtrate on a fresh cell monolayer and record this as a repeat passage. Other viruses such as adenovirus have a distinctive CPE that causes cell rounding and detachment usually without observation of concomitant lysis (Fig. 2c, f). A putative adenovirus CPE can be passaged from RD to RD cells but observation of adenovirus-like CPE in L20B is usually restricted to a single inoculation that will not recur when passaged to either the RD or L20B cell line. Specific testing (immunofluorescence, virus neutralization, PCR) is required to confirm the presence of adenovirus in the cell lysate.

3.5 Characterization of Poliovirus Isolates

A number of tests can be used to confirm the presence of poliovirus in tube cultures with an enterovirus-like CPE. The classic technique was antisera neutralization. Serotype-specific monoclonal and polyclonal antibodies raised against poliovirus are available commercially, as well as pooled antisera against the three serotypes. The harvested cell lysate, or isolate, is incubated with antiserum known to be capable of binding to the poliovirus virion, interfering with the process of cellular infection, and thus, preventing the development of viral CPE when subsequently incubated with a cell suspension. An isolate incubated in parallel with and without anti-poliovirus type 1 antiserum that does not exhibit CPE in the presence of antiserum, but does in its absence, is reported as poliovirus serotype 1. This technique has now been supplanted by RT-PCR methodology due to its rapid turnaround, capacity for high throughput and convenience, not being dependent upon the immediate availability of sufficient cells. Oligonucleotide primers, fluorescent probes and thermal cycler conditions have been published to detect the individual poliovirus serotypes by RT-PCR and all three serotypes (pan-poliovirus). The following protocol describes an indirect immunofluorescence test to rapidly confirm the presence of poliovirus upon observation of enterovirus-like CPE in culture that, unlike antisera neutralization, does not require additional handling of cell lines. Commercial preparations of poliovirus serotype specific antibodies are available for this method but the use of a blend of antibodies to all three poliovirus serotypes will be described.

1. Upon observing enterovirus-like CPE in tube cultures but prior to harvesting the cell lysate by freeze-thawing (Subheading 3.4, **step 10**), pellet the cellular debris at $250 \times g$ for 10 min at ambient temperature (*see* **Note 22**). Remove the supernatant and store in a separate tube.

2. Add 250 μl of complete PBS with a transfer pipette and gently resuspend the cell pellet. Using the same pipette, transfer a single drop of suspension to each well of a three-well hydrophobic-coated microscope slide and let air-dry in a BSC. The remaining lysate may be stored at 4 °C pending the outcome of the immunofluorescence test or be harvested as described in Subheading 3.4 **step 9** (**a**)–(**d**).

3. Prepare three negative control microscope slides in the same manner as described in **steps 1** and **2** using an uninoculated tube culture.

4. Fill a microscope slide staining jar with cold acetone. Fix the cells to the slides by placing the slides in the jar for 10 min at 4 °C and air-dry.

5. During the incubation, dilute the concentrated mouse anti-poliovirus blend antibody preparation, stored at −20 °C, with PBS. Following the manufacturer's recommendation, prepare a range of dilutions to optimize the fluorescent staining; for example, 1:100, 1:200, and 1:400.

6. Add 15 μl of each dilution of anti-poliovirus antibody to just cover the fixed cellular material on a test and control slide.

7. Lay slides on a tray and place in a plastic container lined with paper towel moistened with water. Seal the humidified chamber with a lid and incubate at 37 °C for 30 min.

8. Remove the slides from the humidified chamber and rinse in a microscope staining jar filled with complete PBS for 5 min. Repeat the 5 min rinse with fresh complete PBS.

9. During the rinsing step, dilute the concentrated anti-mouse IgG-FITC labeled secondary antibody preparation, stored at −20 °C, with complete PBS. Follow the manufacturer's

recommendation or dilute the secondary antibody 1:100 with complete PBS.

10. Add 15 µl of diluted anti-mouse IgG-FITC labeled secondary antibody to just cover the fixed cellular material on each of the test and control slides.

11. Lay slides on a tray and place in a sealed humidified chamber and incubate at 37 °C for 30 min.

12. Remove the slides from the chamber and rinse in a microscope staining jar filled with complete PBS for 5 min. Repeat the 5 min rinse with fresh complete PBS.

13. Mount the slides with glycerol.

14. Observe the stained microscope slides at 10–100× using a fluorescence microscope with a filter suitable for FITC. A positive result for the putative poliovirus test slides should appear as a yellow-green fluorescence in the nucleus and/or cytoplasm of the cellular material. The negative control slides should be dull with only punctate or diffuse fluorescence and may have a red appearance if a rhodamine-red counter stain was included.

4   Notes

1. While short-term exposure of complete cell culture medium to fluorescent lighting during usage is acceptable, stocks should be stored in the dark to avoid production of toxic chemicals due to photoactivation of HEPES, riboflavin and tryptophan.

2. Batch-to-batch variation of serum can occur. An aliquot of FBS should be requested from the manufacturer for pretesting with the cell lines in routine usage. Acceptance of the batch of FBS can be based on cell morphology, growth rate, and sensitivity to poliovirus infection and other enteroviruses in parallel with the serum currently used. FBS must be heated at 56 °C for 30 min to inactivate complement proteins prior to addition to cell culture media. Gentle swirling of the serum at 5–10 min intervals during the inactivation will ensure even heat distribution. FBS should not be thawed more than twice and aliquots can be prepared after heat inactivation.

3. Cell sensitivity is measured by routinely testing the cell lines with standard poliovirus preparations of known titre for each serotype. WHO recommend testing is performed at least once per cell revival from liquid nitrogen [12].

4. Cell lines are usually shipped to laboratories in one or two 25 cm² flasks. The cells should be revived from liquid nitrogen one or two days before sending. On the day of shipment, the confluency of the cells is assessed and fresh media added to the

flask. The percentage of FBS in the media will depend upon the confluency of the cells; >50 % confluency use media with 2 % FBS, between 25 and 50 % confluency use 6 % FBS and for cells <25 % confluency use media with 10 % FBS, although it may be preferable to delay the shipment until the cells are more established. The culture media is added to the brim of the flask while in an upright position. Laboratory sealing film is wound tightly around the lid in a clockwise direction to prevent leakage and avoid loosening while in transit. Sufficient absorbent material should be included in the shipment package in case of leakage. Cell lines are shipped at ambient temperature and the courier should be informed that the contents are fragile and must avoid extreme temperatures but does not need to be refrigerated if there is a delay in transit.

5. The action of EDTA is to chelate divalent cations involved with cell attachment to other cells and the surface of the flask. Trypsin is a protease that cleaves amino acids involved with cell attachment and is inhibited by magnesium and chloride ions. Trypsin is often sourced from pig pancreases although recombinant forms are available.

6. With routine handling the period of incubation required to detach a particular cell line will be determined and may proceed in a single step. Some laboratories incubate the cells with trypsin–EDTA only until observing the initial rounding of the cells and prior to detachment. The trypsin–EDTA solution is removed from the flask in a BSC and returned to the incubator for a further period until cell detachment occurs. The authors have not noticed any untoward effects of leaving the trypsin–EDTA solution in the flask throughout the detachment procedure as it will become inactivated upon addition of the growth media at the next step.

7. Performing a cell count each time a cell line is passaged will ensure accurate and consistent results with all laboratory staff. However, sometimes it is sufficient to serial passage a cell line by performing a simple dilution or split ratio of the cell suspension. The growth pattern of a particular cell line is determined by routine handling as well as the dilution required to achieve a confluent cell monolayer in a specified number of days. For a split ratio of 1:5, 1 ml of cell suspension is diluted with 5 ml of growth media.

8. Trypan blue is a toluidine based dye that can traverse the membrane of nonviable cells but is excluded by intact cell membranes. Trypan blue is a hazardous chemical and suspected carcinogen that should be handled with gloves. Only discard Trypan blue waste or unused dye solution via a professional chemical waste disposal company.

9. Place the hemocytometer on a microscope stage and observe the grid pattern etched on each side of the chamber. Check the manufacturer's instructions for performing cell counts as the grid pattern and volume of the counting chamber can vary. The cell concentration is determined by counting viable (non-stained) cells in a specified number of squares on each side of the counting chamber. Nonviable cells that have taken up the Trypan blue dye are excluded from the count. A hand tally counter is useful to keep count. Only single cells should be counted, not clumps. If many clumps or aggregates of cells are observed the cell suspension requires further dispersal with a transfer pipette prior to dilution with Trypan blue. Individual cells that overlap the top and right hand side grid lines are included in the count but not those overlapping the bottom and left hand side grid lines.

10. The seeding concentration for ampoules stored in the cell bank can vary from 1 to 5×10^6 cells per ml depending on the intended usage and how quickly a confluent cell monolayer is needed after revival. Ampoules can be stored at 1×10^6 cells per ml if they are routinely revived into 25 cm^2 flasks whereas a 75 cm^2 flask requires a higher cell concentration at 5×10^6 cells per ml. The cell concentration in the cell bank should be higher than that for a normal passage to allow for loss of cell viability upon revival.

11. Liquid nitrogen is highly dangerous and should be handled with caution at all times. A SOP should be developed for training, routine work and emergency procedures. At atmospheric pressure the liquid is at −196 °C and can cause cold burns if splashed on exposed skin. Personal protective equipment including full-face mask, protective full-length gown and full-length gloves suitable for cryogenic work and covered shoes must be worn by all persons handling liquid nitrogen. As liquid nitrogen evaporates it reduces the oxygen concentration and asphyxiation can occur rapidly if storage containers are opened in enclosed spaces; 1 l of liquid nitrogen can expand to 168 l of nitrogen gas. For this reason liquid nitrogen containers must be stored in a well-ventilated area and a "buddy" assigned to keep watch on procedures outside of the store room to respond in an emergency. A personal oxygen meter can be attached to the protective gown of the lab worker handling liquid nitrogen to monitor the immediate area and sound an alarm if the oxygen concentration falls below a preset limit. In the event of a possible asphyxiation the buddy should not enter the liquid nitrogen store room unless wearing portable oxygen breathing apparatus.

12. Plastic cell culture tubes treated for adherent cell lines should be used. Tubes are approximately 16×125 mm with a screw capped lid that is kept closed during all incubations. Once

inoculated, tubes should not be opened to change the media of negative virus cultures due to the potential for contamination. Tubes with a built-in filter membrane in the lid are available for use with CO_2 incubators. Completely rounded culture tubes require a holder that fits on the microscope stage to assist viewing of the cell monolayer. Alternatively, tubes with a flattened side on the lower surface can be placed directly on the microscope stage. Glass tubes, while providing a clearer field when viewed under the microscope compared to plastic, are not recommended due to biosafety concerns if broken while containing live culture material. Cultures grown in 24 well plates are not recommended for poliovirus isolation due to the potential for contamination from adjacent wells.

13. The number of times a cell line has been passaged after revival from liquid nitrogen should be recorded with each passage. This can be indicated with a permanent marker on the cell culture flask or on a rack for tube cultures with a piece of removable tape labeled with the passage number of the cell line revived from liquid nitrogen followed by the number of times the cell line has been passaged since revival. For example, 20-7 indicates passage 20 is stored in liquid nitrogen and it has been passaged seven times out of a recommended maximum of 15 since revival.

14. The culture tubes need to remain stationery after seeding until the cell monolayer has formed. The rack holding the tubes should be tilted slightly backwards (approximately 5°) so the cells settle towards the end of the tube. The rack can be slanted by attaching a short piece of wooden rod, approximately 5 mm diameter, under the front surface of the rack. If the culture tubes were manufactured with an orientation indicator, turn the tubes in the rack so the indicator is at the top, otherwise indicate the orientation of the tube by drawing a line using a permanent marker. With the orientation indicator turned to the top, the cell monolayer will grow on the lower surface. Once the cell monolayer has formed in the tubes, the L20B cells remain in a stationery rack while the RD tubes are transferred to a tubeholder roller drum assembly that turns at approximately 0.2 rpm.

15. The waste container should be autoclavable and have a separate lid that can be secured prior to removal from the BSC. Line the container with a paper autoclave bag large enough to completely fill the container and be wrapped over the edge. Line the paper bag with a plastic bag of similar size as the paper bag. Place a large wad of absorbent cotton wool in the base of the container and spray the entire lining with disinfectant. Do not over fill with waste; a new container may need to be prepared if many specimens are processed at the same time. When the work procedure is completed, fold the plastic lining

inwards, followed by the paper bag lining, to cover the contents. Secure the lid, decontaminate the outer surface, and remove from the BSC. After autoclaving, carefully remove the paper bag as it may be soggy and dispose in a biological waste bin. Rinse the waste container and leave to dry.

16. As part of training for this procedure, it is recommended new staff weigh out 2 g of stool specimen to visually estimate the amount involved.

17. The purpose of chloroform is to extract lipid material, remove bacteria and fungi and to assist with disaggregation of non-enveloped virus particles. Note that if the specimen extract is to be also tested for enveloped viruses, chloroform treatment will denature the lipid membrane and an aliquot should be removed prior to the addition of chloroform.

18. The mechanical shaker must have a vigorous action that completely mixes the specimen extract either by vortex or horizontal agitation. The specimen extract may not be properly processed to fully separate the poliovirus particles from the organic matter if the mechanical shaker has a feeble mixing action.

19. It is important to record the passage history of virus cultures. Primary inoculation of a cell culture tube with specimen extract can be labeled as R_1 and L_1 on the RD and L20B cell line, respectively. Passage of R_1 to a fresh monolayer of RD cells would be indicated as R_2 and to L20B would be R_1L_1. Passage of L_1 to L20B would be indicated as L_2 and to RD would be L_1R_1.

20. Specimen extracts, especially of stool specimens, can be toxic when in direct contact with the cell monolayer. When inoculating culture tubes pipette the specimen extract down the inner face of the tube opposite the monolayer (same side as the orientation indicator) or directly into the culture media, cap the tube and gently swirl to mix and dilute the extract with the media.

21. At least two sequential incubations of cell monolayers are required for virus culture. The primary inoculation of the specimen extract will enable virus attachment, internalization, and replication to occur within cells sensitive to poliovirus infection but this can occur in competition with cell degeneration due to material present in specimen extracts that are toxic to the cells. Passage of the primary inoculation to a fresh monolayer will assist to identify that the CPE observed is due to poliovirus rather than toxicity; virus titre should increase when passaged to a fresh monolayer whereas toxic material will be diluted. The WHO algorithm for poliovirus isolation recommends passaging between the alternate cell lines for the rapid isolation of poliovirus rather than non-polio enteroviruses (NPEVs); a positive L20B isolate (L_1) is passaged to a fresh monolayer of RD cells and vice versa, R_1 to L20B. If R_1L_1 is

subsequently positive, a further passage back to the RD cell line is recommended as poliovirus usually grows to a higher titre in the RD cell line than L20B to conduct confirmatory tests. An additional passage of positive RD isolates (R_1) back to the RD cell line can help identify the isolation of NPEVs in the R_2 isolate. Since most NPEVs do not grow in the L20B cell line, the other passages (R_1L_1 and L_1R_1) will be negative. However, the L_1R_1 passage can be positive due to growth of an NPEV in the RD cell line from remnant NPEV particles present in the original stool extract inoculation of L20B culture media and carried over with the cell lysate. All positive virus cultures should be confirmed with specialized protocols used for putative poliovirus isolates that can also differentiate between wild and Sabin vaccine strains of poliovirus, known as intratypic differentiation.

22. The immunofluorescence test requires some cellular debris to remain for optimal sensitivity as poliovirus is more easily detected when attached to cellular membranes and fragments rather than floating in culture media. The CPE should be between 25 and 90 % to avoid false negatives due to either too little virus present or lack of cellular material, respectively.

References

1. Enders JF, Weller TH, Robbins FC (1949) Cultivation of the Lansing strain of poliomyelitis virus in cultures of various human embryonic tissues. Science 109:85–87

2. Robbins FC, Enders JF, Weller TH (1950) Cytopathogenic effect of poliomyelitis viruses *in vitro* on human embryonic tissues. Proc Soc Exp Biol Med 75:370–374

3. Robbins FC, Enders JF, Weller TH et al (1951) Studies on the cultivation of poliomyelitis viruses in tissue culture. V The direct isolation and serologic identification of virus strains in tissue culture from patients with nonparalytic and paralytic poliomyelitis. Am J Hyg 54:286–293

4. Robbins FC, Weller TH, Enders JF (1952) Studies on the cultivation of poliomyelitis viruses in tissue culture. II The propagation of the poliomyelitis viruses in roller-tube cultures of various human tissues. J Immunol 69:673–694

5. Freshney RI (2010) Culture of animal cells. A manual of basic techniques and specialized applications, 6th edn. Wiley, New Jersey

6. Dietz V, Andrus J, Olivé J-M et al (1995) Epidemiology and clinical characteristics of acute flaccid paralysis associated with non-polio enterovirus isolation: the experience in the Americas. Bull World Health Organ 73:597–603

7. Laxmivandana R, Yergolkar P, Gopalkrishna V et al (2013) Characterization of the non-polio enteroviruses infections associated with acute flaccid paralysis in South-Western India. PLoS One 8(4):e61650. doi:10.1371/journal.pone.0061650

8. Geraghty RJ, Capes-Davis A, Davis JM et al (2014) Guidelines for the use of cell lines in biomedical research. Br J Cancer 111:1021–1046

9. Pipkin PA, Wood DJ, Racaniello VR et al (1993) Characterisation of L cells expressing the human poliovirus receptor for the specific detection of polioviruses *in vitro*. J Virol Methods 41:333–340

10. Wood DJ, Hull B (1999) L20B cells simplify culture of polioviruses from clinical samples. J Med Virol 58:188–192

11. Nadkarni SS, Deshpande JM (2003) Recombinant murine L20B cell line supports multiplication of group A coxsackieviruses. J Med Virol 70:81–85

12. World Health Organization (2004) Polio laboratory manual WHO/IVB/04.10, 4th edn. WHO Department of Immunization, Vaccines and Biologicals, Geneva

Chapter 5

Molecular Characterization of Polio from Environmental Samples: ISSP, The Israeli Sewage Surveillance Protocol

Lester M. Shulman, Yossi Manor, Musa Hindiyeh, Danit Sofer, and Ella Mendelson

Abstract

Polioviruses are enteric viruses that cause paralytic poliomyelitis in less than 0.5 % of infections and are asymptomatic in >90 % infections of naïve hosts. Environmental surveillance monitors polio in populations rather than in individuals. When this very low morbidity to infection ratio, drops drastically in highly vaccinated populations, environmental surveillance employing manual or automatic sampling coupled with molecular analysis carried out in well-equipped central laboratories becomes the surveillance method of choice since polioviruses are excreted by infected individuals regardless of whether or not the infection is symptomatic. This chapter describes a high throughput rapid turn-around time method for molecular characterization of polioviruses from sewage. It is presented in five modules: (1) Sewage collection and concentration of the viruses in the sewage; (2) Cell cultures for identification of virus in the concentrated sewage; (3) Nucleic acid extractions directly from sewage and from tissue cultures infected with aliquots of concentrated sewage; (4) Nucleic Acid Amplification for poliovirus serotype identification and intratypic differentiation (discriminating wild and vaccine derived polioviruses form vaccine strains); and (5) Molecular characterization of viral RNA by qRT-PCR, TR-PCR, and Sequence analysis. Monitoring silent or symptomatic transmission of vaccine-derived polioviruses or wild polioviruses is critical for the endgame of poliovirus eradication. We present methods for adapting standard kits and validating the changes for this purpose based on experience gained during the recent introduction and sustained transmission of a wild type 1 poliovirus in Israel in 2013 in a population with an initial IPV vaccine coverage >90 %.

Key words Environmental surveillance, RNA purification, RT-PCR, qRT-PCR, Sequence analysis, Intratypic differentiation, Poliovirus, Wild poliovirus, Vaccine-derived poliovirus (VDPV), Oral polio vaccine (OPV), Tissue culture, Plaque assay

Abbreviations

AFP	Acute flaccid paralysis
AR	Analytic reagent grade chemicals
bp	Base pairs
BSL-2	Biological safety level two
CDC	Centers for Disease Control and Prevention, Atlanta, GA, USA
CPE	Cytopathic effect—morphological changes in cells resulting from infection

Javier Martín (ed.), *Poliovirus: Methods and Protocols*, Methods in Molecular Biology, vol. 1387, DOI 10.1007/978-1-4939-3292-4_5, © Springer Science+Business Media New York 2016

Ct	Cycle threshold, the cycle at which specific signal from the probe is first detected above the threshold of detection in qRT-PCR
FBS	Fetal bovine serum
GPEI	Global Poliovirus Eradication Initiative
IPV	Inactivated polio vaccine (trivalent contains all three Salk serotypes)
ITD	Intratypic differentiation, e.g., determining whether the isolate is vaccine-like, VDPV, or wild
MOI	Multiplicity of infection
OPV	Live oral poliovirus (trivalent contains all three Sabin serotypes bivalent contains Sabin serotypes 1 and 3, monovalent contains either serotype 1, 2, or 3, exclusively)
PCR	Polymerase chain reaction
PFU	Plaque forming units
qRT-PCR	Quantitative reverse transcription polymerase chain reaction
RT-PCR	Reverse transcription polymerase chain reaction
SIA	Supplementary Immunization Activities
TD	Typic differentiation determining the serotype of the virus of interest
VDPV	Vaccine-derived polio virus
VP1	Viral capsid protein 1
VP2	Viral capsid protein 2
VP3	Viral capsid protein 3
VP4	Viral capsid protein 4
WHO	World Health Organization
5′ UTR	5′ untranslated region of polio and non-polio enteroviruses

1   Introduction

Polioviruses like other member of the Picornaviridae are encapsidated in an icosahedral structure formed from 60 capsomeres containing one copy each of viral capsid proteins 1 through 4, (VP1, VP2, VP3, and VP4 [1]). There are three serotypes of poliovirus [1, 2]. Poliovirus can cause irreversible paralysis of the infected host and even death, however, most poliovirus infections are asymptomatic [1]. The most common and most efficient route of host-to-host transmission is fecal–oral but poliovirus can also be transmitted by oral–oral transmission [3]. Excretion, duration of excretion, and the amount excreted are dependent on host factors and on vaccination history of the infected individual [2, 3]. A poliovirus is designated as polio vaccine-like, vaccine derived poliovirus (VDPV), or wild poliovirus based on whether the molecular sequence of its VP1 differs from that of the homologous Sabin serotype in live oral polio vaccine (OPV) by ≤1 % (or <0.6 % for serotype 2), from 1 to 15–18 %, or >20 %, respectively [4].

The Global Poliovirus Eradication Initiative (GPEI), launched at the World Health Assembly (WHA, resolution WHA41.28) in 1988, is the single largest, internationally coordinated public health project the world has ever known. Early detection of highly pathogenic infectious organisms such as polio is essential for containing the spread of these organisms. Early detection and measurement of efficacy of response requires high quality surveillance over extended periods of time. Acute flaccid paralysis (AFP) surveillance, is based on monitoring infections in individuals and is the Gold Standard for poliovirus surveillance. AFP surveillance is based on investigating all cases of AFP in children under 15 to rule-in or rule-out poliovirus etiology. Two trivalent vaccines (trivalent since they contain vaccine strains for each or the three poliovirus serotypes) have been used for eradication; IPV formulated from inactivated neurovirulent polioviruses and OPV a vaccine formulated from attenuated live polioviruses. A region is considered to be poliovirus free when over a period of 3 years, the number of AFP cases is equivalent to the expected incidence due to non-poliovirus causes (1 per 100,000 children) and none of the investigated cases were due to poliovirus.

In contrast to AFP surveillance, environmental surveillance for poliovirus is based on the monitoring of poliovirus transmission in human populations by examining environmental samples that contain human feces. The rationale for environmental surveillance is based on the fact that all poliovirus-infected individuals shed large amounts of poliovirus in their feces for several weeks whether or not they have symptoms [5]. The protein capsid is relatively stable enabling poliovirus viruses to remain viable in the environment at ambient temperatures [3, 5]. The length of time when viruses remain viable is extended when the virus adsorbs to solids in sewage [3, 5]. The probability of detecting poliovirus in environmental samples [6] depends on the duration and amount of poliovirus excreted by one or more infected individuals, the effect of physical and mechanical factors on the dilution and survival of poliovirus in the sewage system (reviewed by Dowdle [3]), the frequency of collection and laboratory processing of the environmental samples [7] and the location of the excretor relative to the sample site [5].

Poliovirus may be recovered quantitatively from the environment [8–10]. Two complementary assays for quantifying poliovirus are presented here, plaque assay and quantitative (or semiquantitative) reverse transcription polymerase chain reaction (qRT-PCR). Decreasing the distance between the excretor or excretors and the sample site is usually more effective in increasing the probability of detection, the amount of virus detected and is less labor intensive and more cost efficient than increasing the frequency of sampling [11].

When the very low morbidity to infection ratio (<1:200) of poliovirus infections in naïve populations drops drastically in highly vaccinated populations, environmental surveillance employing manual or automatic sampling coupled with molecular analysis carried out in well-equipped central laboratories becomes the surveillance method of choice. In fact environmental surveillance has been used [5]: to determine extent of a poliovirus outbreak in a population; pinpoint the putative reservoirs and/or epicenters of wild or vaccine-derived polioviruses; to calculate the risk for emergence and transmission of vaccine-derived viruses after immunization campaigns switch from inclusion of live vaccines to exclusive use of inactivated poliovirus vaccine; to screen for unidentified persistently infected asymptomatic individuals in a given population; to monitor for introduction of wild or vaccine-derived viruses into poliomyelitis-free regions; and to determine whether there is sustained person-to-person transmission after introduction of a non-vaccine poliovirus. Environmental surveillance will play an increasingly important role in quality assurance as the amount of wastewater reclamation for agriculture, recreation and drinking purposes grows [5]. Finally, environmental surveillance will also play a critical role in post-eradication surveillance in demonstrating the absence of wild and vaccine strains and for detecting reemergence of poliovirus (*see* **Note 1**).

Standard protocols and algorithms for determining the serotype of a poliovirus, e.g., typic differentiation (TD) and whether the isolate is vaccine-like, VDPV, or wild, e.g., intratypic differentiation (ITD) of poliovirus from tissue culture by molecular means are provided in detail in the WHO Polio Laboratory Manual 4th edition, 2004 WHO/IVB/04.10 [12]. The different steps must be performed in appropriate designated Molecular Areas that are defined in **Note 2**. The molecular assays are based on qRT-PCR using specific primers and labeled TaqMan probes that recognize the 5' untranslated region (5' UTR) of enteroviruses and the VP1 genes of polioviruses in RNA extracted from supernatants of infected tissue cultures (*see* **Note 3**). These standard algorithms and reagents (*see* **Note 4**) are in the process of revision and late draft versions of these revisions were discussed at the 20th Informal Consultation of the Global Polio Laboratory Network held at WHO Headquarters in Geneva Switzerland, on June 26–27, 2014. Thus it would not be productive to provide step-by-step instructions for molecular characterization of the poliovirus RNA extracted in Subheading 3.4 at this time. For the latest routine procedures and algorithms we recommend that the reader contact the Polio Laboratory at the Centers for Disease Control and Prevention in Atlanta, GA, USA for information on what latest kits, reagents, and methods are available; and the WHO Global Polio Laboratory Network (GPLN) Regional Laboratory Coordinator for the region in which the laboratory is situated to obtain the latest SOP including algorithms for molecular characterization. A degenerate primer is a mixture of oligonucleotide primers where the bases at one or more nucleotide positions differ or a primer that contains a universal nucleotide at one or more position, such as inosine, that can complement all four nucleotide bases. Sequence variations among wild poliovirus still in circulation have made it necessary to use degenerate primers especially for wild polioviruses for some of the reactions in these kits.

What we will present is a protocol based on these general protocols that can be used to build a qRT-PCR assay with high sensitivity and specificity for a poliovirus of interest once a wild polio or VDPV has been identified based on experience gained during a silent outbreak (persistent transmission without cases) of type 1 wild poliovirus (WPV1) in Israel in 2013–2014 [9, 13]. After its viral capsid protein 1 (VP1) has been sequenced by the protocol in Subheading 3.4 below, it is possible to develop and validate specific nondegenerate primers and probe for qRT-PCR for that strain [9]. Nondegenerate primers such as those used to identify vaccine strains, usually have 100- to 1000-fold lower limits of quantitation (LOQs) and limits of detection (LODs) than degenerate primers. The qRT-PCRs that will be described are multiplex, e.g., each reaction tube includes primers and probes for more than one target sequence. In PCR, the amount of target is doubled during each cycle. The amount of nucleic acid in the solution being assayed is inversely proportional to the Ct or cycle at which specific signal from the probe is first detected above the threshold of background noise. It is important to correlate the Ct for the specific qRT-PCR results with the number of plaque forming units (PFU) of virus [9]. This will enable the lab to provide quantitative information about the load of the virus of interest even when it is present a mixture of heterotypic or homotypic polioviruses (*see* **Note 5**). Finally the nondegenerate qRT-PCR primers that have been designed to specifically identify the virus of interest can also be used to specifically sequence the VP1 of the virus of interest in RNA extracted from mixtures of heterotypic and homotypic polioviruses without the need for first isolating the virus of interest from these mixtures. Next-generation sequencing (NGS) procedures for RNA extracted from concentrated sewage with or without enrichment for polioviruses by affinity precipitation using anti-poliovirus antibodies or poliovirus receptors are being developed as this chapter goes to press and may supplement or replace some of the procedures described above at specialized molecular laboratories equipped with the necessary machinery and expertise.

2   Materials and Equipment

All solutions use autoclave-sterilized ultrapure water (prepared either by purifying deionized water to attain a sensitivity of 18 MΩ at 25 °C or by distillation) and analytical (AR), molecular biology, or tissue culture grade reagents. Stocks and working solutions were aliquoted before freezing to limit the number of freeze–thaw cycles and preserve component reactivity and uniformity of results. The solutions should be clearly labeled and the label should contain the date of preparation and the expiration date. Prepare and store all reagents at room temperature (unless indicated otherwise). Sewage and all materials derived from sewage must be treated as if contaminated with pathogens. Therefore, appropriate personal protective equipment (PPE) must be worn at all times; lab coats, protective sleeves, splashguard for face or safety glasses, gloves that must be changed frequently, and laboratory approved footwear (no open shoes or sandals) and used biological reagents must be decontaminated before disposal. Appropriate materials to contain and decontaminate spills must be near at hand in accordance with your institutions safety regulations. All stocks of cells should be tested periodically to ensure that they are mycoplasma-free and sensitive to infection by poliovirus. Do not eat, drink, or smoke in laboratory work areas and never pipette by mouth.

2.1 Sewage Processing Components

1. Sterile 0.5, 1, and 2 L screw cap bottles for storage of unprocessed sewage at 4 and at –20 °C.

2. Wide-mouthed screw cap 2 L plastic bottles: 12 cm diameter × 22.5 cm tall, mouth >6 cm in diameter.

3. Magnetic stirrers and magnetic stirring bars (*see* **Note 6**).

4. Vortex mixers.

5. Rotary shakers for containers ≤500 and ≤50 mL.

6. Test tubes: 1.5, 2.0, 15, 50, 250, and 500 mL tubes suitable for centrifugation and for freezing.

7. Centrifuge with head suitable for spinning 500 mL or 250 mL centrifuge bottles at $10,000 \times g$.

8. Centrifuge suitable for spinning tubes ≤50 mL at $1400 \times g$.

9. Microcentrifuges for 1.5 and 2.0 mL microcentrifuge test at up to $14,000 \times g$.

10. Chemical fume hood or biological–chemical biosafety level 2 (BSL-2) laminar flow hood for initial stages of processing.

11. Biosafety level 2 (BSL-2) laminar flow hood.

12. +4 °C refrigerator(s) for temporary storage of large bottles containing volumes up to 2 L.

13. –20 °C freezers.

14. –70 °C freezers.

15. Polyethylene glycol 6000 (Sigma or Merck) (*see* **Note 7**).

16. AR grade NaCl.

17. AR grade chloroform.

18. Chlorine for disinfection: Chlorine tablets (Actichlor Plus (EchoLabs, Northwich, Cheshire CW8 4DX, UK) containing 1.7 g sodium dichloroisocyanurate or the equivalent to prepare fresh solutions containing 1000 ppm of chlorine solution weekly (one tablet per liter of tap water); and/or 3 % 30,000 ppm) or 12 % (120,000 ppm) concentrated chlorine

stock solutions which can be diluted to 1000 ppm for decontamination (*see* **Note 8**).

19. PBS: Dulbecco's calcium- and magnesium-free phosphate-buffered saline (NaCl 8 g, KCl 0.2 g, Na_2HPO_4 1.15 g, KH_2PO_4 0.2 g, 1 L water; pH = 7.3–7.4).

20. PBS–Tween 80: PBS containing 0.1 % Tween 80.

21. PSF stock: Penicillin 50 mg/mL, Streptomycin 50,000 U/mL, and Fungizone 0.5 mg/mL.

22. Mycostatin stock [6250 U/mL].

23. PSMY stock: Penicillin G 50,000 U/mL, Dihydro-Streptomycin 50 mg/mL, and Mycostatin 6250 U/mL.

24. 3 % DIFCO (Becton, Dickinson and Company, NJ, USA) beef extract in water (pH 7.2).

2.2 Cell Cultures Components

1. World Health Organization (WHO) certified cell lines: human rhabdomyosarcoma (RD) currently available form the National Institute of Standards and Controls (NIBSC UK; Blanche Lane, South Mimms, Potters Bar, Hertfordshire, EN6 3QG) and transgenic mouse cells expressing the human CD155 receptor for polioviruses (L20B) and Buffalo Green Monkey Cells (BGM: ATCC CCL-161) and human epithelial cancer cells (HEp2C: ATCC CCL-23) from the ATCC collection (American Type Culture Collection (ATCC), P.O. Box 1549, 10801 University Boulevard, Manassas, VA 20110, USA) (*see* **Note 9**).

2. PSMY stock: Penicillin G 50,000 U/mL, Dihydro-Streptomycin 50 mg/mL, and Mycostatin 6250 U/mL.

3. PSF: Penicillin G 50,000 U/mL, Dihydro-Streptomycin 50 mg/mL, and Fungizone 0.5 mg/mL.

4. Fetal bovine serum (FBS) that has NOT been heat inactivated.

5. L-glutamine 3 % (w/v) in water.

6. Complete M199 medium: M199 medium, 10 % FBS (v/v), 2 mL per 500 mL PSMY antibiotics (v/v). Store at 4 °C.

7. 2× concentrated M199 Medium: 2× concentrated M199 medium, 2 % FBS (v/v), 1 mL per 200 mL PSMY antibiotic (v/v). Store at 4 °C.

8. Complete Eagle's MEM-NAA medium: Eagle's MEM-NAA medium, 10 % fetal bovine serum (v/v), 2 mL per 500 mL PSMY antibiotics (v/v), 4 mL per 500 mL L-glutamine (v/v). Store at 4 °C.

9. Complete Viral Growth Medium: Eagle's MEM-NAA medium, 2 % FBS (v/v), 4 mL per 500 mL PSMY antibiotics (v/v), 4 mL per 500 mL L-glutamine (v/v). Store at 4 °C.

10. Plaque plating medium: M199 medium, 2 mL per 500 mL PMSY (v/v) *without* FBS.

11. Plaque suspension medium: M199, 4 mL per 500 mL PMSY (v/v) *without* FBS.

12. Trypsin-Versene solution: 0.25 % trypsin, 0.05 % EDTA.

13. 1.8 % DIFCO (Becton, Dickinson and Company, NJ, USA) Bacto agar in sterile deionized water.

14. Agar overlay: a solution of 1.8 % DIFCO Bacto Agar in water melted in a microwave (approximately 1 min at the maximum setting).—**CAUTION**: the solution is very hot and may boil vigorously and spill out of the bottle when the bottle is removed from the microwave—Keep melted agar at 50 °C in a water bath until just before use when an equal volume of 2× concentrated M199 medium (brought to 50 °C) is added.

15. Plaque staining solution: M199 medium, 1 mL per 100 mL 1 % neutral red in H_2O (v/v).

16. Tissue culture plates, tubes, and flasks: 10 cm style disposable, tissue culture grade petri dishes; 25, 75 of 150 cm^2 style tissue culture flasks; Greiner Bio-One CELLSTAR 12 mL Cell Culture Tubes or their equivalent (Greiner Bio-One International AG, 4550 Kremsmünster Austria).

17. Separate incubators for uninfected and infected cell cultures (*see* **Note 10**).

18. Sterile pipettes and/or transfer pipettes suitable for volumes of 1, 2, 5, 10, and 25 mL.

19. Micropipettes with adjustable volumes with the following ranges up to 0.5 to 10, ≤2 to 20, 10 to 100, 20–200, and 200–1000 μL.

20. Sterile micropipette tips for volumes up to 10, 20, 100, 200, 1000 μL with aerosol barriers.

21. Disposable test tubes: sterile 1.5, 2.0, 15, and 50 mL tubes suitable for centrifugation and frozen storage with appropriate test tube racks.

22. Reusable centrifuge tubes for 250 and 500 mL.

23. Biosafety level 2 laminar flow hood.

24. +4 °C refrigerator(s).

25. –20 °C freezers.

26. –70 °C freezers.

27. Liquid nitrogen freezing and storage facilities for maintenance of frozen cell stocks.

28. Regular or inverted (phase contrast optional) light microscopes.

29. Decontaminating solutions for wiping down work areas before and after work.

30. Optional but preferred—Lab mat (absorbent paper with a nonpermeable bottom layer) to place on working surfaces before starting and discarded in a container for biological decontamination afterwards.

31. Vacuum line, tubing, and flasks for removing tissue culture medium from tissue cultures (*see* **Note 11**).

2.3   RNA and cDNA Extraction

1. QIAamp Viral RNA Mini Kit (QIAGEN Inc, Valencia, CA, USA) or its equivalent.

2. KingFisher Purification System (Thermo Scientific, Waltham, MA, USA) using the viro_totRNA_KFmini extraction protocol with NucleoMag 96 RNA extraction Kits (Macherey-Nagel GmbH Duren, Germany) or Thermo Fisher Total RNA Kits (Thermo Scientific, Waltham, MA, USA) or their equivalents.

3. NucliSENS easyMag semiautomatic extractor (bioMérieux, Marcy l'Etoile, France) using the Specific B extraction protocol with easyMag extraction kits or their equivalent.

4. High Pure PCR Product Purification Kit (Roche Diagnostics, Indianapolis, IN, USA) or its equivalent.

5. QIAquick Gel Extraction Kit (QIAGEN Inc, Valencia, CA, USA) or its equivalent.

6. Big Dye Terminator Purification Kit (Life Technologies, Foster City, CA, USA).

7. Test tubes: 1.5, 2.0, 15, 50 mL suitable for centrifugation and frozen storage with appropriate test tube racks.

8. Storage boxes (9×9) for storing tubes ≤2 mL.

9. +4 °C refrigerator(s).

10. –20 °C freezers.

11. –70 °C freezers.

12. Centrifuge for spinning tubes ≤50 mL at ≤1500×g.

13. Microcentrifuge for spinning 1.5 and 2.0 tubes up to 15,000×g.

14. Spin-down centrifuges for 0.2 mL tubes or strips.

15. BSL-2 laminar flow hood.

16. External RNA control: shielded RNA or MS2 coliphage as an external control for the presence of inhibitors of PCR or RT-PCR that might be co-extracted with the RNA.

2.4   Nucleic Acid Amplification

1. QIAGEN one-step RT-PCR kit (QIAGEN Inc, Valencia, CA, USA) or its equivalent.

2. ABI Prism 7500 sequence detection system (Life Technologies, Foster City, CA, USA) for qRT-PCR using AgPath-ID™ One-Step RT-PCR Kit (Life Technologies/Rhenium LTD).

3. PCR machines for 0.2 mL tubes or 8-tube strips.

4. Test tubes: 0.2, 0.5, 1.5, 2.0, 15, and 50 mL suitable for centrifugation and frozen storage with appropriate test tube racks.

5. Axygen 0.2 mL 8-test tube strips compatible with ABI 7500 instruments or their equivalent.

6. Centrifuge for spinning tubes ≤50 mL at ≤1500×g.

7. Microcentrifuge for spinning 1.5 and 2.0 tubes up to 15,000×g.

8. Centrifuges for spin down for 0.2 mL tubes or strips.

9. Metal cooling blocks for holding twelve, 8-tube strips (0.2 mL volume) and maintaining the tubes at cold temperatures.

10. Test tube strip transfer holders for transporting filled test tube strips (can be sterilized, nucleic acid-free (chlorine treated), empty aerosol barrier tip boxes).

11. Micropipettes with adjustable volumes with the following ranges up to 0.5 to 10, ≤2 to 20, 10–100, 20–200, and 200–1000 μL.

12. Sterile micropipette tips for volumes up to 10, 20, 100, 200, 1000 μL with aerosol barriers.

13. Aliquots of stock solutions of primers and probes. Sequences and stock concentrations are listed in Table 1.

14. E buffer (for 10× concentrated stock: 48.44 g Trizma Base, 16.4 g Na-acetate, and 362 mg EDTA per liter water brought to pH = 7.2 with acetic acid).

15. Ethidium bromide (1 mg/mL in water).

16. 2 % agar for gels: (1 % agarose; 1 % NuSieve agarose in E buffer).

2.5 Sequencing and Preliminary Sequence Analysis

1. PCR machines for 0.2 mL tubes or 8-tube strips.

2. Automatic Sanger Sequence Detection System: 3100 Genetic Analyzer; 3500 Genetic Analyzer (Life Technologies, 850 Lincoln Centre Drive, Foster City, CA 94404, USA) or their equivalent.

3. Test tubes: 0.2, 0.5, 1.5, 2.0, 15, and 50 mL suitable for centrifugation and frozen storage with appropriate test tube racks.

4. Corning Axygen 0.2 mL 8-test tube strips compatible with ABI 7500 instruments.

5. Centrifuge for spinning tubes ≤50 mL at ≤1500×g.

6. Microcentrifuge for spinning 1.5 and 2.0 tubes up to 15,000×g.

7. Centrifuges for spin down for 0.2 mL tubes or strips.

8. Metal blocks for holding twelve 8-tube strips and maintaining the tubes at cold temperatures.

Table 1
Sequences and stock concentrations

Primer name	Sequence (5′ → 3′)[a,b]	Position[c]	qRT-PCR (pmol per μL)	qRT-PCR Sequencing	Sequencing (pmol per μL)	Strains	Ref.
Y7	GGGTTTGTGTCAGCCTGTAATGA	2419–2441	–	√	10	Generic	[28]
Y7Rc	GGTTTTGTGTCAGCTTGYAAYGA	2419–2441	–	√	40	Generic	[28]
PV1,2S	TGCGIGAYACIACICAYAT	2459–2477	–	√	80	Generic	[28]
246S-S1	CGAGATACCACACATATAGA	2461–2480	–	√	10	Sabin 1	[28]
247S-S2	CGAGATACAACACACATTAG	2461–2480	–	√	10	Sabin 2	[28]
248S-S3	CGAGACACCACTCACATTTC	2461–2480	–	√	10	Sabin 3	[28]
255S-S1	GGGTTAGGTGCAGATGCTTGAAAGCATG	2500–2526	–	√	10	Sabin 1	[28]
256S-S2	GGAATTGGTGACATGATTGAAGGG	2500–2523	–	√	10	Sabin 2	[28]
257S-S3	GGTATTGAAGATTTGATTTC	2500–2519	–	√	10	Sabin 3	[28]
P1-SENSE	GGICARATGYTTGARAGIATG	2506–2526	–	√	80	Serotype 1	[28]
P2-SENSE	CTIGGIGAYATIMTIGARGG	2503–2522	–	√	80	Serotype 2	[28]
P3-SENSE	ACIGARGTIGCICARKGYGC	2515–2535	–	√	80	Serotype 3	[28]
249A-S1	CACTGTAAATAGCTTATCC	2820–2802	–	√	10	Sabin 1	[28]
250A-S2	AACCGAAAACAATCTGCTG	2820–2802	–	√	10	Sabin 2	[28]
251A-S3	CATGGCAAATAGTTTCTGT	2820–2802	–	√	10	Sabin 3	[28]
PanPV4S	ACITAYAARGAYACIGTICA	2830–2849	–	√	40	Generic	[28]
PanPV2S	CITAITCIMGITTYGAYATG	2876–2895	–	√	80	Generic	[28]
PanPV1A	TTIAIIGCRTGICCRTTRTT	2954–2935	–	√	80	Generic	[28]
SeroP3A	CCCCIAIPTGRTCRTTIKPRTC	3176–3157	–	√	80	Generic	[28]
252A-S1	ATATGTGGTCAGATCCTTGGTG	3407–3387	–	√	10	Sabin 1	[28]

(continued)

Table 1
(continued)

Primer name	Sequence (5′ → 3′)[a,b]	Position[c]	qRT-PCR	qRT-PCR (pmol per µL)	Sequencing	Sequencing (pmol per µL)	Strains	Ref.
253A-S2	ATAAGTCGTTAATCCCTTTTCT	3407–3387	–	–	√	10	Sabin 2	[28]
254A-S3	ATATGTGGTTAATCCTTTCTCA	3407–3387	–	–	√	80	Sabin 3	[28]
PV8A	GCYTTRTTTGTTGICCRAA	3431–3412	–	–	√	80	Generic	[28]
PV10A	GTRTAIACIGCYTRTTYTG	3440–3421	–	–	√	80	Generic	[28]
Q8	AAGAGGTCTCTRTTCCACAT	3527–3508	–	–	√	10	Generic	[28]
AFRO P1W 1A	ACTGARAAYARYTTRTCYTTKGA	–	√	80	±		Generic	[29]
AFRO P1W 2S	ACTGARAAYARYTTRTCYYTKGA	–	√	80	±		Generic	[29]
AFRO P1W probe	FAM-CARCAYAGRTCIMGNTCAGARTCYAG-BHQ1	–	√	15	–		Generic	[29]
SOAS P1W 1A	ACTGARAAYARYTTRTCYYTKGA	–	√	80	±		Generic	[29]
SOAS P1W 2S	ACRGGRGCYACRAACCNTT	–	√	80	±		Generic	[29]
SOAS P1W probe	FAM-CARCAYAGRTCIMGNTCAGARTCYAG-BHQ1	–	√	15	–		Generic	[29]
AFRO P3W 5A	TCYTTRTAIGTRATGCGCCAAG	–	√	80	±		Generic	[29]
AFRO P3W 6S	GTYRTACARCGRCGYAGYAGRA	–	√	80	±		Generic	[29]
AFRO P3W probe	FAM-TTCTTYGCAAGIGGRGCRTGYGT-BHQ1	–	√	15	–		Generic	[29]
SOAS P3W 5A	TCYTTRTAIGTRATGCGCCAAG	–	√	80	±		Generic	[29]
SOAS P3W 6S	GTYRTACARCGRCGYAGYAGRA	–	√	80	±		Generic	[29]
SOAS P3W probe	FAM-TTCTTYGCAAGIGGRGCRTGYGT-BHQ1	–	√	15	–		Generic	[29]
panEV PCR1	GCGATTGTCACCATWAGCAGYCA	603–581	√	10	na		panEV	[28]
panEV PCR2	GGCCCCTGAATGCGGCTAATCC	458–480	√	10	na		panEV	[28]
panEV probe	FAM-CCGACTACTTTGGGWGTCCGTGT-BHQ1	546–568	√	5	na		panEV	[28]

Primer	Sequence	Position[b]				Target	Ref
panPV/PCR-1[d]	AYRTACATIATYTGRTAIAC	2978–2956	√	80	±	Pan-Poliovirus	[28]
panPV/PCR-2	CITAITCIMGITTYGAYATG	2876–2895	√	80	±	Pan-Poliovirus	[28]
panPV Probe 21A	FAM-TGRTTNARIGCRTGICCRTTRTT-BHQ1	2957–2935	√	15	−	Pan-Poliovirus	[28]
seroPV1A[d]	ATCATIYTPTCIARPAITYTG	2528–2509	√	40–80	±	Serotype 1	[28]
seroPV1,2S	TGCGIGAYACIACICAYAT	2459–2477	√	40–80	±	Serotype 1	[28]
seroPV1 Probe 16A	FAM-TGICCYAVICCYTGIGMIADYGC-BHQ1	2510–2488	√	15	−	Serotype 1	[28]
seroPV2A	AYICCYTCIACIRCICCYTC	2537–2518	√	40–80	±	Serotype 2	[28]
seroPV1,2S	TGCGIGAYACIACICAYAT	2459–2477	√	40–80	±	Serotype 2	[28]
seroPV2 Probe 5S	FAM-CARGARGCIATGCCICARGGIATNGG-BHQ1	2482–2507	√	15	−	Serotype 2	[28]
seroPV3A[d]	CCCIAIPTGRTCRTTIKPRTC	3178–3157	√	40–80	±	Serotype 3	[28]
seroPV3S	AAYCCITCIRTITITYTAYAC	3037–3056	√	40–80	±	Serotype 3	[28]
seroPV3 Probe 11S	FAM-CCRTAYGTNGGITTRGCVAAYGC-BHQ1	3091–3113	√	5	−	Serotype 3	[28]
Sab1/PCR-1	CCACTGGCTTCAGTGTTT	2600–2583	√	5	±	Sabin 1	[28]
Sab1/PCR-2[d]	AGGTCAGATGCTTGAAAGC	2505–2523	√	5	±	Sabin 1	[28]
Sab1/Probe[d]	CY5-TTGCGCGCCCCACCGTTTCACGGA-BHQ3	2563–2540	√	5	−	Sabin 1	[28]
Sab2/PCR-1[d]	CGGCTTTGTGTCAGGCA	2595–2579	√	5	±	Sabin 2	[28]
Sab2/PCR-2	CCGTTGAAGGGATTACTAAA	2525–2544	√	5	±	Sabin 2	[28]
Sab2/Probe[d]	FAM-AITGGTTCCCCGACTTCCACCAAT-BHQ1	2550–2572	√	5	−	Sabin 2	[28]
Sab3/PCR-1[d]	TTAGTATCAGGTAAGCTAIC	2591–2572	√	5	±	Sabin 3	[28]
Sab3/PCR-2	AGGGCGCCCTAACTTT	2537–2552	√	5	±	Sabin 3	[28]
Sab3/Probe[d]	ROX-TCACTCCCGAAGCAACAG-BHQ2	2554–2571	√	5	−	Sabin 3	[28]

[a]IUB ambiguity codes; I, inosine; P, pyrimidine base analog (T + C). Probes: Reporters Cy5, FAM, Rox; Quenchers BHQ1 and BHQ3

[b]Position relative to the positions reported by Toyoda et al. [27]

[c]Inosine-containing primers, use degenerate PCR conditions (lower annealing and extension temperatures)

[d]Modification of previously published primer sequence

9. Spectrophotometer for measuring the amount of nucleic: NanoDrop ND-1000 Spectrophotometer (Thermo Fisher Scientific, 3411 Silverside Road, Bancroft Building, Suite 100, Wilmington, DE 19810, USA) or equivalent.

10. Micropipettes with adjustable volumes with the following ranges up to 0.5 to 10, ≤2 to 20, 10 to 100, 20–200, and 200–1000 μL.

11. Sterile micropipette tips for volumes up to 10, 20, 100, 200, 1000 μL with aerosol barriers.

12. Computer with connection to the internet and loaded with computer programs for analysis of nucleic acid sequences. Highly recommended: Sequencher version 5 or later (GeneCodes, Ann Arbor, MI, USA).

3 Methods

3.1 Sewage Collection and Concentration with High Throughput with Minimal Turnaround Time

Composite sewage samples (1–2 L) are collected by pooling aliquots of sewage collected at timed intervals over a 24 h period by in-line automatic collectors at the inlet to sewage treatment facilities. Composite sewage samples are collected at upstream sites and at sewage treatment facilities lacking in-line collectors by using automatic portable, computerized composite sewage collectors (Sigma SD900 portable samplers, HACH, Loveland, CO, USA) (*see* **Note 12**). Samples must be transported under cold chain conditions (kept below 4 °C) to the processing laboratory to avoid loss of viability of any poliovirus in the samples. Surveillance sites are chosen according to WHO recommended Guidelines for environmental surveillance of poliovirus circulation [7]. A revised version of these WHO guidelines is available on-line from the WHO (http://www.polioeradication.org/Portals/0/Document/Resources/GPLN_publications/GPLN_GuidelinesES_April2015.pdf last accessed 17-05-2015).

Method for concentrating virus in sewage by 30- to 40-fold (*see* Fig. 1 for timing and time course for performing each step):

1. The starting volume of sewage for concentration is 500 mL (±50 mL). Volume in excess of 1 L is discarded after decontaminating with chlorine solution. Pour approximately 500 mL of the sewage plus sediment into a 2 L wide mouth plastic bottle. The remaining half liter is poured into a 500 mL disposable plastic bottle and stored at 4 °C for backup if needed. Sewage samples may be stored at 4 °C for up to 2 weeks until processing is begun without significant loss of virus viability and for up to a month for recovery of genomic RNA for molecular assay but this should be validated for conditions relevant for your laboratory.

2. Add 40 g of polyethylene glycol 6000 and 8.8 g of AR grade NaCl (*see* **Note 13**).

3. Add a sterile magnetic bar (*see* **Note 6**) and stir for ≥60 min at room temperature in a chemical hood.

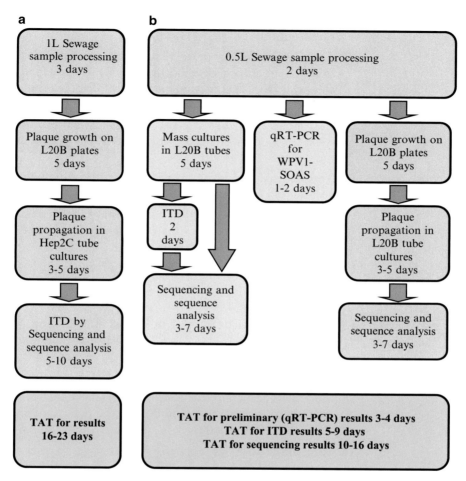

Fig. 1 Poliovirus isolation and identification working algorithms employed during the routine environmental surveillance (*a*) and after discovery of the WPV1-SOAS (*b*). The working algorithm shown in A allows isolation of any poliovirus type by sequence analysis within 16–23 days and allows processing of 6–12 samples/week. The quantitative evaluation of plaque number is done after the final identification. The working algorithm shown in *b* incorporating additional steps (*blue shapes*) allows primary detection and quantification of WPV1-SOAS by qRT-PCR within 3–4 days, confirmation by ITD and quantitative evaluation of plaque number within 7–9 days and full sequence analysis within 10–16 days. It also allows processing of up to 50 samples per week

4. Precipitate the virus by letting the sample sit overnight at 4 °C.

5. Gently spill off some of the uppermost liquid, mix the sediment and remaining liquid, and add this suspension to one 500 mL centrifuge tube or two 250 mL centrifuge tubes (*see* **Note 14**). Decontaminate any excess liquid with chlorine solution.

6. Centrifuge the samples at $10,000 \times g$ for 1 h at 4 °C (*see* **Note 15**). **Caution**: Use a transfer pipette to balance tubes before

centrifugation and decontaminate any excess liquid with chlorine solution.

7. The virus is in the pellet. Therefore, carefully pour off the supernatant into a new bottle into which you add chlorine solution (for decontamination) and dispose the chlorine-treated liquid after 30 min.

8. Resuspend the pellet in 7 mL of PBS–Tween 80. Pipette and wash the sides of the centrifuge tube to resuspend all pelleted material.

9. Transfer this suspension to a 50 mL centrifuge tube and add ½ volume of chloroform.

10. Agitate vigorously for 20 min on a rotary shaker at room temperature to break apart small clumps and inactivate bacteria (*see* **Note 16**).

11. Centrifuge the suspension at $1400 \times g$ for 20 min at 4 °C.

12. Transfer the supernatant that contains most of the virus to a fresh 50 mL test tube.

13. Reextract any virus trapped in the pellet by adding 2–3 mL of 3 % beef extract (pH 7.2) to the pellet and mixing vigorously for 5 min (*see* **Note 16**).

14. Centrifuge at $1400 \times g$ for 20 min at 4 °C.

15. Add the supernatant to the liquid from **step 12**. Decontaminate (with chlorine solution) and discard the pellet and any chloroform that was carried over.

16. Add antibiotics for concentrated sewage to the combined supernatants (for each 50–60 mL of supernatant, add 0.5 mL of PSF, 0.5 mL of Mycostatin stock, and 0.8 mL of PSMY stock). The final volume should be approximately 15 mL. If it is less, bring the volume to 15 mL by adding medium containing PSF and PSMY. **Important**: Record the volume of the concentrated sewage for calculating the fold concentration (*see* **Note 17**).

17. Incubate the processed sewage from **step 16** for 20 min at room temperature before proceeding or freezing the sample at −20 °C (*see* **Note 18**).

3.2 Cell Cultures for Identification of Virus in Concentrated Sewage

3.2.1 Preparation and Maintenance of Stocks of Uninfected L20B, RD, HEp2C and BGM Cells

Cell cultures are essential for testing for the presence of viable virus in sewage surveillance samples. Cell lines are passed once a week on the fifth workday into flasks for maintaining the cell lines and into tube cultures for viral challenge on the second workday of the following week. Each week there must be enough flasks of cell cultures on hand to maintain continuous passage of each cell line and provide enough extra for preparing the necessary number of uninfected tube cultures and plate cultures anticipated for characterization of the virus in the samples during the coming week. Cell cultures are renewed periodically from frozen stocks. Confirm the sensitivity of L20B and RD to polioviruses midway through the expected use of 15 passages or every 3 months. The other cell lines should also be checked but the interval between sensitivity testing can be longer (*see* **Note 19**). Uninfected and infected cell culture tubes should be maintained in separate incubators.

1. Growth medium and passage ratios for maintaining uninfected cell cultures: Grow cultures at 37 °C in closed 75 cm² flasks (*see* **Note 10**).

   - **L20B** cells stock cultures are passage at a ratio of 1:3 or 1:4.
   - **RD** at a ratio of 1:5; **HEp2C** at a ratio of 1:6 or 1:7.
   - **BGM** at a ration of 1:4. Complete M199 medium is used for L20B, RD, and BGM.
   - **HEp2C** cells are cultured in complete Eagle's MEM-NAA. L20B can also be cultured in complete Eagle's MEM-NAA.

2. Preparations of cell suspensions for maintenance and for tube and plate cultures.

   - Pour off medium.
   - Add 20–30 mL of Trypsin-Versene solution (*see* **Note 20**).
   - Pour off all but a few milliliters of the Trypsin-Versene solution.
   - Incubate the culture at 37 °C for >5 min.
   - Tap the bottle to release cells.
   - Add 5 mL of complete tissue culture medium (with FBS).
   - Pipette up and down a few times to break apart any clumps.
   - Either aliquot the cell suspensions in the ratios indicated above for continuing passage or add complete medium in the amounts indicated in the next paragraph for plating the cells in tube cultures or petri dishes for viral challenge and isolation (*see* **Note 21**).

3. Tube cultures per sample to be tested: Four **L20B** and four **RD** tube cultures are prepared 1 week before use. You will also need to prepare at least one L20B culture tube for repassage of supernatant from RD cells that develop cytopathic effects (CPE) during the previous week (*see* **Note 22**). Prepare at least one additional tube for each cell line to serve as uninfected control.

   - Dilute cell suspensions from one 75 cm² flask diluted into a final volume of 85 mL of complete M199 medium.
   - Add 2 mL of this is suspension to each 12 mL cell culture tubes (*see* **Notes 23** and **24**).
   - Close the cap tightly and incubate at 37 °C in test tube racks that keep the tubes at a 20° slant (*see* **Note 10**).

4. Tube cultures for **HEp2C** cells. Prepare one HEp2C culture tube for each plaque isolate to be amplified. Prepare at least one additional tube to serve as uninfected control (*see* **Note 24**).

   – Resuspend cells from one 75 cm² flask in a final volume of 85 mL of complete MEM-NAA medium and

   – Add 2 mL of this is suspension to 12 mL cell culture tube.

   – Close the cap tightly and incubate at 37 °C in test tube racks that keep the tubes at a 20° slant (*see* **Note 10**).

5. **L20B monolayers for plaque assay and plaque isolation**. Prepare three 10 cm petri dishes of L20B cells for each sewage sample and one for negative challenge control 4 days before intended use.

   – Dilute suspended cells from one 150 cm² flask in 100 mL of complete M199 medium.

   – Plate 12.5 mL of this suspension into each 10 cm style tissue culture petri dishes.

   – Incubate plates in a 5 % CO_2 incubator at 37 °C.

6. **BGM and RD monolayers for plaque assay and plaque isolation** (*see* **Note 25**). Prepare one 10 cm petri dishes of RD cells for each sewage sample and one for negative challenge control 4 days prior to viral challenge.

   – Dilute suspended cells from one 150 cm² flask in 100 mL of complete Eagle's MEM-NAA medium—Plate 12.5 mL of this suspension into each 10 cm style tissue culture petri dishes 4 days before intended use.

   – Incubate plates in a 5 % CO_2 incubator at 37 °C.

3.2.2 Viral Challenge of Uninfected Cell Cultures

Viral challenge of tube cultures

1. Thaw concentrated sewage in room temperature water and gently vortex to ensure homogeneous resuspension of the virus.

2. Replace complete medium with viral growth medium in the tube cultures.

   – Pour off or remove complete medium by vacuum.

   – Replace with 2 mL of virus growth medium.

   – Tightly close the cap tubes and incubate tubes at 37 °C.

3. Add concentrated sewage (200 μL) to each of four replicate tube cultures of L20B cells and four replicate tube cultures of RD cells.

4. Maintain the tube cultures at 37 °C for 5–7 days (*see* **Note 10**).

5. Visually inspect for CPE.

- CPE-positive L20B cultures most likely contain polioviruses. The presence of poliovirus is confirmed and the
serotype and type determined by molecular TD and ITD
assays described in Subheading 3.4 below.

- RD CPE-positive cultures may contain human enteroviruses including poliovirus.

6. Confirm whether CPE is caused by poliovirus by characterizing the virus present in CPE-positive cultures (*see* **Notes 26
and 27**, and especially **Note 28**).

- Perform molecular ITD (Subheading 3.4) on RNA
extracted (Subheading 3.3) from CPE-positive supernatants of L20B cells.

- Perform molecular ITD (Subheading 3.4) on RNA
extracted (Subheading 3.3) from CPE-positive supernatants of L20B cells. Significant amounts of work and
molecular reagents can be saved by repassing 200 μL of
supernatant from CPE-positive RD tube cultures (or
pooled supernatant from replicate RD tube cultures) on
L20B tube cultures. These cultures are called RL cultures
to indicate that the L20B tube culture was infected with
supernatant from and RD tube cultures (*see* **Note 27**).

- If CPE develops on RL tube cultures, the presence of
poliovirus is confirmed by molecular ITD assay.

Plaque assays and plaque purification

1. Infect the cell monolayer.

- Dilute 4 mL of concentrated sewage with 12 mL of plaque
plating medium.

- Remove medium by vacuum from three petri dishes of
L20B and one petri dish of BGM plated 4 days earlier

- Add 4 mL of diluted sample from **step 1** to each plate (*see*
Note 29).

- Rock the plates gently to ensure that the liquid is evenly
distributed over the entire surface.

- Incubate for 1 h at 37 °C. Rock the plates gently every
15 min to keep all of the surface area of the monolayer
moist and all cells alive.

- Prepare the agar overlay and keep it at 50 °C in a water
bath during this incubation period in the previous step.

- Remove the 4 mL of diluted sample by vacuum

- Gently overlay the monolayer with 12.5 mL of the agar
overlay. Add the agar to the lower sidewall of the plate that
is held at a shallow angle rather than directly onto the cell
monolayer to avoid tearing the monolayer. Make an effort
not to create bubbles.

- Let the overlay solidify at room temperature (approximately 5–10 min).
- Invert the petri dishes.
- Incubate them at 37 °C for a minimum of 2 days.

2. Inspect the monolayers for plaques and count the number of plaques.
   - Invert the petri dishes so the cells are on the bottom.
   - Add 8.5 mL of plaque staining solution.
   - Incubate at 37 °C for 1 h.
   - Remove the plaque staining solution with vacuum.
   - Incubate the petri dishes in an upright position at 37 °C overnight.
   - Count the number of plaques on each 10 cm-style petri dish. **CAUTION**: never hold the plates above your face when counting plaques to reduce the chance of bio-contamination by liquid falling from the plates onto your face or body (*see* **Note 30**).
   - Calculate the number of plaques in 1 mL of unconcentrated sewage (PFU/mL) by multiplying the average of the number of plaques on the three plates by the dilution factor (starting volume approximately 500 mL divided by the volume of concentrated sewage recorded) (recorded in Subheading 3.1).

3. Pick individual plaques of virus.
   - Pick virus in a plaque by stabbing through the agar overlay using a disposable plastic volume transfer pipette (*see* **Note 31**).
   - Transfer the agar plug from the stab into a 2 mL microcentrifuge test tube containing 200 μL of plaque suspension medium (*see* **Note 32**). **Caution**: Take precaution so that the transfer of the agar plug does not cause the liquid to splash.
   - The tube with the agar plug can be store at 4 °C for up to 4 days or at −20 °C until primary stocks of the virus are prepared in tube cultures (HEp2C or RD preferred).

4. Primary amplification of virus in the plaque—preparation of a primary stock.
   - Vortex the plaque suspension.
   - Spin down the liquid by a brief centrifugation.
   - Add 150 μL of the supernatant HEp2C or RD tube culture containing viral growth medium.
   - Incubate cultures for up to 5 days at 37 °C.

- Inspect the cultures for CPE daily starting on the third day.

- Transfer tubes with full CPE to 4 °C or freeze them at –20 °C or at –70 °C (preferred) (*see* **Note 33**).

5. Prepare high-titer secondary stocks (*see* **Note 34**).
High-titer secondary stocks better suited for some types of molecular analysis such as sequencing can be prepared as follows:

- Add 150–200 μL of primary stock to a fresh HEp2C or RD tube culture.

- Incubate for 3–5 days at 37 °C.

- Transfer tubes with full CPE to 4 °C or freeze them at –20 °C or at –70 °C (preferred) (*see* **Note 34**).

3.3 RNA Extraction from Tissue Culture Supernatants and Concentrated Sewage

RNA is extracted from supernatants of individual CPE-positive tube cultures, from pools supernatants of tube cultures (*see* **Notes 27** and **35**) and from concentrated sewage. The volume used for extraction varies between 50 and 1000 μL depending on the extraction procedure used. Most commercial extraction systems provide RNA of more or less equivalent quality (for example *see* [14]). Two alternatives are provided here: (1) total nucleic acid (NA) extraction using the EasyMag system (bioMérieux, Marcy l'Etoile, France); and (2) RNA extraction using a KingFisher System (Waltham, MA, USA). We recommend adding protected non-polio RNA such as shielded RNA or MS2 coliphage (ATCC 15597-B1) either to the lysis buffer to serve as an external control for the presence of inhibitors of PCR or RT-PCR chemistry that might be co-extracted with the RNA or adding them to each sample before extraction to serve as a control for extraction and inhibitors. We add MS-2 to the lysis buffer at a concentration sufficient to produce 28–32 Cts in a multiplex qRT-PCR. Each extraction run should also include extraction of (1) a negative control that also serves as positive for MS2 in the absence of inhibitors; and (2) an extract positive control (*see* **Note 36**), a sample containing an amount of enterovirus that will give a consistent Ct in the range of 28–32 after extraction for quality control for extraction.

1. Thaw stored samples for samples that have been frozen.

- Thaw in room temperature water.

- Vortex to prepare a uniform suspension.

2. Extract the RNA in the sample (*see* **Note 28**):

EasyMag extraction system

- Add up to 1 mL of tissue culture supernatant or concentrated sewage to 2.0 mL of lysis buffer containing MS2 coliphage at a concentration of 10,000 PFU/mL.

- Vortex.
- Extract according to the manufacturer's instructions on the NucliSENS EasyMAG system.
- Elute the RNA in 55 μL elution buffer (supplied with the kit) or sterile nuclease-free water.

KingFisher Extraction System

- Add 50 μL of tissue culture supernatant or concentrated sewage to the lysis well on the 8-well strip of the RNA extraction kit.
- Extract using the recommended viro_totRNA_KFmini protocol that can be downloaded and installed on the KingFisher extractor using the Thermo Scientific BindIt Software program (Waltham, MA, USA).
- RNA is eluted in 55 μL elution buffer (supplied with the kit) or sterile nuclease-free water.

3. Store the extracted RNA.

- Store RNA at –70 °C pending analysis.

3.4 Nucleic Acid Amplification for Poliovirus Molecular Serotype Identification (TD) and Intratypic Differentiation (ITD)

The latest WHO recommended standard kits for poliovirus TD and determining whether the poliovirus is vaccine-like, vaccine-derived, or wild, e.g., ITD differentiation (and the SOPs to perform them can be obtained from the Polio laboratory at the CDC, Atlanta, Georgia). These kits can directly identify all wild and vaccine polioviruses from sewage and indirectly identify VDPVs by a process of elimination when all of the reactions are run (pan-entero; pan-polio; generic type 1, 2, and 3; Sabin 1, 2, and 3; and optionally VDPV 1 and 2 and SOAS wild type 1 and 3 and WEAF wild type 1 and 3). Since the kits are currently under revision, we will not include their protocols here. Instead we will provide detailed instructions on how to provide critical rapid, high through-put information necessary for making evidence-based decisions during outbreaks [9, 15, 16] by modifying the kits, primers, and/or probes and/or procedures for a virus of interest once that virus has been detected and identified. [Caution: the procedures outlined here are for detecting the virus of interest. Unless the other standard assays are also applied, other viruses of interest may be overlooked.] All specific assay modifications should conform to guidance for assay design and optimization, analytical validation, and environmental field validation in Saunders et al. [17]. For these procedures, plaque purified poliovirus stocks need to be prepared for designing specific primers and probes to identify the virus of interest. The steps that follow were designed for amplification and detection by TaqMan technology using an Life Technologies 7500 sequence detection systems (Life Technologies, NY, USA) (*see* **Note 37**).

1. **Prepare primary and secondary virus stocks of the virus of interest for preparing specific reagents for detection and characterization.**

   - Plaque-purify the virus of interest as described in Subheading 3.2 above (*see* **Note 38**).

   - Prepare a **primary stock**, from one plaque.

   - Determine the PFU/mL of the primary stock.

   - Prepare a **secondary stock** at a multiplicity of infection (MOI) of one PFU per cell (*see* **Note 38**).

   - Incubate at 35 °C for 48 h.

   - Freeze-thaw the tube at –70 °C to increase viral yield (*see* **Note 33**).

   - Clarify the cell debris by centrifugation at $3000 \times g$ for 10 min.

   - Determine the PFU/mL of the secondary stock as above.

   - Adjust the virus concentration of the secondary stocks to 2×10^8 PFU/mL.

   - Aliquot the secondary stock and store at –70 °C.

2. **Design primers and probes specific for the virus of interest.**

   - Sequence the complete viral capsid protein 1 (VP1) of the twice plaque-purified secondary stock or stocks using generic primers Y7 and Q8 (Table 1) and the procedure outlined in Subheading 3.5 below (*see* **Note 38**).

   - Align VP1 sequences from more than one related virus isolate of interest (*see* **Note 38**).

   - Use a computer program (*see* **Note 39**) to design at least two different nondegenerate pairs of primers and probe suitable for qRT-PCR that will not react with the corresponding OPV strain [9] (*see* **Note 40**).

3. **Optimize reagents for a standard qRT-PCR assay protocol.**

   It is assumed that the operator is familiar with the use of the qRT-PCR machine. Remember to set the qRT-PCR machine to read the dyes used on the probe. We optimize all qRT-PCR reactions to the same run conditions so that we can test different viruses or combinations of viruses in the same run (*see* **Note 41**). The procedures here are for qRT-PCR on an ABI 7500 instrument (Life Technologies, Thermo Fisher Scientific, Waltham, MA, USA) using 8-well microtiter strips for maximum flexibility (*see* **Note 42**). Modifications must be validated if different qRT-PCR machines, enzymes, and buffers are used.

Molecular Area 1

- Prepare a work sheet listing all RNAs to be tested and the position they will appear on each strip before starting any qRT-PCR run. Include positive and negative controls for amplified target and the extraction positive control (*see* **Note 36**).

- Take cold metal blocks kept at 4 °C in the refrigerator to keep reagents and mixtures cold throughout preparation.

- Remove an aliquot of Mastermix from the AgPath-ID™ One-Step RT-PCR Kit (Life Technologies/Rhenium LTD) kit that had been kept at –20 °C, thaw it by spinning the tube in a microcentrifuge, vortex it gently for 5 s, and then respinning it for a few seconds in a centrifuge.

- Place the Master mix tube in the cold block and transfer the calculated amount into a 1.5 mL test tube. Add a black dot on the top of the Master mix tube each time it is thawed and immediately return the Master mix to the freezer. Do not use after five freeze–thaw cycles.

- Once the amount of Mastermix has been transferred to the 1.5 test tube, remove the Reverse Transcriptase/Polymerase enzymes tube from the –20 °C freezer, briefly centrifuge it for 5 s, and transfer the required amount of enzyme to the tube with the Master mix. Return the enzyme immediately to the –20 °C freezer to maintain activity of the enzyme.

- Place 0.2 mL tube strips on the cold block. Label the top end of each strip with strip numbers corresponding to those on the work sheet.

- Vortex the tube containing Master mix plus enzyme for 5 s, spin-down for 5 s, and then aliquot 20 μL to each reaction tube.

- Gently place a cover on top of the 0.2 mL strip. Hold the strip cap in place by closing only the tubes at both ends of the strip, but do not close them completely (*see* **Note 43**).

- Do NOT remove the cold block from Molecular Area 1. Instead, transfer the strips to an empty chemically and biologically sterile pipette tip box kept in Molecular Area 1 and bring the transfer box to a 4 °C refrigerator in Molecular Area 2.

Molecular Area 2

- Transfer the strips to a metal cold block in a dead-air box in Molecular Area 2.

- Remove the extracted RNA from the –20 °C freezer.

Table 2
Assay results

	Results for RNA extracted directly from environmental samples						
	qRTPCR	Tubes	MS2 ΔCt	Preliminary report	Action	TC[a]	Final interpretation
1	≤37	≥1	≤3	Positive	None	Positive	Positive
2	≤37	≥1	≤3	Positive	Redo TC	Negative	Positive
3	>37 and ≤45	2	≤3	Weak positive	None	Positive	Positive
4	>37 and ≤45	2	≤3	Weak positive	None	Negative	Weak positive
5	>37 and ≤45	1, 1 negative	≤3	Pending	Repeat qRT-PCR in triplicate[a]	Positive	Positive
6	>37 and ≤45	1, 1 negative	≤3	Pending	Repeat qRT-PCR in triplicate[a]	Negative	Positive OR weak positive OR inconclusive[b]
7	>45	2	≤3	Pending	None	Positive	Positive
8	>45	2	<3	Pending	None	Negative	Negative
9	>45	2	>6	Pending	Dilute RNA 1:10 and/or add 1.2 % BSA; repeat qRT-PCRa in triplicate	Positive	Positive
10	>45	2	>6	Pending	Dilute RNA 1:10 and/or add 1.2 % BSA; repeat qRT-PCRa in triplicate	Negative	Indeterminate (inhibition not resolved) OR result of qPCR[c]
	Results for RNA extracted from tube cultures						
1	<30		≤3			L20B positive	Positive
2	≥30[d]		≤3			L20B positive	Negative

[a]TC results will only be available 1 (L20B) or 2 (RD→L20B) weeks after qRT-PCR so samples are repeated before knowing TC results

[b]TC Negative: The sample is "Positive" if at least one of the triplicate repeats is ≤37; "weak Positive" if at least one of the triplicates is 37 < result ≤45; and "inconclusive" if none of the triplicate repeats are ≤45 and the TC is Negative

[c]Ct > 6 inhibition is considered to be completely inhibited and qRTPCR is repeated using conditions for reducing inhibition. TC Negative: If the repeat triplicates are ≤45, results are interpreted as "Positive," "Weak Positive," or "Inconclusive" according to qRT-PCR criteria for TC Negative. If the results are >45 and the sample is still inhibited, then the result is "Indeterminate"

[d]Results normally fall between Ct 14–20. Higher Ct values than this may indicate some incompatibility with PCR chemistry, the presence of inhibitors, or the presence of mixtures of polioviruses. Virus can be repassed to obtain a higher titer and/or virus can be sequenced using algorithms for heterotypic or homotypic mixes

- Thaw the unknowns by centrifuging, vortexing, and respinning the extracted RNA samples in a microcentrifuge.

- Load 8-μL of RNA from the samples using a separate aerosol barrier tip for each RNA sample.

- Place the caps back on the strip and close firmly using a strip cap-closing tool kept in the dead-air box.

- Include additional reactions for control RNAs (positive poliovirus RNA, MS2 RNA, negative control, and extract positive and negative controls) stored in the –20 °C freezer.

- Spin-down the contents of the tubes in a small strip centrifuge.

4. **Optimization of qRT-PCR specific for the virus of interest**.

   Standard qRT-PCR conditions are followed for a reaction mix of 25 μL; 5 μL of RNA is added to a 20 μL reaction mixture containing AgPath-ID One-Step RT PCR reagents (Life Technologies, NY, USA), primers and probe specific for the poliovirus of interest, and MS-2 primers and probes.

   - RT at 48 °C for 30 min.

   - Taq polymerase activation at 95 °C for 10 min.

   - 50 cycles of strand separation at 95 °C for 15 s, and elongation at 60 °C for 1 min. Data collection is ON only for the 60 °C elongation step.

   - Interpretation of the assay results is according to Table 2 and the number of reactions for any given RNA is dependant on these initial results.

   - Evaluate different concentrations of the newly designed VP-1 nondegenerate primers (300, 600, and 900 nM) and probe (200 and 300 nM) to optimize the multiplex qRT-PCR in a multiplex using 150 nM each of MS2 external control primer and 50 nM MS-2 probe.

   - Choose the concentrations that provide the lowest limit of quantitation and detection as described in the next two subsections.

5. **Determine the Analytical sensitivity, the limit of detection (LOD), and precision of the qRT-PCR specific for the virus of interest**.

   - Determine the analytical sensitivity, the limit of detection (LOD), and precision of the qRT-PCR by testing in parallel 4 tenfold serial dilutions in water of RNA extracted from the secondary virus stock starting with 2×10^7 PFU/mL as in Dreier et al. [18].

   - Inspect the multicomponent curve for each qRT-PCR reaction to confirm that a positive Ct is actually the result of a true event.

6. **Determine the analytical specificity of the qRT-PCR specific for the virus of interest**.

    – Determine the analytical specificity of the specific primers and probe in silico using Basic Local Alignment Search Tool (BLAST) search (available at: http://www.ncbi.nlm. nih.gov, last accessed July 25, 2014)

    – Determine the analytical specificity of the specific primers and probe in vitro by testing them on RNA and DNA from polio and non-polio enteroviruses and as wide a range of reference strains of non-enteroviruses as the your laboratory possesses.

7. **Validation of the primers and probe for the qRT-PCR specific for the virus of interest**.

    – Validate the qRT-PCR by extracting 200 µL of pools of poliovirus-negative concentrated sewage collected before first appearance of the poliovirus of interest spiked with tenfold serial dilutions of the secondary stock (starting from 2×10^6 PFU/mL).

    – In parallel, confirm the actual titer in each dilution by plaque assay of 200 µL after bring the volume to 4 mL and using the plaque assay conditions as described above.

    – Finally validate the assay by testing 20–50 concentrated sewage samples known to contain the poliovirus of interest. Reconfirm that they contain polio by retesting on L20B tube cultures in parallel with the qRT-PCR.

8. **Validate your qRT-PCR for Sabin strains**.

    – Validate a multiplex qRT-PCR as above for a qRT-PCR multiplex using your Master mix, buffers and enzymes with CDC-designed [19] primers and probes for Sabin 1, Sabin 2, and Sabin 3 (Table 1).

    – Inspect the multicomponent curve for each qRT-PCR reaction to confirm that a positive Ct is actually the result of a true event.

9. **Important note added in proof: The CDC has started to distribute new ITD v4.0 and VDPV v4.0 Kits in 2015**.

    The reaction protocol of the ITD Kit v4.0 has been simplified and differs from previous versions of ITD kits. There are five not six reaction mixtures: (1) a quadriplex for pan-enterovirus, Sabin 1, Sabin 2, and Sabin 3; (2) a singleplex for pan-poliovirus; (3) a duplex for both currently circulating lineages of wild type 1 polio virus (SOAS and WEAF); and separate singleplex reactions for (4) wild type 3 SOAS; and (5) wild type 3 WEAF polioviruses. The three serotype-specific reactions from the previous ITD kit have been discontinued. In addition, changes have been made in the composition of some of the primers and probes and/or the dyes and quenchers. The sequences are provided in the Kit insert.

The VDPV v4.0 kits are still based identification of a VDPV by ruling out the presence of Sabin strains of the same serotype. The reaction protocol of the ITD Kit v4.0 also has been simplified and differs from previous versions of ITD kits.

As before when using both kits to characterize polioviruses in sewage in areas where OPV is still administered and sewage contains Sabin-like isolates, VDPVs especially highly diverged VDPVs may be missed in environmental samples containing homotypic mixtures of highly diverged VDPVs and Sabin strains of the same serotype.

The CDC has tested the new reaction mixtures and amplification protocols from the v4.0 kits using many available commercial enzymes. The efficiency of the reactions varied widely. For example for one of the Kit reaction components, Cts for the same RNA sample ranged from 12 to 36 and the quality of the amplification plots ranges from poor to excellent. The CDC recommends two commercial enzyme preparations that provided the best results for all of v4.0 kit reaction components. It is possible to purchase the primers and probes separately, formulate mixes with different buffers than these in the CDC kits, and use different enzymes. However given the extensive validation needed and the requirement for uniformity of analysis conditions for all the laboratories of the Global Poliovirus Laboratory Network, we now strongly recommend using the kits supplied by the CDC and the enzymes they recommend. The kits also contain positive controls that can be used to evaluate differences between labs when standard protocols are used. Furthermore we recommend validating the assay in your laboratory for both recommended enzymes in case there arise problems in availability for one of them.

3.5 Molecular Characterization of Poliovirus RNA by RT-PCR and Sequence Analysis

The complete VP1 sequence must be determined to confirm the presence of a poliovirus of interest once it has been identified by the qRT-PCR described in Subheading 3.4 above. For this, RNA is amplified by RT-PCR and both strands of the amplicon are sequenced. This section describes how to amplify the poliovirus VP1 gene, purify the amplicon, and sequence both strands on a Sanger-based [20] automatic sequencer. It is assumed that the reader either knows how to use a Sanger-based automatic sequencer or can send the processed cDNA to a sequencing laboratory. In the future it will be possible to obtain full poliovirus genomes using next-generation sequencing. Subheading 3.5 concludes with instructions on how to correct the raw sequence data generated by the automatic sequencer and how to easily and rapidly perform TD and ITD on the sequence.

As throughout this section, the work flow is unidirectional from Molecular Area 1 to Molecular Area 4 for both RT-PCR and Sequencing. The RT-PCR reaction mix (50 µL final volume with nucleic acid) is prepared in **Molecular Area 1**; the RNA is added to the RT-PCR mix in **Molecular Area 2**; the RT-PCR is performed in **Molecular Area 3**; RT-PCR results are analyzed in **Molecular Area 4**; and the amplicon is purified and labeled in **Molecular Area 4**. Sequencing reaction mix is prepared in **Molecular Area 1**; the amplicon is added and labeled in **Molecular Area 4**, the labeled amplicon is purified in **Molecular Area 4**; and the labeled amplicon is sequenced in **Molecular Area 3 or 4**. **RT-PCR** conditions are described for RT-PCR using the QIAgen one-step RT-PCR Kit (QIAGEN GmbH, Chatsworth, CA, USA). Conditions need to be validated and optimized when equivalent RT-PCR kits are used.

1. **Preparing specific amplicons for sequencing.**
   - Prepare the reaction mix in **Area 1**: 28 µL sterile distilled water, 10 µL of 5× buffer, 2 µL of 10 mM dNTP mix, 2 µL of RT-PCR enzyme mix, 0.5 µL of RNase inhibitor, 0.5 µL of 100 mM dDTT per reaction and 1 µL each of forward and reverse primers chosen by the following three algorithms based on whether or not the RNA contains mixtures of polioviruses and the serotypes of the viruses in the mixtures. Stock concentrations of the primers are listed in Table 1 and were calculated to provide the correct final concentration when 1 µL is added to the mix.
   - Add 5 µL of RNA in Area 2 (*see* **Note 44**).
   - Run the RT-PCR using the following conditions: 42 °C for 45 min; 94 °C for 10-min; 35 cycles of 94 °C for 30 s, 42 °C for 45 s, ramp to 60 °C T 0.4 °C/s, and 60 °C for 2 min; 60 °C for 5 min; and 4 °C hold.

Algorithm 1: the RNA contains only the virus of interest or mixtures of sequence variants of this virus.
   - Generate an amplicon containing the complete VP1 sequence by RT-PCR using generic forward primer Y7 or Y7R with generic reverse primer Q8.
   - Purify the amplicon.
   - Confirm that it is of the correct size (as described below).
   - Label the amplicons for sequencing in separate reactions with the same primers used to prepare the VP1 amplicons.
   - Sequence both strands of the amplicon in separate reactions using primers Y7 and Q8.

Algorithm 2: Heterotypic mixes - the RNA contains the virus of interest or mixtures of sequence variants of this virus together with polioviruses of different serotypes.

-   Generate two overlapping amplicons specific for the virus of interest. Specificity is obtained by running two hemi-nested RT-PCRs. One hemi-nested RT-PCR is with generic forward upstream primer Y7 (or Y7R) and the reverse primer from the specific primer pair designed for qRT-PCR in **step 2** of Subheading 3.4 above. The second hemi-nested RT-PCR is with the forward primer from the specific primer pair designed for qRT-PCR in **step 2** of Subheading 3.4 above and generic primer Q8 (*see* **Note 45**).

-   Purify the amplicon as described below.

-   Confirm that it is of the correct size by gel electrophoresis.

-   Label the amplicons for sequencing in separate reactions with the same primers used to prepare each amplicon.

Alternative for Algorithm 2:

-   Use generic primers Y7 or Y7R with Q8 to amplify the VP1 from all viruses in the mix.

-   Purify the amplicon as described below.

-   Confirm that it is of the correct size by gel electrophoresis.

-   Specifically label the amplicons for the virus of interest in separate reactions using the specific qRT-PCR primers for the virus of interest as sequencing primers (*see* **Note 46**).

-   When Sabin 2-like viruses may need to be sequenced from heterotypic mixtures of polio to rule out emergence or silent circulation of type 2 VDPVs, first try RT-PCR with primer Q8 and 247 s (1061 nt) and then Y7 with 253A (989 nt), and use the same primers for sequencing (*see* **Note 47**).

Algorithm 3 Homotypic mixes: The RNA contains the virus of interest or mixtures of sequence variants of this virus together with polioviruses of the same serotype, for example the vaccine strain.

The most difficult situation is to specifically sequence the genome of interest from a mixture of viruses of the same serotype.

-   Produce Amplicons using any of the methods in Algorithm 2.

-   Purify the amplicon.

- Confirm that they are of the correct size by gel electrophoresis.
- Selectively label the amplicons using qRT-PCR specific primers or any serotype specific primers that will not cross hybridize with the other viruses in the homotypic mix (*see* **Notes 46** and **47**).

2. Confirm that the amplicons generated from algorithms 1, 2, or 3 above are of the correct size by gel electrophoresis.
   - Load one tenth of the reaction mix onto a lane on a 2 % agarose gel containing 1 μL of a 1 mg/mL aqueous Ethidium bromide solution per 10 mL of agarose solution in a bath containing 1× E buffer.
   - Visualize the results with UV light, and record the results with a camera or other permanent recording device.
   - Estimate the size by comparison with fragments in a 100 base pair ladder DNA size marker loaded onto an adjacent lane and run at the same time.

3. Purify the specific amplicon from the mixture of buffers, enzymes and primers according to the whether the electrophoresis pattern in the previous step indicated a single band of expected size or more than one band, including one of the anticipated size.

Algorithm 1: a single band of anticipated size.

Purify the amplicon directly from the mix remaining in the RT-PCR tube using a High Pure PCR product Purification Kit (Roche Diagnostics, Indianapolis, IN, USA).

- Add binding buffer (green cap) in the amount of five times the volume of the reaction mix not including the volume of the mineral oil (if any). It is not necessary to remove the mineral oil.
- Mix by pipetting gently to avoid spilling.
- Transfer all of the liquid to the High Pure filter column (*see* **Note 48**).
- Centrifuge for 1 min at $8000 \times g$ at room temperature.
- Transfer the column to a clean 2 mL test tube. Discard the flow through.
- Wash the column by adding 500 μL of wash buffer (blue cap) and centrifuging for 1 min at $8000 \times g$.
- Transfer the column to a clean 2 mL test tube. Discard the flow through.
- Wash the column by adding 200 μL of wash buffer (blue cap) and centrifuging for 1 min at $8000 \times g$.

- Transfer the column to a clean 1.5 mL test tube. Discard the flow through.
- Centrifuge for 1 min at $8000 \times g$.
- Transfer the column to a new clean 1.5 mL test tube labeled on the lid with the name of the sample.
- Add 50 µL of elution buffer (vial #3), or sterile distilled water. If the band is faint elution can be in a minimum of 25 µL.
- Let stand for 1 min at room temperature.
- Centrifuge for 1 min at $8000 \times g$. Discard column.
- Save the flow through liquid containing the amplicon at –20 or –70 °C.

Algorithm 2: Multiple bands, one of which is a band of anticipated size.

- Load the remaining volume of reaction mix onto a 2 % agarose gel (*see* **Note 49**).
- Load a 100 base pair size marker in a separate lane.
- Electrophorese the sample and marker.
- After electrophoresis, excise the band of the correct size for the amplicon from the agarose gel with a clean sharp disposable scalpel blade and place it in a pre-weighed 15 mL test tube. Re-weigh the tube to determine the net weight of the agarose gel slice by subtracting the initial weight from the final weight (*see* **Note 50**).
- Add 3 volumes of Buffer QG to 1 volume of 1 % agarose gel or 6 volumes of Buffer QG to 1 volume of 2 % agarose gel using the approximation that 100 µL is equivalent to 100 mg (*see* **Note 50**).
- Incubate at 50 °C for 10 min with flicking every 2–3 min. Ensure that pH remains OK (solution yellow) otherwise correct with 3 M sodium acetate (pH = 5.0).
- Add 1 volume of isopropanol and mix.
- Load sample on column (≤800 µL per run) and centrifuge at $10,000 \times g$ for 1 min on a microcentrifuge.
- Place the QIAquick column into new 2 mL eppendorf centrifuge tube, not into same tube as specified in manufacturer's protocol. Discard tube and flow through (*see* **Note 51**).
- Add 500 µL of Buffer QG and centrifuge at $10,000 \times g$ for 1 min on a microcentrifuge centrifuge.
- Transfer the QIAquick column into new 2 mL eppendorf centrifuge tube, not into same tube as specified in manu-

facturer's protocol (*see* **Note 51**). Discard tube and flowthrough.

- Wash the column by adding 750 μL of Buffer PE, incubate at RT for 5 min, and centrifuging at $10,000 \times g$ for 1 min.

- Transfer the column into new 2 mL tube and centrifuge at $10,000 \times g$ for 1 min rather than discarding flowthrough and reusing same tube as in manufacturer's protocol (*see* **Note 51**). Discard the original tube with flowthrough.

- Transfer the spin column into 1.5 mL microcentrifuge tube and elute the DNA by applying 30–50 μL of water directly to the center of the QIAquick spin column membrane, letting the spin column stand for 2 min at RT, and centrifuging at $10,000 \times g$ for 1 min.

- Discard the spin column.

- Store DNA in the flow through at –20 or –70 °C.

4. **Determine the concentrations of the amplicon DNA using a NanoDrop ND1000 spectrophotometer** (Thermo Scientific, Waltham, MA, USA).

- Turn the machine on, and then open the computer program for the NanoDrop.

- Clean the spectrophotometer by lifting the sample arm and adding sterile distilled water to the measurement pedestal, closing and opening the sample arm and wiping both surfaces with a soft clean absorbent tissue (*see* **Note 52**).

- First blank the NanoDrop machine by adding 1.8 μL of sterile water to the pedestal of the sample arm, close the sample arm, and then press the blank (F3) button on the computer program.

- Open the sample arm and wipe both surfaces with a soft clean absorbent tissue.

- Add an aliquot of sterile water to the lower measurement pedestal.

- Close the sample arm and initiate a spectral measurement by pressing the measure button (F1) on the operating software on the computer.

- After the measurement wipe both surfaces of the sample arm with a soft clean absorbent tissue.

- Clean and repeat until the water blank gives a flat spectrum.

- Finally, add samples and measure by pressing the measure button (F1) on the operating software on the computer (*see* **Note 53**).

- Enter the name for each sample on the computer program and finished, print out a report.

- Clean the spectrophotometer at the end of use.

5. **Label the dsDNA amplicon for sequencing with the Z reaction**.

- Prepare the Z reaction mix for $n + 0.5$ reactions in **Molecular Area 1**: (2 µL Big Dye Terminator Ready Reaction Premix (Life Technologies, Foster City, CA, USA), 1 µL of 5× buffer Tris–HCl, pH = 9.0 and $MgCl_2$), 5 µL of primer diluted to give 3 pmol/reaction, and 5 µL of sterile deionized water: the final volume with cDNA is 10 µL.

- Transfer the tube with the Z mix to **Molecular Area 3**.

- In **Molecular Area 3**, add 1 µL the cDNA from **step 11** above to the Z mix after first diluting the cDNA to a concentration of 3–10 ng per 100 nucleotides per 1 µL water and add.

- Incubate the mix in a PCR machine using the following settings: 25 cycles of 94 °C for 20 s, 42 °C for 15 s, ramping to 60 °C at 0.4 °C/s, and 60 °C for 4 min, followed by a hold at 4 °C.—Purify the end-labeled cDNA using a BigDye X Terminator Purification Kit (Life Technologies, Foster City, CA, USA).

- Transfer the 10 µL reaction mix into a well of a 96-well microtiter plate.

- Add 45 µL of SAM Solution and 10 µL of BigDye X Terminator solution to the well. Use a wide bore tip for the BigDye X Terminator solution.

- Cover the palate with adhesive film.

- Thoroughly mix plate on a vortex or rotary shaker for 30 min.

- Centrifuge the plate at $1000 \times g$ for 2 min in a centrifuge with buckets suitable for 96-well microtiter plates.

6. **Sequencing the labeled VP1 amplicons**

- Transfer 15 µL of the Z reaction supernatant to the 96 well microtiter plate sample holder of the ABI 3100 or 3500 Genetic Analyzer (*see* **Note 54** for suggestions on how to name the sample).

- Sequence the amplicons.

- Save the graphic sequence output files (they end in .AB1) from the automatic sequencing machine on a CD, DVD, disc-on-key, or equivalent.

7. **ITD and sequence analysis of VP1**.

The protocol provided here is for the commercial Sequencher program v5.0 or later (GeneCodes, Madison, WI,

USA) because we feel offers a very user-friendly platform for correcting primary sequence data and for immediately resolution of ITD. It can also highlight amino acid substitutions or differences and aligned sequences can be easily exported in a number of different formats including FASTA NEXUS and Phylip for more advanced phylogenetic analysis using other computer programs. **Bold italicized text** refers to the menu commands and a forward slant "**/**" indicate that the command following the forward slant is in a submenu. There are other commercial, share-ware, and free-ware programs for sequence correction and preliminary analysis.

1. Prepare a Generic Sequencher project file for correction and preliminary ITD analysis of all polioviruses.

    - This file should contain text files for the complete genome of Sabin 1, Sabin 2, and Sabin 3 (DDBJ/GenBank/EMBL database access numbers V01150, AY184220, and X00925, respectively)

    - The file should contain text files of the complete VP1 sequence of all three Sabin strains extracted from the complete sequence files (906, 903, and 900 nt, respectively, for Sabin 1, Sabin 2, and Sabin 3).

    - This file should contain three copies of Y7 and Q8 primer sequences (*see* **Note 55**).

    - The file may contain graphic sequence files (files ending in .AB1) for amplicons of all three Sabin strains amplified using Y7 and Q8 primers.

    - Select all sequences (with the exception of the primers) and assemble automatically using the default values for Maximum Match Percentage (85 %) and minimum overlap (20 bp) in the Assembly Parameters to produce three separate contigs, one for each vaccine serotype. A contig is a sequence generated in silico by de novo assembly of sequences that partially or fully overlap.

    - Select each contig and a single copy of Y7 and Q8 primers and assemble automatically using the default values for Maximum Match Percentage (85 %) and minimum overlap (20 bp) in the Assembly Parameters.

    - Save this file with the name "proj polio".

    - Duplicate this "proj polio" file and use the duplicate COPY for ITD analysis of all poliovirus sequences.

2. Prepare a separate folder for each virus that will be sequenced and name it after the isolate. This folder should contain all raw (uncorrected) graphic sequence files from the given isolate even if from different regions of the genome and a copy of the "proj polio" file. The folder and project file should be renamed to reflect the name of the isolate (*see* **Note 56**).

3. Transfer the .AB1 raw sequence files to the computer with the Sequencher program.

   Add the name of the isolate to the beginning of the .AB1 file name (*see* **Note 57**).

4. Load the raw data .AB1 file into the project file by using the File/import and one of the options in the next submenu commands or by dragging and dropping the icons of the .AB1 sequence files onto the icon of Sequencher or the open Sequencher window.

5. Activate all newly loaded sequences (place the curser over the file name and for Mac: **Command click**; PC: **control click**) and trim the ends of the sequence (**Sequences/Trim ends without preview**).

6. Preliminary correction of individual sequences in the individual sequence windows.

   – Open all of the newly trimmed sequences, e.g., **double-click** on the name or icon. This will open separate windows for each of the trimmed sequences.

   – Click on the button **Show Chromatogram**. The computer generated base calls are located below the line above the graphic peak for each nucleotide. These computer generated base calls cannot be edited, however there is an editable copy of each base call that appears above the line. It is these above-line base calls that are used by the Sequencher program for comparisons.

   – Correct the bases where the program cannot unambiguously call a base and returns a degenerate IUB base code (examples: "R" for a pu*R*ine, an "A" or a "G", "N" for a *N*y base) by placing the curser on the base call.

   – Visually inspect the beginning and end of the chromatogram. **Trim ends without preview** removes bases above the line at the 5′ and 3′ ends. You can restore as many trimmed bases that are readable by placing the cursor on the computer generated bases below the line and choosing **Sequence/Revert to Experimental Data**.

   – Correct ambiguous base calls of the restored sequences after visual inspection of the graphic data (*see* **Note 58**).

   – Use the **Backspace** key NOT the **Delete** key to remove extra bases.

7. Prepare a contig of the corrected sequences from the previous step:

   – Activate the all the sequences form the same isolate (place the curser on the file name and Mac: **Command click**; PC: **control click**).

- Press the **Assembly Automatically** button.
- Place the cursor on the temporary text name of the contig, **double-click** to open the name for editing.
- Rename the contig with the name of the viral isolate.

8. Correct the sequences in the contig.

- Open the contig by **double-clicking** on the icon to visualize the graphic results (change from **Overview** to **Bases**). While in overview, drag window to the region you wish to correct before using bases.
- Correct the mismatches between overlapping sequences based on the graphics (*see* **Note 58**).
- Use degenerate bases if there are more than one graphic trace at a given nucleotide position and the graphic traces are approximately equal on all overlapping strands at that nucleotide position. Otherwise choose the major peak if it is the same on all overlapping strands.
- Save file at frequent intervals.

9. Prepare a text file containing the consensus sequence of the corrected contig.

- Select all bases in the consensus sequence (place cursor on consensus line and press Mac: **Command A**; PC: **control A**).
- Copy the consensus (Mac: **Command C**; PC: **control C**).
- Open a new sequence **Sequence/Create a new sequence**.
- Type in the name of the new sequence in the small window that opens (**isolate name**; space or underline; **VP1**).
- Click **enter**.
- Paste the consensus sequence into the new window that opens (Mac: **Command V**: PC: **Control V**); closing the window and selecting "**Record as Experimental data**" from the new window that opens (*see* **Note 59**).
- Close all open windows.
- Add the newly created consensus sequence to the contig by activating the contig and the newly created consensus text sequence and clicking on "**Assemble automatically**".
- Open the contig file and switch to **Bases**.
- Place the cursor on the name of consensus sequence and drag the newly created text sequence to the top of the list of sequences.
- Activate all sequence names in the Sequence Name window and click on **Sequence/compare bases to/"top sequence"**. The position of all mismatches will be indi-

cated in the window that opens. Go to each nucleotide position and correct any mismatches.

- **Save** the file. This is critical so that if there is a problem in the next steps before the next **save**, work will not be lost. if there is a problem it allows the option of closing the file without saving and then restarting from the saved version

10. Final correction and Automatic ITD determination. Determine the serotype and whether the virus is vaccine, a minimally diverged VDPV, a highly diverged VDPV, or wild by performing the following steps in order.

    - Activate the contig file containing the corrected consensus file and all three Sabin serotype files and form a new contig containing graphic and text files of the new sequence and one of the three Sabin prototypes by clicking on "**Assemble automatically**" (*see* **Note 60**).

    - Bring the consensus file of the sequence of interest to top of the list of sequences and the Sabin VP1 sequence immediately below it. If all has gone well, the consensus sequence of interest will be longer than the Sabin VP1 file.

    - Trim the consensus text file to the length of the Sabin VP1 sequence. If the sequence is shorter, it is possible to restore more of the trimmed sequences by placing the cursor on the position of the consensus sequence that includes the last base in the consensus sequence of the sequence of the virus of interest in **Base view mode** (bottom line), activating **Show Chromatogram**, placing the cursor over the lower window and clicking to activate the chromatogram window, going to the chromatogram of the graphic file of the sequence of interest in the lower window, moving the cursor to bases below the line on the graphic display of the graphic file of the sequence of interest and restoring trimmed bases as in **step 6**. Once again review and correct the restored segment (*see* **Note 61**).

11. **Save** the proj file. **Copy/paste** or **export** the corrected and trimmed consensus sequence for the virus of interest to your sequence data base.

12. If the unknown does not form a contig with one of the Sabin contigs using the default value for **Assembly parameters**, open the **Assembly Parameters** and successively increase the **Minimum Match Percentage** in steps of 2 % until a match occurs or you reach 81 %, since at 80 % all Sabin strains will form a single contig. If no match occurs at the 81 % setting, BLAST search the text consensus sequence of the unknown against the nt library at DDBJ/GenBank/EMBL either directly from Sequencher or by copy pasting the sequence into the BLAST window at NCBI (http://www.ncbi.nlm.nih.gov/

last accessed July 2014) to determine whether it is a poliovirus and which polioviruses serotype of wild or highly diverged VDPV it most closely resembles. Return to the Sequencher project file, activate the contig of the unknown and the Sabin serotype corresponding to the serotype of the closest matches in the BLAST search, and continue to increase the **Minimum Match Percentage** in steps of 2 % until a match occurs. Correct the file by comparison with the corresponding Sabin VP1 after removing paired gaps (*see* **Note 60**). **Save** the proj file. **Copy** or **export** the consensus sequence of your unknown isolate into your data base file. Change the "n" in "pvn" for the name of the folder and proj file to the serotype of the unknown virus.

4   Notes

1. In practice, most poliovirus-positive environmental samples contain one or at best a few polioviruses of interest [2, 5]. Under these conditions environmental surveillance protocols operate at the lower limits of detection. Thus negative findings cannot rule out the presence of polioviruses at levels below detection. However negative findings gain significance when they are part of a long sequence of negative results from frequent routine surveillance at the given site. It also requires that the limits of quantitation (LOQ) and limits of detection (LOD) of your molecular assays are well documented.

2. The procedures of reagent preparation; nucleic acid extraction; nucleic acid amplification; detection of amplified products, and other post-amplification manipulation should be carried out in separate rooms designates as Molecular Areas 1 through 4 using equipment and even personal protective equipment such as color coded lab coats dedicated for exclusive use in each area. (Use clearly separated work areas if it is not possible to use separate rooms.) The work schedule must be planned in advance so that there is unidirectional movement from Area 1 to Area 4 on the same workday in order to avoid molecular or biochemical contamination. **Molecular Area 1** is the area dedicated to the preparation of reagents for amplification by PCR. Use gloves and color-coded disposable lab coats at all times when working in this room. Do not use lab coats that have been outside of the room. Use a biological cabinet or dead-air box equipped for UV irradiation for handling all reagents using pipettors, pipettes tips, etc. that are kept in this room at all times. The only nucleic acids allowed in the room are primers and probes. It is prohibited to bring RNA or DNA, and especially amplified DNA into this room. After finishing work, the pipettors and any test tube racks should be decon-

taminated using chemical means or a UV Cross linker. **Molecular Area 2** is for work with unamplified nucleic acids for example nucleic acid extraction from sewage and for the addition of extracted nucleic acids to the master mix. The room should contain a BSL-2 laminar flow hood for working with the samples and a dead air box for addition of the nucleic acids to the reaction mix. The biological safety hood and the pipettes should be cleaned with bleach/PCR decontamination spray All pipettors, pipettes, tips, etc. used in Molecular area 2 should be kept in this room at all times. These pieces of equipment and this area should not be used for pipetting or processing amplified DNA. **Molecular Area 3** is an area used for molecular amplification and detection of amplified DNA. Infected tissue cultures that have reached full CPE, contain very high titers of virus and probably should be manipulated in Area 3 rather than Area 2. If Area 2 is used, work with the tissue cultures should be done last and the area thoroughly disinfected using agents that inactivate the virus and destroy any residual nucleic acids. Dedicated pipettors, pipettes and equipment should be kept in this area at all times. Use a color-coded lab coat in this area. Do not bring amplified DNA or any supplies or equipment from Area 3 into Areas 1 or 2. **Molecular Area 4** is the area where amplified nucleic acids are manipulated. Manipulation includes but is not limited to gel electrophoresis, purification and labeling of amplified nucleic acids, labeling for sequencing and cloning,

3. Initial test results and downstream molecular tests that are used depend on the types of poliovirus that may be present in sewage and in some cases on which WHO region of the Global Polio Laboratory Network a specific laboratory is located. For example in OPV-free areas (vaccinated exclusively with IPV) all viruses recovered from sewage are viruses of interest, whereas in OPV areas where viruses of interest may be a minority among OPV strains, selective tests must be used to pick out these viruses of interest. Histories of VDPVs from persistently infected individuals and population movements from areas with circulating VDPVs and wild polioviruses also play some role. The full battery of positive rule in tests recommended by the WHO/CDC will cover all these possibilities. The negative or rule out molecular tests for vaccine derived viruses may fail to pick up VDPVs if there are other viruses of the same serotype present in the sample. Rule-in tests are currently under development. *See* also **Note 4**.

4. The primers and probes for semiquantitative or quantitative reverse transcript polymerase chain reaction (qRT-PCR) that have been developed and distributed by the CDC for the GPLN are based on positive identification of poliovirus in gen-

eral, and wild poliovirus lineages still in circulation and for vaccine strains in particular [19, 21–23]. There are three CDC/WHO molecular kits for characterization of polioviruses based on positive results (rule in). (Kit 1) The CDC/WHO poliovirus diagnostic rRT-PCR Kit contains reagents for six qRT-PCR tests for each RNA preparation: pan enterovirus, pan-poliovirus, type 1, type 2, and type 3 poliovirus, and a multiplex for Sabin 1, 2, and 3 (Kits 2 and 3). There are two separate kits, one for identifying wild type one or wild type three polioviruses for lineages circulating in South Asia and the other for lineages circulation western Africa. Additional qRT-PCR Kits test for the presence of vaccine derived polioviruses (VDPVs); however, they are based on ruling out that a poliovirus is vaccine or wild, and thus it is possible to obtain false-negative results when there are mixtures of VDPVs with vaccine or wild poliovirus strains. Protocols reagents and kits are currently under development by the CDC for positive identification of VDPVs.

5. When the virus of interest is the only virus in the environmental sample, for example in sewage from a population vaccinated exclusively with IPV, plaque assay for correlating the number of viable viruses with RNA can be performed on concentrated sewage. Once the correlation has been made, it is possible to continue to follow the amount of virus of interest in a sample even after the introduction of the corresponding OPV strain in Supplementary Immunization Activities (SIAs) when plaque-based assays lose meaning because environmental samples contain homotypic (if monovalent OPV is used) or homotypic and heterotypic polio viruses (when bivalent or trivalent OPV is used) ([9] and Shulman unpublished). When the poliovirus of interest is already in homotypic or heterotypic mixtures, such as in sewage from a population vaccinated with OPV or combined program using both OPV and IPV, the correlation must be done using purified virus of interest and or by spiking poliovirus-negative environmental samples with the virus of interest.

6. It is preferably that the magnetic bars be oval or egg shaped to minimize mechanical failure by entrapment in sediment in the sewage.

7. Polyethylene glycol is hygroscopic and will adsorb moisture—therefore store it in closed containers.

8. Chlorine solutions are corrosive. Wash surfaces after 15 min to remove chlorine residues with water or 70 % ethanol followed by water.

9. Obtain stocks with the lowest passage numbers and freeze aliquots of early passages for secondary laboratory stocks. The WHO recommends that fresh cultures should be prepared

from frozen secondary stocks after 15–20 passages. Sensitivity of cells must be demonstrated at frequent intervals for cultures that have been passed more than 20 times after restarting from the secondary stock. Cell sensitivity results should br reported to the WHO Regional Polio Laboratory Coordinator within 48 h of test completion.

10. It is better if the incubators are 5 % CO_2 incubators. However non-CO_2 incubators can be used when plating tube cultures or after adding small volumes of virus to tube cultures as long as the tubes are tightly closed. The high number of cells that adhere to the wall and their metabolism in complete medium will keep the culture form turning alkaline. Tube cultures that are alkaline should be discarded. It is better to use CO_2 incubators after viral challenge. Cells in viral growth medium do not metabolize as fast as in complete medium and are much more sensitive to pH changes. Cells in alkaline medium will undergo cytotoxic CPE that is difficult to distinguish from CPE caused by virus infection

11. With the exception of plaque assays, liquids from uninfected cell cultures can be carefully poured off into collecting bottles that are disinfected afterwards.

12. When it is not possible to use automatic composite samplers, collect 4–6 grab samples (about 400–200 mL each) during 2–3 h of the peak flow rate of the sewage. Mix the grab samples and keep 1 L for the concentration procedure.

13. It is possible to pre-weigh 40 g portions of polyethylene glycol 60000 and add 8.8 g of NaCl as long as the portions are stored under conditions that prevent hydration of the hygroscopic polyethylene glycol.

14. Most of the virus is in the sediment so pouring off a does not result in significant losses. The liquid is poured off primarily to avoid spillage during centrifugation using fixed angle centrifuge heads when centrifuge tubes are completely filled. When the head spins, the top of the liquid will turn perpendicular to the radius of the centrifuge head. Make sure that either before or during centrifugation no liquid will enter the capped region of the centrifuge tube.

15. We use one 500 mL reusable centrifuge bottles per sample, a Sorval SLA 3000 Superlite rotor, and a refrigerated Sorval RC5C Plus centrifuge. Smaller rotors can be used for 250 mL centrifuge tubes

16. This is important to minimize loss of virus during processing in general and for quantitation by decreasing micro-clumping of viruses (viruses trapped together by organic matter).

17. In some cases the volume may >15 mL. Record the volume and proceed. This is important for determining the quantity of PFU per mL in unconcentrated sewage.

18. This delay is very important to ensure the uptake and effectiveness of the antibiotics and antimycotics.

19. Cell Sensitivity Testing of L20B and RD cells must be performed routinely (at least midway through their expected use of 15 passages or every 3 months) according to written WHO Standard Operating Procedures using Laboratory Quality Control (LQC) standards prepared according to the WHO recommended procedure. Sufficient quantities of appropriately labeled aliquots of the standards are stored at $-20\,^{\circ}C$. Permanent records of results must be kept in a format that allows analysis of trends. Corrective action must be taken (e.g., replacement of cell lines, retesting of samples) in response to evidence of reduced cell sensitivity.

20. Some people prefer to wash the cell monolayer with PBS or a few mL of the Trypsin–Versene solution, pour it off and then add the rest of the 20–30 mL of Trypsin–Versene as indicated. This removes FBS and increases the efficiency of the Trypsin–Versene and may yield more single cells and fewer clumps.

21. In cases when it is important to know the number of cells in the tube culture, place a drop of the suspension on a Neubauer hemocytometer counting chamber ($1/400$ mm^2 and $1/10$ mm deep). The counting chamber contains a grid that forms 9 large squares containing 16 smaller squares. The smaller squares in the center and middle large outer squares on all sides are further subdivided into 16 even smaller squares. Count the number of cells in any of the nine largest squares (or average the number of cells in two or more of the large squares) and multiply this by 10^4 to get the number of cells per mL. When counting choose two adjacent sides of the large cell and include all cells that overlap these lines. Ignore any cells that overlap the lines forming the remaining two adjacent sides.

22. The WHO protocol currently requires that CPE-negative cultures be repassed once as well to increase the chance of isolating poliovirus.

23. Pipette the cell suspensions up and down a few times before plating in order to break up any clumps. Do not let the cells sit too long before distribution or else they will start reattaching to the sides or bottom of the plastic flask. Reattachment may take longer if the cells are transferred to a sterile glass bottle or non-tissue culture plastic 50 mL centrifuge tube.

24. To save on medium and serum, only prepare as many tubes needed. Remember, 5 mL of suspension diluted in medium is

enough for twenty 2 mL-tube cultures, so 1 mL of this suspension diluted to 8 mL is sufficient for four test tubes.

For tube culture placed on test tube racks with a 20° slant, it is also important to place a dot or equivalent indicator on the upper side of the tube cultures after adding cell suspensions so that it is easy to locate the lower surface of the tube where the cells have settled and adhered to the surface. The dot should be on the bottom when using a regular microscope and on the top when visualizing the cells with an inverted microscope. We recommend adding a string of linked rubber bands through the holes on the slanted test tube rack to keep the tubes from turning once the cells have adhered since this will dry and kill the cells. More cells are needed if a roller apparatus is used instead of a slanted rack since more surface area of the tube will need to be covered by the monolayer. To economize, round culture tubes can be used instead of tubes with a plat culture surface.. Cells on the inner surface of the round tubes can be viewed on standard or inverted microscopes if tubes are kept from rolling with a simple angled stand made from two parallel rods with a leg at one end.

25. BGM and RD plaque assay plates are used to determine the number of enteroviruses in the sewage as a quality control for viral recovery from concentrated sewage. According to WHO guidelines is enough to challenge tube cultures or small flask cultures of these cell lines and score the culture as CPE positive or negative. The WHO recommends a minimum positivity rate of 30 % for sewage as an indication that conditions have been maintained that would enable isolation of polio from the sample if it were present. We count plaques to obtain information on the level of non-polio enterovirus infections in a community that could potentially compete with and reduce the efficiency of vaccination with OPV. The number of non-polio enterovirus sewage samples will vary for each country. In Israel, 80 % of the sewage samples normally contain enteroviruses.

26. When there are very low numbers of poliovirus particles and/or viable particles in a sewage sample it is not unusual to receive positive results for either tissue culture or molecular assay of RNA extracted from concentrated sewage, but not both. It is also possible to find different viruses or combinations of viruses in replicate tube cultures.

27. The WHO recommends a blind passage of CPE negative L20B cultures on L20B and a second culture on L20B for each RD culture without regard to CPE. During outbreaks and especially after outbreak response with OPV, many RD cultures are positive. For rapid results, any poliovirus in the RD CPE-positive cultures can be characterized directly by molecular ITD assay (Subheading 3.4) from RNA extracted from the

supernatant (Subheading 3.3) when molecular methods are sensitive enough to detect 0.2 PFU per RT-PCR reaction for RNA extracted from concentrated sewage [9]. This saves 1 week.

It is possible to pool aliquots from all replicate CPE-positive RD cultures from a single sewage sample (or to pool supernatants from all replicate RD cultures regardless of CPE when at least one tube culture is positive). It is critical to change micropipette aerosol barrier tips between tubes when pooling to prevent cross-contamination of the RD tubes. Unpooled RD tubes should be stored at −20 °C. To save on costly reagents it is possible to characterize the poliovirus directly from pooled aliquots by molecular ITD assay (Subheading 3.4) using RNA extracted from the pooled supernatants (Subheading 3.3) and/or to challenge a single L20B tube culture with the pooled supernatant. The reason for storing unpooled RD samples is that each of the tubes may contain different mixtures of poliovirus and sometimes this makes downstream processing involving sequencing and characterization of a virus of interest from mixture easier to perform (*see* Subheading 3.4). Characterizing the poliovirus in individual RD tubes may also provide a crude indication of viral load.

28. During an outbreak and especially as the outbreak winds down and samples become negative it is crucial to be as careful as possible to avoid cross contamination of samples with intact virus and/or purified nucleic acids. It is also crucial when the laboratory also performs tests on clinical samples. Molecular assays and sequencing works best with RNA extracted from tissue cultures with full CPE. However these cultures contain very high titers of virus. To prevent cross contamination from intact virus and/or subsequently purified nucleic acids, we strongly recommend that these tissue culture samples be processed in an area completely separate from other activities in the lab where environmental samples are concentrated, stool samples processed, and especially where CSF is processed. Thus, whenever L20B cultures are positive, the presumption is that poliovirus caused the CPE and that these tubes contain very high titers of poliovirus. These tubes should only be opened in a separate BSL-2 laminar flow hood designated for this purpose and situated in a separate room. We used the KingFisher RNA extraction procedure for extracting RNA from these test tube cultures for cost considerations (the extractor is less expensive than the easyMAG extractor), convenience (a dedicated machine is in the same BSL-2 laminar flow hood as used to open the CPE-positive tube cultures) and for cross-contamination control (we extract clinical samples on the easyMAG extractor).

29. Recommendation: add the sample to the sidewall of the plate rather than directly on the cell monolayer to avoid disturbing the monolayer.

30. Unopened plates can be counted when held in the upright position or after inverting them. Don't keep them at an angle or inverted for too long as the monolayer may tear and slide under the agar overlay and ruin the plaque assay or prevent picking plaques for amplification and characterization. For convenience when counting or designating plaques to pick, mark the positions of the plaques on the underside of each dish with a dot from a marker.

31. Alternatives: use disposable glass Pasteur pipettes or 0.5 mL reusable glass pipettes that are cleaned with chlorine solutions and autoclaved for repeated use.

32. There is a much lower chance of splashing if larger tubes (3–15 mL) are used to collect the agar plug; however, using larger tubes presents a disadvantage for short-term and long-term storage.

33. Highest titers can be recovered after one pr two cycles of freeze thawing of the tube cultures. For ease in storage and conservation of storage space, supernatants may be transferred to small 1.5 or 2.0 mL tubes. RNA may be extracted immediately, after up to 4 days of storage at 4 °C, or at any time after storage at −20 or −70 °C.

34. Preparation of secondary stocks (repassage) is necessary to ensure high enough titers for successful sequencing. Incubate the tube cultures at 37 °C until culture reaches full CPE, usually 1–3 days but may take as long as 5 days for some plaques.

35. Pooling before extraction will reduce by up to fourfold the number of RNA extractions and the number of diagnostic rRT-PCR reactions that need to be performed without significantly reducing Ct for positive samples. Specifically if four samples are pooled the Ct of a semiquantitative qRT-PCR reaction will be increased by two cycles. If the workload is not excessive and sufficient storage space is available, store separate unpooled 1 mL samples from each tube that had CPE, otherwise store the pool (also *see* **Note 27**).

36. Storing small single-use aliquots of the extraction positive control at −70 °C provides the best reproducibility between tests so that test performance can be evaluated using Westgard Rules [24, 25] and Levy Jennings plots [26]

37. We recommend trying the primers designed by the CDC (Table 1) using your enzymes and buffers and compare results with the buffers in the kits supplied by the CDC. In some cases this may lower limits of detection by 3–4 orders of magnitude

due to incompatibilities between the CDC buffer and your enzymes. In the future to avoid any such problems, CDC-designed primers and probes may be sent without buffer.

38. Infecting cells at an MOI of >5 may produce secondary stocks with poorer ratios of infectious to noninfectious particles, while infection at a MOI < 0 may take more than 48 h for the virus in the culture to reach a high enough concentrations of virus for downstream molecular analyses. The ratio of infectious to noninfectious particles is important for the correlation of PFU per mL with qRT-PCR results since both infective and noninfective particles will produce positive signals in qRT-PCR. Two rounds of plaque purification are recommended since the original plaques occasionally contain more than one virus due to micro-clumping or overlapping of adjacent plaques. Prepare primary and secondary stocks from the original sample or if possible from different contemporary samples. Design of primers and probes using related samples makes it less likely that there will be sequence variation within the area chosen for the primers or probe. A miss-match in the 3′ base of primers will render the primer useless, and as variability increases in the target sequence complementary to the probe specific binding of the probe will decrease until detection is no longer reliable.

39. We used Primer Express Software v3.0 (Life Technologies, Life Technologies, Thermo Fisher Scientific, Waltham, MA, USA) for designing compatible primers and probes for our ABI 7500 instrument. For our assay the specific probes for the poliovirus strain of interest were labeled with 6-carboxyfluorescein (FAM) and black hole quencher 2 (BHQ-2; Metabion, Planegg/Steinkirchen, Germany) so that we could use them in multiplex with our MS-2 primers and VIC labeled probe (Table 1) based on sequences previously published by Dreier et al. [18] that we purchased from Life Technologies (Life Technologies, NY, USA).

40. If possible, design one primer pair and probe from the first third of VP1 and the other in the last third of the VP1 sequence. Validate both. Remember the primer sequence is nondegenerate. The virus of interest may mutate and render one of the primers or the probe ineffective. Two validated sets will also allow for longer overlapping sequences when using the forward primer form the first third with reverse generic primer Q8 and the reverse primer from the last third with generic forward primer Y7 (see Table 1 and the discussion on sequencing a virus of interest from RNA extracted form a homotypic mix in Subheading 3.5).

41. We optimize results for this standard assay condition for all reactions with nondegenerate primers and probes so that we can run RT-PCR and even PCR reactions for many different viruses in the same qRT-PCR run. The only time we may need to vary run conditions is when degenerate primers are used such as for the Pan poliovirus qRT-PCR described here.

42. Take all precautions to prevent contamination. Handle the 0.2 mL PCR strips for the qPCR machines by the lids, being careful not to contaminate the tubes. To decrease the chance for cross contamination, place a fresh sheet of absorbent lab mat on the work area and distribute reagents in a dead-air hood. Expel used aerosol barrier tips into a disposable waste cup or bag.

43. If the cap is firmly closed there is an increased chance for splashing and contamination when the cap is briefly removed in Molecular Area 2 for addition of the RNA template.

44. The final reaction volume is 50 µL. Add water to compensate for any decrease in volume of primers. The choice of primers depends on the identity of polioviruses in the tissue culture or concentrated sewage sample. The Samples may contain (A) just the virus of interest or mixtures of sequence variants of this virus, (B) a mixture of the virus of interest and viruses of other serotypes (heterotypic mix), or (C) mixtures of vaccine, vaccine-derived, and/or wild, polioviruses of the same serotype as the virus of interest (homotypic mixtures). qRT-PCR results from molecular ITD tests described in the previous section indicate which of these which of the three possibilities is the case and this information determine the choice of primers from among those listed in Table 1 for both amplification and sequencing. RNA. Amplification conditions may differ depending on whether the primers are nondegenerate or degenerate. For example the final concentration of primers in the reaction mix depends on whether or not they are degenerate and on how many degenerate bases they contain. For convenience this is taken into account (Table 1) when preparing the concentration of primer stock by adjusting the concentration so that the necessary amount of primer is present in 1 µL of stock.

45. To create two specific amplicons with a long overlap from the two the hemi-nested RT-PCRs, choose as the forward specific primer, the forward specific qRT-PCR primer from the pair designed for the 5′ third of VP1 and the reverse specific qRT-PCR primer designed for the 3′ third of VP1. Specific primers from either one of the specific qRT-PCR primer pairs can also be used in the hemi-nested reactions. This will produce two amplicons with a much a shorter overlap and difficulty in reading sequences immediately adjacent to the sequencing primer may result in gaps where there is no overlap.

One or both hemi-nested reactions may not work. Try RT-PCRs with generic primers paired with type specific internal VP1 primers or just pairs of type specific primers from Table 1 (for example P1W2s with Q8 (900 bp) and Y7 with P1W1a (250 bp) for wild type one).

If it is not possible to obtain sequences from heterotypic mixes and it is important to characterize the virus of interest, the virus of interest may be enriched or be isolated by one or two rounds of tissue culture of dilutions of the mixture grown in the presence of neutralizing antibodies to the other virus serotypes in the mixture.

46. If two sets of specific primers were prepared for the qRT-PCR than use the forward primer from the first third of VP1 and the reverse primer from the last third of VP1 to label amplicons with the longest overlap. Specific primers appropriate for the wild serotype (Table 1) can be substituted for the specific qRT-PCR.

    An additional alternative is to use specific degenerate primers designed to be the CDC to identify SOAS type 1 or 3 poliovirus from currently endemic to south Asia or WEAF type 1 or 3 polioviruses endemic to western Africa.

47. Sabin 2 is the OPV serotype that causes most live vaccine poliomyelitis outbreaks. If all else fails and it is important to characterize the type two sequence, isolate or enrich for type two virus by growing limiting dilutions of the mixture in the presence of neutralizing antibodies to the other virus serotypes in the mixture. Specific kits for rule-in type 1 and type 2 VDPVs by positive identification rather than rule-out are in the last stages of testing by the CDC. If it is not possible to obtain sequences from heterotypic mixes and it is important to characterize the virus of interest, the virus of interest may be enriched or be isolated by one or two rounds of tissue culture of dilutions of the mixture grown in the presence of neutralizing antibodies to the other virus serotypes in the mixture.

48. Use the same tip a few times without resetting the volume. The maximum amount of RNA that can be loaded on a column is 400 mg.

49. When forming the agarose gel rather than running the sample in two adjacent lanes, use cellophane tape to close the space between adjacent teeth of the slot (lane) to make a single slot long enough to accommodate all of the remaining volume of the reaction mix.

50. When purifying a number of bands at the same time it is easier to bring all of the gel slices to the same weight by adding a volume of 1× E buffer equal to the difference in weight between each sample and the heaviest sample. If you do this you do not need to reset pipette volumes for each sample when dispensing reagents form the kit.

51. Modification of manufacturer's instructions to reduce chance of contamination with amplified PCR template.

52. Cleaning between samples and especially after a sample with a high concentration of nucleic acids reduces the chance of carryover to a sample containing low levels. If necessary the sample arm can be decontaminated from potential biologically active materials with a freshly prepared solution of 0.5 % solution of sodium hypochlorite

53. The ratio of sample absorbance at 260 and 280 nm assess the purity of the DNA and RNA a ratio of ~1.8 is generally accepted as "pure" for DNA and ~2.0 is considered "pure" for RNA. Lower rations may indicate the presence of protein or other contaminants some of which may affect downstream reactions.

54. Suggested naming conventions: Name each sequencing reaction according to the following convention: Date (YY-MM-DD), organism (polio), the ID of the cDNA RT-PCR (example P1002-2), the primer (example Q8), and the position on the microtiter plate (example C10) combined into a single name without spaces by using the underline character. In this case the name is 14_07-08-POLIO-P1002-2-Q8C10.

55. RT-PCR produces a double strand amplicon containing both primers sequences. Sequencing of one strand will read through the primer at the other end if the amplicon is short enough. These primer sequences must be trimmed off from the end of the sequence since the sequence of nucleotides in the primer may not be 100 % identical with actual genome sequence. Having them aligned on the sequence makes trimming easier.

56. We use the following file naming convention before the serotype of the virus has been identified: pvn_isolate ID number that may include a hyphen and letter to indicate the sample type_three letter country code and two digits for the year of isolation (for example: pvn_8066-e_ISR13 identifies a poliovirus isolated from environmental sample 8066 collected in Israel in 2013). The n in pvn is changed to 1, 2, or 3 when the serotype becomes known. The word "polio" in the name of the copy of the generic "proj polio" file should be changed to reflect the ID of the sequence, e.g., "proj pv_8066-e".

57. Each raw sequence file should have the ID of the virus added at the beginning of the full name. This is helpful because the original raw data file name indicates the date of sequencing reaction, the amplicon sequenced, and the primer used for sequencing is preserved. Putting the ID name first enables the computer to use alphanumeric criteria to easily sort together raw sequence files from the same isolate using different sequencing primers especially when sequenced on different days.

58. Correct ambiguous calls by the computer based on a clear call for a complementary strand or sequence from the same strand

produced using different primers since in almost all cases the sequences should be identical, e.g., usually there is a unique base at each position of the sequence. Normally when sequences do not match 100 %, this is an indication of a mistake in identification, requiring reviewing all steps from sample to sequence. However multiple peaks can be observed at one or more position when there are mixtures of closely related viruses and a minority species accounts for >10 % of the population. In addition sequences may differ when overlapping amplicons are sequenced for closely related mixtures of viruses using nondegenerate primers and there is a mismatch in some of the sequences at the 3′ base of one of the primers. When this occurs the two amplicons are no longer equivalent as a sub-member of the mix may be overrepresented or underrepresented. See discussion on homotypic mixtures in environmental samples i.e. when environmentl samples contain mixtures of the same serotype.

The position of mismatches is indicated by a dot under the bottom line containing the consensus and in the consensus itself by a degenerate base or semicolon (for gaps after alignment). Place the cursor on the first base of the consensus sequence and press command N (Mac) or control N (PC) and the position will shift to the next mismatch to the right. After doing this twice the space bar becomes a shortcut. If the shift is depressed at the same time the shift is used, the program will scan for mismatches 5′ (to the left). Pressing Command 4 or Control 4 with automatically reverse complement all of the sequences in the contig without affecting alignment. This is convenient since in Sequencher, restoring and deleting sequences in an alignment is easier when done at the 3′ end of a sequence.

At the 5′ end there may be problems with uniform distances between peaks (an electrophoresis artifact resulting from the influence of neighboring bases on secondary structures of short sequences) that results in two partially overlapping peaks with one computer base call. If so correct this by placing the cursor on the base to the right of the missing base above the line and pressing the **tab** key to introduce a semicolon, ":". Move the cursor to the semicolon and type in the missing base. **At the 3′ end,** the periodicity of peaks and especially the delineation between peaks and troughs for repetitive bases may be disrupted resulting in stretched out peaks for which the program will insert an extra base.

Usually there is a unique base at each position of the sequence. However multiple peaks can be observed at one or more position when there are mixtures of closely related viruses and a minority species accounts for > 10 % of the population.

59. Alternative: select all of one sequence and paste into the new sequence, return to the contig and highlight and copy additional nonoverlapping sequences from other strands starting with the first nonoverlapping base.

60. Note when you use the default Assembly Parameters for overlap (20 nt) and minimum match Percentage the Sabin contigs for each serotype remain separate and the unknown will match with one of them if it is vaccine or vaccine-derived with <15 % sequence difference. The sequence will not form a contig with one of the Sabin serotypes if it is from a highly diverged (>15 %0) VDPV or form a wild poliovirus.

61. When correcting, look especially for gaps, paired gaps between the sequence of interest and the Sabin sequence, and at all bases that are different from the Sabin sequences. The graphic file of the Sabin strain can help to resolve problems at the beginning and end of sequences where the periodicity of peaks may be nonuniform and lead to incorrect automatic base calls. During alignment, the penalty for gaps for a short segment may be less than the penalty for the number of mismatched bases and the program will introduce a pair of gaps one in the prototype and another in the unknown. When this occurs activate the segment of aligned sequences between and including the matched gaps, gather the gaps to the left or right (Sequence/gather gaps/right) and delete the aligned gap by putting the cursor on the gap on the consensus line and pressing the backspace key.

References

1. Racaniello VR (2001) Picornaviridae: the viruses and their replication. In: Fields BN, Knipe N, Howley P (eds) Virology, 4th edn. Lippincott Williams & Wilkins, Philadelphia, PA, pp 685–722

2. Shulman LM (2012) Polio and its epidemiology. In: Meyers RA (ed) SpringerLink (Online service). Encyclopedia of sustainability science and technology. Springer, New York, pp 8123–8173

3. Dowdle W, van der Avoort H, de Gourville E, Delpeyroux F, Desphande J, Hovi T, Martin J, Pallansch M, Kew O, Wolff C (2006) Containment of polioviruses after eradication and OPV cessation: characterizing risks to improve management. Risk Anal 26(6):1449–1469

4. Kew OM, Sutter RW, de Gourville EM, Dowdle WR, Pallansch MA (2005) Vaccine-derived polioviruses and the endgame strategy for global polio eradication. Annu Rev Microbiol 59:587–635

5. Hovi T, Shulman LM, van der Avoort H, Deshpande J, Roivainen M, De Gourville EM (2012) Role of environmental poliovirus surveillance in global polio eradication and beyond. Epidemiol Infect 140(1):1–13

6. Ranta J, Hovi T, Arjas E (2001) Poliovirus surveillance by examining sewage water specimens: studies on detection probability using simulation models. Risk Anal 21(6):1087–1096

7. WHO (2003) Guidelines for environmental surveillance of poliovirus circulation: WHO, Dept of Vaccines and Biologicals. http://apps.who.int/iris/handle/10665/67854. Last accessed 23 July 2014

8. Hovi T, Stenvik M, Partanen H, Kangas A (2001) Poliovirus surveillance by examining sewage specimens. Quantitative recovery of virus after introduction into sewerage at remote upstream location. Epidemiol Infect 127(1): 101–106

9. Hindiyeh MY, Moran-Gilad J, Manor Y, Ram D, Shulman LM, Sofer D, Mendelson E (2014) Development and validation of a quantitative RT-PCR assay for investigation of wild type poliovirus 1 (SOAS) reintroduced into Israel. Euro Surveill 19(7):pii: 207103800

10. Lodder WJ, Buisman AM, Rutjes SA, Heijne JC, Teunis PF, de Roda Husman AM (2012) Feasibility of quantitative environmental surveillance in poliovirus eradication strategies. Appl Environ Microbiol 78(11):3800–3805

11. Shulman LM, Manor Y, Sofer D, Mendelson E (2012) Bioterrorism and surveillance for infectious diseases—lessons from poliovirus and

enteric virus surveillance. J Bioterr Biodef S4:004

12. WHO (2004) Polio laboratory manual, 4th ed. WHO/IVB/04.10, WHO

13. Anis E, Kopel E, Singer SR, Kaliner E, Moerman L, Moran-Gilad J, Sofer D, Manor Y, Shulman LM, Mendelson E, Gdalevich M, Lev B, Gamzu R, Grotto I (2013) Insidious reintroduction of wild poliovirus into Israel, 2013. Euro Surveill 18(38):pii: 20586

14. Shulman LM, Hindiyeh M, Muhsen K, Cohen D, Mendelson E, Sofer D (2012) Evaluation of four different systems for extraction of RNA from stool suspensions using MS-2 coliphage as an exogenous control for RT-PCR inhibition. PLoS One 7(7):e39455

15. Manor Y, Shulman LM, Hindiyeh M, Ram D, Sofer D, Moran-Gilad J, Lev B, Grotto I, Gamzu R, Mendelson E (2014) Intensified environmental surveillance supporting the response to wild-type poliovirus type 1 silent circulation in Israel, 2013. Euro Surveill 19(7):20708

16. Shulman LM, Gavrilin E, Jorba J, Martin J, Burns CC, Manor Y, Moran-Gilad J, Sofer D, Hindiyeh MY, Gamzu R, Mendelson E, Grotto I, Genotype-Phenotype_Identification_(GPI)_Group (2014) Molecular epidemiology of silent introduction and sustained transmission of wild poliovirus type 1, Israel, 2013. Euro Surveill 19(7):pii=20709, Available online: http://www.eurosurveillance.org/ViewArticle.aspx?ArticleId=20709

17. Saunders N, Zambon M, Sharp I, Siddiqui R, Bermingham A, Ellis J, Vipond B, Sails A, Moran-Gilad J, Marsh P, Guiver M, Division HPAMS (2013) Guidance on the development and validation of diagnostic tests that depend on nucleic acid amplification and detection. J Clin Virol 56(3):260–270

18. Dreier J, Stormer M, Kleesiek K (2005) Use of bacteriophage MS2 as an internal control in viral reverse transcription-PCR assays. J Clin Microbiol 43(9):4551–4557

19. Kilpatrick DR, Yang CF, Ching K, Vincent A, Iber J, Campagnoli R, Mandelbaum M, De L, Yang SJ, Nix A, Kew OM (2009) Rapid group-, serotype-, and vaccine strain-specific identification of poliovirus isolates by real-time reverse transcription-PCR using degenerate primers and probes containing deoxyinosine residues. J Clin Microbiol 47(6):1939–1941

20. Sanger F, Coulson AR (1975) A rapid method for determining sequences in DNA by primed synthesis with DNA polymerase. J Mol Biol 94(3):441–448

21. Kilpatrick DR, Ching K, Iber J, Campagnoli R, Freeman CJ, Mishrik N, Liu HM, Pallansch MA, Kew OM (2004) Multiplex PCR method for identifying recombinant vaccine-related polioviruses. J Clin Microbiol 42(9):4313–4315

22. Kilpatrick DR, Nottay B, Yang CF, Yang SJ, Da Silva E, Penaranda S, Pallansch M, Kew O (1998) Serotype-specific identification of polioviruses by PCR using primers containing mixed-base or deoxyinosine residues at positions of codon degeneracy. J Clin Microbiol 36(2):352–357

23. Kilpatrick DR, Nottay B, Yang CF, Yang SJ, Mulders MN, Holloway BP, Pallansch MA, Kew OM (1996) Group-specific identification of polioviruses by PCR using primers containing mixed-base or deoxyinosine residue at positions of codon degeneracy. J Clin Microbiol 34(12):2990–2996

24. Westgard JO, Groth T, Aronsson T, Falk H, de Verdier CH (1977) Performance characteristics of rules for internal quality control: probabilities for false rejection and error detection. Clin Chem 23(10):1857–1867

25. Westgard JO, Barry PL, Hunt MR, Groth T (1981) A multi-rule Shewhart chart for quality control in clinical chemistry. Clin Chem 27(3):493–501

26. Levey S, Jennings ER (1950) The use of control charts in the clinical laboratory. Am J Clin Pathol 20(11):1059–1066

27. Toyoda H, Kohara M, Kataoka Y, Suganuma T, Omata T, Imura N, Nomoto A (1984) Complete nucleotide sequences of all three poliovirus serotype genomes. Implication for genetic relationship, gene function and antigenic determinants. J Mol Biol 174:561–585

28. Kilpatrick DR, Iber JC, Chen Q, Ching K, Yang SJ, De L, Mandelbaum MD, Emery B, Campagnoli R, Burns CC, Kew O (2011) Poliovirus serotype-specific VP1 sequencing primers. J Virol Methods 174:128–130

29. Centers for Disease Control and Prevention, Atlanta, GA., Kit Instructions In: Poliovirus rRT-PCR ITD 4.0 Kit. A kit for serotyping of L20B positive cell cultures and intratypic differentiation of polioviruses in support of the Global Polio Eradication Initiative. Distributed by The WHO Collaborating Centre for Enteroviruses and Polioviruses, Centers for Disease Control and Prevention, 1600 Clifton Road NE, Mailstop G-10, Atlanta, Georgia 30333 USA, +1-404-639-1341, Fax: +1-404-639-4011. Email: MMandelbaum@cdc.gov, HSun@cdc.gov, SOberste@cdc.gov. November 20, 2014

Chapter 6

Quality Assurance in the Polio Laboratory. Cell Sensitivity and Cell Authentication Assays

Glynis Dunn

Abstract

The accuracy of poliovirus surveillance is largely dependent on the quality of the cell lines used for virus isolation, which is the foundation of poliovirus diagnostic work. Many cell lines are available for the isolation of enteroviruses, whilst genetically modified L20B cells can be used as a diagnostic tool for the identification of polioviruses. To be confident that cells can consistently isolate the virus of interest, it is necessary to have a quality assurance system in place, which will ensure that the cells in use are not contaminated with other cell lines or microorganisms and that they remain sensitive to the viruses being studied.

The sensitivity of cell lines can be assessed by the regular testing of a virus standard of known titer in the cell lines used for virus isolation. The titers obtained are compared to previously obtained titers in the same assay, so that any loss of sensitivity can be detected.

However, the detection of cell line cross contamination is more difficult. DNA bar coding is a technique that uses a short DNA sequence from a standardized position in the genome as a molecular diagnostic assay for species-level identification. For almost all groups of higher animals, the cytochrome c oxidase subunit 1 of mitochondrial DNA (CO1) is emerging as the standard barcode region. This region is 648 nucleotide base pairs long in most phylogenetic groups and is flanked by regions of conserved sequences, making it relatively easy to isolate and analyze. DNA barcodes vary among individuals of the same species to a very minor degree (generally less than 1–2 %), and a growing number of studies have shown that the COI sequences of even closely related species differ by several per cent, making it possible to identify different species with high confidence.

Key words Cell authentication, Cell sensitivity, Quality assurance

1 Introduction

The Global Poliovirus Eradication Initiative (GPEI) is close to achieving its goal of eliminating wild type poliovirus; as a result enhanced surveillance protocols are being introduced. False negative or false positive virus isolation reporting can have serious consequences for the surveillance programs. It has been estimated that for each paralytic poliovirus case identified there may be as many as 100,000 people infected, consequently, missing the opportunity to identify a circulating poliovirus can lead to substantial numbers of

Javier Martín (ed.), *Poliovirus: Methods and Protocols*, Methods in Molecular Biology, vol. 1387,
DOI 10.1007/978-1-4939-3292-4_6, © Springer Science+Business Media New York 2016

infections going unreported. False positive reporting can lead to the unnecessary implementation of emergency vaccination programs, and so optimum sensitivity of the entire virus isolation procedure is crucial for enterovirus surveillance.

The use of human cell lines in research and diagnostic laboratories is widespread, so too is the misidentification of cell lines. Improvements in genetic analysis such as Short Tandem Repeat (STR) analysis have enabled custodians of cell banks to analyze and characterize the cell lines they hold. The International Cell Line Authentication Committee (ICLAC) has produced a list of misidentified cell lines [1, 2], which were mainly misidentified at submission to the various cell banks.

More worrying is the inadvertent cross-contamination of cell lines within the laboratory, generally due to poor handling techniques. Although occasionally it may be possible to see that the cells in culture have different morphologies, most commonly, the contamination goes unnoticed, and can have serious consequences for the work carried out, particularly where cell lines are used for diagnostic purposes.

Polioviruses will grow in human cell lines, such as RD, HEp2c,[1] and HeLa cells, as do many other enteroviruses. L20B cells are mouse L cells which have been modified to contain the human poliovirus receptor site [3], and so with very few exceptions, only support the growth of poliovirus. This feature is used in the WHO Global Polio Laboratory Network (GPLN) as a surveillance diagnostic tool to quickly identify poliovirus from stool samples [4, 5]. Contamination of cell lines can confuse the isolation of enteroviruses, particularly if the L20B cells are contaminated with RD or other cells which allow other enterovirus to grow in the L20B culture.

The use of DNA Bar Coding as a tool for species identification was first suggested by Paul Hebert et al in 2003 [6]. Their suggestion was based on the observation that the DNA of the mitochondrial cytochrome *c* oxidase gene subunit 1 (CO1), did not vary by more than 1 or 2 % within a particular animal species, but varied by several per cent between species. The CO1 DNA region used is relatively short, 648 bases, and is flanked by regions which are generally conserved between species, making it a convenient target for bar coding analysis. A consortium, The Bar Code of Life, has been set up which compiles the Barcode of Life Data Systems (BOLD) database—a reference library of DNA barcodes [7] that can be used to assign species identities to unknown specimens.

By using the differences in CO1 sequences it is possible to identify the species derivation of laboratory cell lines, and find any contaminants. However, it is difficult to establish cell identity by Sanger sequencing alone, since the majority DNA sequence may mask any minor contaminating sequences. By using a real-time

---

[1] HEp2c cells have been shown to be HeLa cells by STR analysis.

PCR (qPCR) assay, and melt curve analysis, it is possible to separate two or more genetically distinct DNAs.

qPCR gives the user the ability to monitor the progress of the PCR in real time, as well as measuring the precise amount of amplicon, above a certain threshold, in each tube at the end of each cycle. The progress of the reaction can be measured using dyes which will increase in fluorescence in direct proportion to the DNA present in the tube. The cycle number at which the DNA will exceed the critical threshold of detection is directly related to the concentration of the template DNA and is termed the C_t value (cycle threshold).

Melt curve analysis measures the dissociation of double stranded DNA when heated. This temperature dependant process can be monitored as it occurs using a real-time PCR machine and DNA intercalating fluorophores such as SYBR Green or EvaGreen. SYBR Green and EvaGreen fluoresce strongly when bound to the minor groove of DNA, but as the temperature increases, the DNA strands will dissociate or "melt." As there is no longer any double stranded DNA in the dissociated amplicons, there is nothing for the SYBR or EvaGreen to bind to and fluorescence decreases rapidly [8].

The energy required to break the hydrogen bonds between two strands of DNA, that is the temperature at which the strand melts, is dependent on the base composition of the DNA and its length. Therefore it is possible to design primers which will discriminate between different DNA sequences in the CO1 region of the mitochondrial gene. Although the melting temperature can also be affected by salt concentration, standardizing the PCR reaction will ensure that the amplified species specific CO1 DNA will melt at a predictable temperature.

By using a quantitative PCR (qPCR) method, it is also possible to detect and distinguish very low levels of contaminating cells. There are between 100 and 1000 mitochondria per cell [9], depending on the tissue the cells originate from, consequently the qPCR assay can detect a single cell contaminant in a culture of 10^5 cells per ml.

In addition to monitoring cell line authenticity, it is important to routinely monitor the sensitivity of cell lines for virus isolation, to provide assurance that a cell line retains the ability to detect poliovirus and other enteroviruses, even if present at low titer. A well-characterized reference virus preparation of known and reproducible titer should be used to evaluate cell line sensitivity at least once throughout the expected number of cell passages, generally 15 passages for L20B and RD cell lines in the GPLN [10].

Many factors can affect cell-line sensitivity including: mycoplasma contamination, quality of growth media, fetal calf serum, as well as growth conditions such as temperature of incubation. The cell sensitivity test enables these variations to be assessed in a standardized way and identify any components and culture conditions which may have an adverse effect on virus isolation.

2   Cell Authentication

<table>
<tr><td>

2.1   Materials

</td><td>

Pipette Aids and 10 ml pipettes.

</td></tr>
<tr><td>

2.1.1   Nonsterile

</td><td>

- −20 °C freezer.
- −70 °C freezer.
- Pipettes in calibration with tips, 10, 20, 200, and 1000 μl.
- Spectrophotometer or fluorimeter.
- Centrifuge.
- Microcentrifuge.
- Proteinase K (20 mg/ml).
- Phenol–chloroform–isoamyl alcohol (in the ratio 25:24:1).
- 3 M sodium acetate pH 5.2.
- 100 and 70 % Ethanol.
- Real-time PCR thermal cycler.
- Disposal containers for disposal of pipettes and tips.
- 70 % alcohol for cleaning work area.
- Paper towels.
- Marking pens.
- Inverted microscope.
- Ice bucket with ice.
- Liquid waste container.
- Biological safety cabinet.
- Culture flasks with confluent monolayers of cells for testing.
- Neubauer hemocytometer or electronic cell counter.
- Heating block cable of being set at 56–60 °C.

</td></tr>
<tr><td>

2.1.2   Sterile

</td><td>

- Phosphate buffered saline without calcium and magnesium.
- 0.25 % Trypsin solution (Sigma, Cat. no. T4049).
- Trypan blue (0.04 % w/v) (Life Technologies, Cat. no. T10282).
- Cell culture tubes and flasks.
- Pipettes.
- 2× QuantiTect SYBR Green PCR Master Mix (Qiagen Cat No. 204141).
- DNA extraction kit, e.g., Qiagen Blood and Tissue Kit (Qiagen Cat. no. 69504/6).
- Sense primer (VFQ) 5′-TTC TCA ACC AAC CAC AAR GAY ATY GG (10 μM).
- Antisense primer (VRQ) 5′-GGR ACW AGT CAG TTG KCA AAG CCT CC-3′ (10 μM).

</td></tr>
</table>

- Sense primer (VFHu) 5′-GCT GGG CCA GCC AGG CAA C-3′ (10 µM).

- Sense primer (VFMo) 5′-ATT AGG TCA ACC AGG TGC A-3′ (10 µM).

- Antisense primer (HuREV) (5′-GTT GTT TAT GCG GGG AAA CGC C-3′) (10 µM).

- Antisense primer (MoREV) (5′-CTA TTG ATG ATG CTA GGA GAA GGA GAA AT-3′) (10 µM).

- DNase-free water.

- Thin walled PCR tubes—suitable for real-time PCR thermal cycler.

2.2 Method

2.2.1 Harvesting of Monolayer Cultures

1. Remove flasks of cell cultures from a 37 °C incubator. Using an inverted phase contrast microscope, select flasks with confluent monolayers. Decant spent medium from culture into a suitable discard container.

2. Rinse the cell monolayers with Sterile Phosphate Buffered Saline: 10 mM Na_2HPO_4, 1.8 mM KH_2PO_4, 2.7 mM potassium chloride KCl, and 137 mM sodium chloride NaCl, pH 7.4, at 25 °C, autoclaved at 121 °C for 20 min (*see* **Note 1**). Use 5 ml of PBS for 25 cm² and 20 ml for 75 cm² flask or 175 cm² flasks, and discard the buffer into a suitable container.

3. Add approximately 3–5 ml of trypsin solution 0.25 % trypsin solution to the flask. Wash over cells briefly. Decant the solution and replace with fresh trypsin (1 ml for 25 cm² flask; 3 ml for 75 cm² or 175 cm² flasks). Wash over cells moving the flasks to ensure the trypsin covers all the cells.

4. Incubate the flask (37 °C) until the cell sheet is dislodged. Check flasks within 5 min and then again within 10 min if the cells are not dislodged, this will ensure that the cells are not over trypsinized.

5. Resuspend the cells in 10 ml of sterile PBS; using a 10 ml pipette, gently aspirate the medium to form a uniform cell suspension (*see* **Note 2**).

6. Count the cells using the trypan blue dye exclusion method; remove 200 µl of cell suspension using sterile conditions and add an equal volume of 0.04 % trypan blue. Mix gently.

7. Count the cells using a cell counter or hemocytometer: Clean the hemocytometer and moisten the coverslip with water or exhaled breath. Slide the coverslip over the etched chamber back and forth applying slight pressure until Newton's rings appear under the coverslip (*see* **Note 3**).

8. Fill both sides of the chamber with cell suspension in trypan blue (10–20 µl per chamber). View under a light microscope using ×20 magnification.

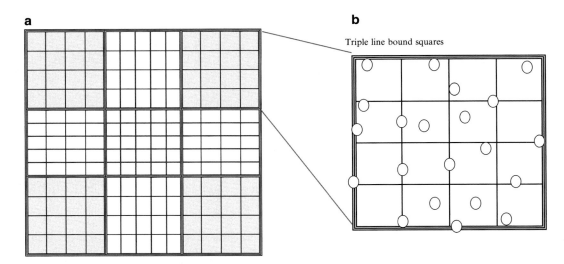

Count the cells in the 16 squares of the four corners. Viable cells will be bright and non-viable cells will be stained blue. Cells touching the top and left side triple lines are counted, but cells touching the bottom and right side triple lines are not.

Mean of viable cells counted $(Mv) = \dfrac{\text{Total number of viable cells}}{\text{Number of triple line bound squares}}$

D = Dilution Factor   = 2 if equal volumes of trypan blue and cells used
C = Correction Factor, given by the manufacturer of the chamber commonly 10^4

Fig. 1 Cell counting using an improved Neubauer hemocytometer

9. For the Neubauer improved hemocytometer, count the viable cells, which will appear to be bright, in the four large corner squares bounded by triple lines (1 mm²) outlined in the diagram below. Average the count for these four squares (Fig. 1a).

10. Cells on the boundary of the area being counted, should only be included if they fall on the top or left hand boundary lines, but not those falling on the bottom and right hand boundary lines (Fig. 1b).

11. Count the four corner squares of the second chamber in a similar way, and average the count.

12. Counts should ideally have cells in the range of 100–200 per chamber. If there are more cells, then the cells should be diluted to make the count more accurate. The counts for the two chambers should be within 20 % of each other. If not, clean the hemocytometer and recount the cells.

13. To calculate the number of viable cells per ml of cell suspension, use the formula below. Remember to correct for any dilution factor, e.g., trypan blue.

$$\text{Viable Cell Count}\,(\text{Live cells per ml})$$
$$= \dfrac{\text{Number live cells counted}}{\text{Number of large corner squares counted}} \times \text{Dilution} \times 10^4$$

14. Centrifuge the appropriate number of cells (maximum 5×10^6) at $1,500 \times g$ for 5 min. Discard the medium and proceed to DNA extraction of the harvested cell pellet. The cell pellet can be disrupted by freezing at -20 °C prior to DNA extraction.

2.2.2 DNA Extraction

1. Remove cell samples from -20 °C freezer and allow to thaw.

2. Take the cell pellet and use a suitable commercial method to extract the DNA such as Qiagen Blood and tissue kit according to the manufacturer's instructions,
    Or treat the cell pellet with 2.5 μl proteinase K (20 mg/ml) at 60 °C for 60 min, mixing every 10 min to ensure equal digestion.

3. Extract the cell/proteinase K mixture with phenol–chloroform–isoamyl alcohol. Actual volumes are not given—this protocol is dependent on the amount of starting material.

4. Add an equal volume of phenol–chloroform–isoamyl alcohol (25:24:1).

5. Vortex for a few seconds to mix well and form an emulsion.

6. Centrifuge at room temperature for 5 min at $14,000 \times g$.

7. Pipette off and keep the top aqueous phase—you MUST avoid taking any precipitated material from the interphase or any phenol.

8. Add 1/10th volume 3 M sodium acetate pH 5.2.

9. Add 2.5 volumes of 100 % ethanol, and mix slowly by inversion—the DNA may be seen to precipitate out of solution as a "ball of fluff."

10. Place the DNA at -20 °C overnight or at -70 °C for at least 1 h. Centrifuge at $14,000 \times g$ for 15 min in a microcentrifuge, a white DNA pellet should be visible at the bottom of the tube.

11. Remove the ethanol with a pipette tip and wash the pellet with 70 % ethanol; make sure all the ethanol is removed—recentrifuge if necessary.

12. Air-dry, but do not over-dry.

13. Take up each DNA pellet in 50 μl water and ensure that they are uniformly mixed by gentle pipetting.

14. Briefly centrifuge the DNA samples. Measure the DNA using a spectrophotometer or fluorimeter. (The typical yield of DNA from 10^6 cells is 7–15 μg.) Dilute DNA with enzyme free H_2O to 10 ng/μl (this will give 50 ng/reaction).

15. Calculate the number of cells in 1 μl of DNA solution.

16. Store the DNA at -20 °C.

2.2.3 PCR Amplification of the Entire COI Region (648 bases) for Species Identification

Prepare the following PCR master mix:	
For each sample you require:	
2× QuantiTect SYBR Green PCR Master Mix	12.5 µl
Sense primer VFQ (10 µM)	0.5 µl
Antiense primer VRQ (10 µM)	0.5 µl
Template DNA	5.0 µl
H₂O	6.5 µl
Total	25.0 µl

Keep all the reagents on ice.

1. Label thin walled PCR tubes (for the real-time thermal cycler to be used) appropriately on the lids rather than the side of the tube.

2. Add 20 µl of master mix to each tube.

3. Add 5 µl of template DNA to each appropriately labeled tube and keep at +4 °C

4. Open the operating software for the real-time thermal cycler and complete the sample list.

5. Set up the cycling conditions for the real-time thermal cycler (*see* **Note 4**).

6. Example cycling conditions for the Qiagen Rotor Gene Q Real-Time PCR machine:

   Step 1: *Hold* @ 95 °C for 15 min.

   Cycling (40 repeats):

   Step 2: 94 °C, 15 s.

   Step 3: 50 °C, 30 s.

   Step 4: 72 °C, 30 s (Acquiring to SYBR Green channel).

   Step 5: *Melt* from 72 to 95 °C. Hold for 10 s on the first step, hold for 5 s on each next step, acquiring data to the SYBR Green channel.

7. Make sure that all safety features are in place and start the machine.

8. Once the machine has completed the run, save the data, remove the samples from the machine, and store frozen (*see* **Note 5**).

Each real-time thermal cycler will differ in the specifics of the analysis programs, but will generally perform the same processes.

The raw data is analyzed to show the amplification profiles of the samples under test (*see* Figs. 2 and 3).

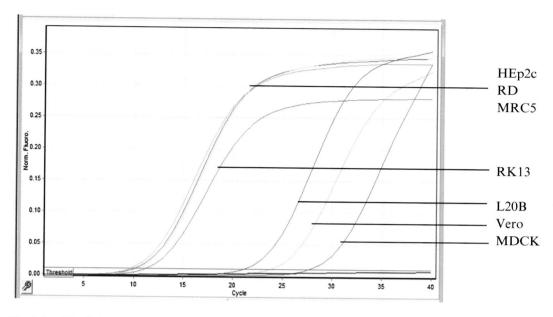

Fig. 2 Amplification curves using VRQ and VRQ primers

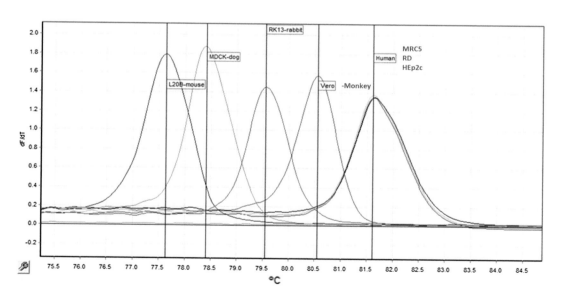

Fig. 3 Melt curve analysis showing that mitochondrial CO1 DNA from different species melt at different temperatures. Three human cell lines were analyzed, MRC5, RD, and HEp2c(HeLa), all derived from different individuals, but whose mitochondrial DNA within the CO1 gene melted at the same temperature

2.2.4 PCR Amplification Using Human and Mouse Specific Primers of the COI Region

Prepare the following PCR master mix:

For the specific mouse primers you require for each sample:

2× QuantiTect SYBR Green PCR Master Mix	12.5 µl
Sense primer VFMo (10 µM)	1.0 µl
Antiense primer Mo REV (10 µM)	1.0 µl
Template DNA	5.0 µl
H₂O	5.5 µl
Total	25.0 µl

For the specific human primers you require for each sample:

2× QuantiTect SYBR Green PCR Master Mix	12.5 µl
Sense primer VFHu (10 µM)	1.0 µl
Antiense primer HuRev (10 µM)	1.0 µl
Template DNA	5.0 µl
H₂O	5.5 µl
Total	25.0 µl

Keep all reagents on ice.

1. Label thin walled PCR tubes (for the real-time thermal cycler to be used) appropriately on the lids rather than the side of the tube. (*NOTE*: The labels are placed on the lid of the tube because the fluorescence is measured through the side of the tubes.)

2. Add 20 µl of master mix to each tube.

3. Add 5 µl of template DNA to each appropriately labeled tube and keep at +4 °C.

4. Open the operating software for the real-time thermal cycler and complete the sample list.

5. Set up the cycling conditions for the real-time thermal cycler (*see* **Note 4**).

   Example cycling conditions for the Qiagen Rotor Gene Q real-time PCR machine:

   Step 1: *Hold* @ 95 °C for 15 min.

   Cycling (40 repeats):

   Step 2: 94 °C, 15 s.

   Step 3: 50 °C, 20 s.

   Step 4: 72 °C, 20 s (Acquiring to SYBR Green channel).

   Step 5: *Melt* from 72 to 95 °C. Rising by 1° at each step, hold 90 s on the first step, hold for 5 s on each of the next steps (acquire to SYBR Green channel).

Make sure that all safety features are in place and start the machine.

Once the machine has completed the run, save the data, remove the samples from the machine and store frozen (*see* **Note 5**).

Each real-time thermal cycler will differ in the specific of the analysis programs, but will generally perform the same processes; an example is given below:

2.2.5 Analysis of Real-Time PCR

1. Check that filters Dynamic Tubes and Slope Correct are active. Check Auto-Scale (to fit the scale to minimum and maximum readings for the data obtained) and Linear Scale. Adjust the threshold manually above the nonspecific background. The threshold should cross a sample's curve in the exponential phase (~0.01). The cycle threshold value (C_t) is the cycle number where a PCR product is detected.

2. Before reading the data check that the negative control is negative and the positive controls are positive to validate the assay.

3. The amplification C_t data estimates the quantity of cellular DNA. The standard curve estimates the amount of DNA equivalent to that from a known number of cells for each L20B and RD cells.

4. The melting curve analysis allows differentiating DNA amplified from human or mouse cells (depending on the primers used) and false positive reactions due to primer-dimers or nonspecific amplification.

The raw data is analyzed to show the amplification profiles of the samples under test.

The results below (Fig. 4) show that the RD (human) cell culture under analysis has been contaminated with mouse cells. The DNA has been amplified by both sets of specific primers and the melting temperature of the contaminating cell line is consistent with that of mouse cells, probably L20B cells.

3 Cell Sensitivity Test

Validation of a Laboratory Quality Control (LQC) preparation for routine use in cell sensitivity assays.

3.1 Materials

- Class II Biological Safety Cabinet.
- Vortex mixer.
- 100, 200, 1000 μl calibrated pipettes with aerosol-resistant tips (ART).
- Incubator set at 36 °C or CO_2 incubator set at 36 °C.

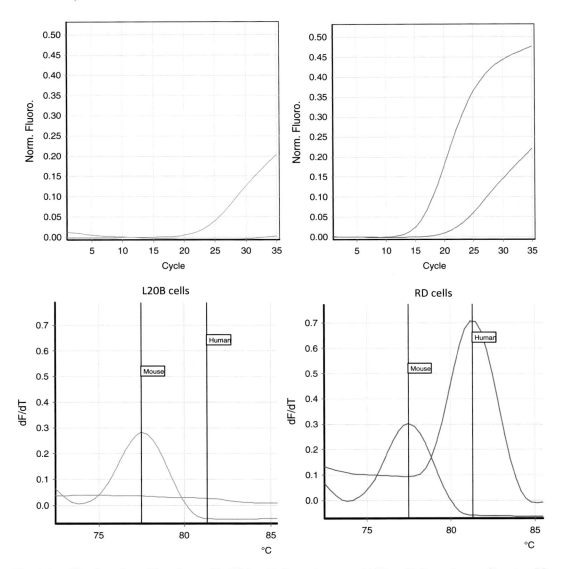

Fig. 4 Amplification of an RD culture with VFMo + MoRev primers and VFHu + HuRev primers: Showing RD (human) cells contaminated with mouse cells

- Microcentrifuge.
- Multichannel or multidropper pipette with appropriate ART tips.
- 1, 5, and 10 ml plastic disposable pipettes and electronic pipetting aid or multidispenser with appropriate tips.
- 5 ml sterile tubes with caps for making dilutions.
- Tube rack.
- Sterile 96-well flat-bottomed cell-culture microtiter plates with lids.

- Sterile, nontoxic plate sealers (if a CO_2 incubator is not to be used).

- Ice bucket with ice.

- Weighing balance.

- Magnetic stirrer and stirrer magnets.

- Sabin poliovirus Laboratory Quality Control (LQC) virus preparations for the three poliovirus serotypes: 1, 2, and 3.

- Standard Reference Preparation (SRP) of known and validated titer for the three poliovirus serotypes: 1 (01/528), 2 (01/530), and 3 (01/532) (available from NIBSC, Potters Bar, Herts. UK. htpp://www.nibsc.org).

- Flasks with a confluent layer of healthy L20B cells.

- Flasks with a confluent layer of healthy RD cells.

- Maintenance medium: Minimum Essential Medium (Sigma, Cat. no. M5650) containing sodium bicarbonate and phenol red (MEM), 2 % fetal calf serum, 1 % penicillin and streptomycin, 1 % 1 M HEPES Buffer, 1 % 200 mM l-glutamine.

- Phosphate Buffered Saline without calcium and magnesium : Add 8.00 g NaCl, 0.20 g KCl, 0.91 g Na_2HPO_4 (anhydrous) and 0.12 g KH_2PO_4 to 600–800 ml distilled water and stir until dissolved. Add 2 ml of 0.4 % phenol red as pH indicator, then make up to 1000 ml with distilled water and autoclave at 10 psi (70 kPa) for 15 min (110 °C).

- 0.25 % Trypsin (Sigma, Cat. no. T4049).

- Neubauer hemocytometer.

- Inverted light microscope.

- Trypan blue (0.04 % w/v) (Life Technologies, Cat. no. T10282).

- Naphthalene Black staining solution: Add 13.6 g sodium acetate to 800 ml distilled water and stir until dissolved. Add 60 ml acetic acid to the sodium acetate solution. Make up to 1 l with "ultrapure" water. Add 1.0 g naphthalene black to the acetic acid and sodium acetate solution and dissolve the stain by stirring for at least 30 min.

- Divide the naphthalene black stain into 500 ml amounts, label, and store for a maximum of 5 years at room temperature.

- Vacuum aspirator or multichannel pipette.

- Virucidal disinfectant.

- Waste container.

3.2   Method

Working in the biological safety cabinet (BSC):

1. Label six sets of 5 ml dilution tubes 10^{-1} to 10^{-8}, one set of three (for three serotypes) for each reference strain (*see* Fig. 1).

2. Dispense 4.5 ml of maintenance medium into tubes 1–8 for the six sets of tubes, using a 5 ml pipette or a multi dispenser.

3. Rapidly thaw aliquots of a Standard Virus Preparation (SVP), one tube of each poliovirus serotype, and aliquots of a Laboratory Quality Control (LQC), one tube of each poliovirus serotype, vortex and spin the tubes in a microcentrifuge and place in the ice bucket.

4. Add 0.5 ml of type 1 poliovirus to the first tube using a sterile pipette or pipettor with ART tip (*see* **Note 6**).

5. DO NOT immerse the tip in the dilution medium (*see* **Note 7**). Discard the tip.

6. Cover the tube and vortex gently.

7. Take another pipette/pipette tip and transfer 0.5 ml of the 10^{-1} dilution to the second tube (10^{-2}) without immersing the tip into the medium in the dilution tube, and discard pipette/pipette tip.

8. Cover the tube and vortex gently.

9. Repeat the dilution steps, transferring 0.5 ml of the previous dilution each time and always changing pipette/pipette tip between dilutions, up until tube 8 (*see* Fig. 5). Repeat for each dilution series for each serotype and reference.

10. For each serotype and SVP and LQC dilution series:

    Add 100 µl of virus dilution 10^{-5} to wells 1–10 in rows A and B into two sterile 96-well flat-bottomed cell-culture microtiter plates (*see* Fig. 6).

Transfer 0.5ml of virus dilution each time

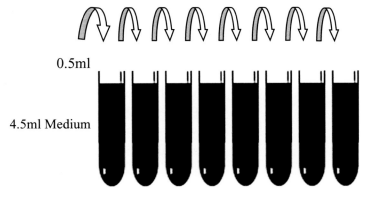

0.5ml

4.5ml Medium

Fig. 5 Preparation of virus dilution of Sabin Poliovirus Reference

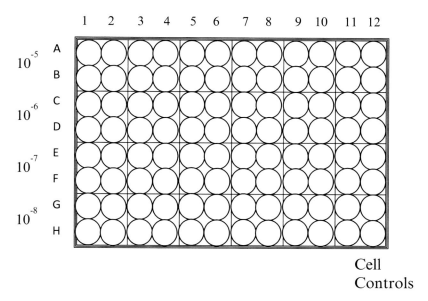

Fig. 6 Plate layout for titration of quality control standard

11. Add 100 μl of virus dilution 10^{-5} to wells 1–10 in rows C and D into two sterile 96-well flat-bottomed cell-culture microtiter plates.

12. Add 100 μl of virus dilution 10^{-5} to wells 1–10 in rows E and F into two sterile 96-well flat-bottomed cell-culture microtiter plates.

13. Add 100 μl of virus dilution 10^{-5} to wells 1–10 in rows G and H into two sterile 96-well flat-bottomed cell-culture microtiter plates.

14. Add 100 μl of maintenance medium to all the remaining empty wells, 11 and 12 in rows A to H.

15. Prepare at least 60 ml of L20B cell suspension containing $1–2 \times 10^5$ cells per ml in maintenance medium (*see* **Note 8**).

16. Add 100 μl of L20B cell suspension to each well of one of the plates for each serotype.

17. Incubate the plates in an incubator at 36 °C for 7 days. (Use lids for the CO_2 incubator and sealing film for non-CO_2 incubator.)

18. Allow the BSC to run for at least 10 min before preparing at least 60 ml of an RD cell suspension at $1–2 \times 10^5$ cells per ml in maintenance medium.

19. Add 100 μl of RD cell suspension to each well of the other plate for each serotype.

20. Incubate the plates in an incubator at 36 °C for 7 days. (Use lids for the CO_2 incubator and sealing film for non-CO_2 incubator.)

21. Examine the call in the microtiter plates using an inverted microscope daily to check the condition of the cells.

22. After 7 days, examine the wells in each plate and score each well as positive or negative for cytopathic effect (CPE), and record the results.

23. Remove the medium from the wells by aspiration and stain with Naphthalene Black staining solution, at least 100 μl per well.

24. Allow to stain for at least 1 h, remove the stain and wash the plates with water, discarding all stain and water into a container of suitable virucidal disinfectant.

3.2.1 To Obtain Valid Test Results

For a valid test, the cell controls should have a complete monolayer of healthy cells.

The lowest dilution (10^{-5}) should have at least 18 CPE positive wells.

The highest dilution (10^{-8}) should have at least 18 CPE negative wells.

If these criteria are fulfilled, calculate the virus titer using the Kärber formula.

3.2.2 To of the Virus Titer by the Kärber Formula

$$\log CCID50 = L - d(S - 0.5), \quad \text{where:}$$

L = log of lowest dilution used in the test.
d = difference between log dilution steps.
S = sum of proportion of "positive" tests (i.e., cultures showing CPE).
In the example shown in Fig. 7:
$L = -5.0$
$d = 1.0$.
$S = 1 + 0.55 + 0.40 + 0 = 1.95$.
$\log CCID50 = -5 - 1(1.95)$
$\log CCID50 = -5 - 1.95 = -6.95$
Virus titer = $10^{6.95}$ CCID50/0.1 ml

The end-point estimation using the Kärber method requires the observation of both a dilution with 100 % and 0 % CPE, or the observation of ≥90 % or ≤10 % CPE at the first or last dilution respectively. Results within 10 % of 100 or 0 % can be accepted, as the next lower dilution is considered as 100 % or the next higher dilution is considered as 0 %.

If the titration range for the plated virus dilutions is too low or too high, either the 100 or 0 % CPE end-points may not be observed. This can occur if the Laboratory Quality Control

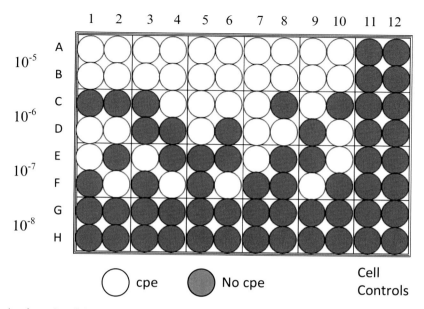

Fig. 7 Example of results of titration of Sabin Poliovirus Reference Standard

preparation has an unusually high or low titer. The recommended dilutions to be plated outlined in this procedure are based on typical reference preparations and it is completely permissible to change the recommended plated dilutions during validation by either higher or lower dilutions as appropriate, to obtain valid test results.

The titration of the Laboratory Quality Control (LQC) and Standard Reference Preparation (SVP) should be performed in parallel and repeated three times in three separate assays using independent titrations of the SVP and LQC.

If the titers obtained for the SVP are within the range stated for the standard in the relevant cell type, and the titration of the LQC is valid, then a virus titer can be assigned to the LQC. The LQC can then be used in routine cell sensitivity assays to monitor the cell lines in use.

3.2.3 Routine Use of the Cell Sensitivity Assay

The values obtained for the cell sensitivity assay LQC should be recorded each time the assay is performed, any variation of ±0.5 log from the established tire of the LQC should be investigated, particularly if media components have been changed, e.g., fetal bovine serum. Repeat the cell sensitivity assay once to eliminate human error in the making the dilutions, if the titers are still outside the expected range, test the cells for mycoplasma contamination and/or perform a cell authentication test to ensure that the cell lines have not become cross contaminated with either L20B or non-primate cells. If the LQC is still outside the expected range, particularly if the titers are lower than expected, obtain a fresh batch of cells and repeat the cell sensitivity assay.

4 Notes

1. Trypsin is inactivated in the presence of serum and so it is essential to remove all traces of serum with PBS without Ca^{2+}/Mg^{2+}. Use phosphate buffered saline without calcium and magnesium, as they will also inhibit the action of trypsin.

2. Adding at least the same amount of PBS as trypsin used should ensure that the trypsin is inactivated.

3. Newton's refraction rings appear as rainbow like rings under the coverslip.

4. The basic features of the PCR program on the thermocycler are shown below.

   Step 1 is an initial 95 °C hold to completely denature the original DNA template.

   Step 2 is the denaturing step.

   Step 3 is the annealing step whose temperature will depend on the sequence of the primers.

   Step 4 is the extension step whose time depends on the length of the product.

   Step 5 are the melting steps which will dissociate the double stranded DNA.

   For other machines it may be necessary to optimize the cycling protocol given here.

5. The melt reaction can be performed again using the same PCR reaction.

6. It is permissible to make the first dilution by adding 100 μl of virus to 900 μl of maintenance medium. This will conserve the stocks of reference preparations.

7. Virus will be on the outer surface of the tip or pipette and immersion of the tip into the medium will introduce additional virus into the dilution tube and will make the dilution inaccurate due to "carryover."

8. You will require 10 ml of cell suspension for each 96-well plate used.

References

1. www.iclac.org
2. http://standards.atcc.org/kwspub/home/the_international_cell_line_authentication_committee-iclac_/Database_of_Cross_Contaminated_or_Misidentified_Cell_Lines.pdf
3. Mendelsohn CC, Wimmer E, Racaniello VR (1989) Cellular receptor for poliovirus molec-
ular cloning, nucleotide sequence and expression of a new member of the immunoglobulin superfamily. Cell 56:855–865
4. Pipkin PA, Wood DJ, Racaniello VR, Minor PD (1993) Characterisation of L cells expressing the human poliovirus receptor for the specific detection of poliovirus in vitro. J Virol Methods 41:333–340

5. Wood DJ, Hull B (1999) L20B cells simplify culture of polioviruses from clinical samples. J Med Virol 58:188–192

6. Hebert PDN, Cywinska A, Ball SL, de Waard JR (2003) Biological identifications through DNA barcodes. Proc R Soc Lond B 270:313–321

7. http://www.barcodeoflife.org

8. Wittwer CT, Reed GH, Gundry CN, Vandersteen JG, Pryor RJ (2003) High-resolution genotyping by amplicon melting analysis using LCGreen. Clin Chem 49(6): 853–860

9. Bogenhagen D, Clayton DA (1974) The number of mitochondrial deoxyyribonucleic acid genomes in mouse L and human HeLa cells. J Biol Chem 249:7991–7995

10. WHO (2004) WHO polio laboratory manual, 4th edn. WHO/IVB/01.10, Geneva, Switzerland

Chapter 7

A Transgenic Mouse Model of Poliomyelitis

Satoshi Koike and Noriyo Nagata

Abstract

Transgenic mice (tg mice) that express the human poliovirus receptor (PVR), CD155, are susceptible to poliovirus and develop a neurological disease that resembles human poliomyelitis. Assessment of the neurovirulence levels of poliovirus strains, including mutant viruses produced by reverse genetics, circulating vaccine-derived poliovirus, and vaccine candidates, is useful for basic research of poliovirus pathogenicity, the surveillance of circulating polioviruses, and the quality control of oral live poliovirus vaccines, and does not require the use of monkeys. Furthermore, PVR-tg mice are useful for studying poliovirus tissue tropism and host immune responses. PVR-tg mice can be bred with mice deficient in the genes involved in viral pathogenicity. This report describes the methods used to analyze the pathogenicity and immune responses of poliovirus using the PVR-tg mouse model.

Key words Poliovirus receptor, Rodent model, Neurovirulence, Intracerebral inoculation, Paralysis, Viral antigen, Motor neuron, Innate immune response, Knockout mouse, Tissue tropism

1 Introduction

Humans are the natural hosts of poliovirus, but chimpanzees and old-world monkeys are also susceptible to experimental infections with poliovirus. In general, other animal species are not susceptible to poliovirus, except for some artificially adapted virus strains [1–3]. This narrow host range means that monkeys have been the main model for studying poliovirus infection in vivo. However, experiments that use monkeys are not only very expensive and labor intensive but also entail ethical problems and a potential risk of zoonoses.

Murine cells are resistant to poliovirus infection; however, transfection of viral RNA into these cells yields infectious particles, which suggests that poliovirus can complete its replication cycle even in mouse cells, provided the virus can circumvent the entry step [4–6]. The viral receptors for picornaviruses play important roles in attachment, internalization, and uncoating of poliovirus [7]. These previous studies, and other experimental evidence,

Javier Martín (ed.), *Poliovirus: Methods and Protocols*, Methods in Molecular Biology, vol. 1387,
DOI 10.1007/978-1-4939-3292-4_7, © Springer Science+Business Media New York 2016

suggest that the poliovirus receptor (PVR) is the primary determinant that restricts the host cell tropism. The "*Poliovirus sensitivity (PVS)*" gene is located on human chromosome 19 [8]. Mendelsohn et al. [9] showed that the lack of susceptibility of mouse L cells was overcome by the ectopic expression of a gene of human origin. This gene was PVR, which encodes an integral membrane protein comprising three immunoglobulin-like domains within the extracellular region [10, 11]. The NH_2-terminal Ig-like domain of PVR binds the virus, whereas the mouse ortholog or paralogs of PVR, i.e., tumor-associated glycoprotein-E4 (Tage4) and poliovirus receptor-related protein-1 and -2 (Prr-1 and Prr-2), cannot [12–16].

Thus, previous studies strongly suggest that transgenic expression of the human *PVR* gene will render mice susceptible to poliovirus infection. Ren et al. [17] and Koike et al. [18] introduced cosmid DNAs, which encoded the whole human *PVR* gene, into mice; PVR was then expressed under the control of its own promoter. The PVR protein was expressed in a wide range of cells in the transgenic mice (tg mice). After inoculating poliovirus via the intracerebral, intravenous, and intraperitoneal routes, the mice developed a paralytic disease that resembled human poliomyelitis. However, the mice were not as sensitive as humans to oral infection. Histopathological examination of the paralyzed mice showed that neurons in the central nervous system were infected and destroyed by poliovirus. No severe pathological changes were observed in other organs.

Oral live Sabin vaccines were developed from parental virulent strains after serial passage in cultured cells [19]. The molecular mechanism underlying attenuation has been one of the most important issues in poliovirus research. Nucleotide sequence analysis showed that the attenuated Sabin 1 strain harbored 57 nucleotide and 21 amino acid substitutions compared with the parental Mahoney strain [20]. The neurovirulence levels of the two viruses and their recombinants were evaluated using monkeys, and several determinants of the attenuation phenotype were mapped in the viral genome [21, 22]. Similarly, determinants of the attenuation phenotypes were identified in type 2 and 3 polioviruses [23, 24]. The most important common genetic determinant in the three serotypes was a nucleotide substitution in stem-loop V of the internal ribosomal entry site (IRES) within the 5′ noncoding region. The substitution in the IRES modifies the stem-loop structure and reduces the translation initiation efficiency of the poliovirus protein [25, 26].

Initially, it was not clear whether the neurovirulence levels of viruses evaluated using the tg mouse model correlated with those obtained using monkey models. Thus, the recombinant viruses were inoculated into PVR-tg mice [27–29] and the neurovirulence level of each virus was evaluated based on the 50 % lethal dose

(LD_{50}) or the 50 % paralysis dose (PD_{50}). The neurovirulence levels determined using the tg mouse model correlated well with those using the monkey model. This suggested that it is possible to use tg mice for neurovirulence tests instead of monkeys. The evaluation of monkey neurovirulence levels requires precise histopathological examination; however, the mice died or were paralyzed in a dose-dependent manner after poliovirus inoculation. Therefore, it is possible to evaluate the neurovirulence levels in tg mice simply by calculating the LD_{50} or PD_{50}, without the need for histological examination [29, 30]. Of the PVR-tg mouse lines, PVR-tg21 was selected for further study because its sensitivity range was suitable for the evaluation of attenuated and virulent virus strains [28, 31]. The tg mice are more sensitive to intraspinal inoculation of poliovirus than to intracerebral inoculation, which means that intraspinal inoculation is suitable for the validation of oral live polio vaccine lots. This method is described in detail in previous studies [29, 32]. The intracerebral inoculation route is suitable for evaluating relatively virulent strains. PVR-tg mice have been used to evaluate the circulating vaccine-derived polioviruses, which caused recent outbreaks [33–37].

PVR-tg mice have also been used to study the tissue tropism and host immune responses of poliovirus. Mechanisms of host innate immune response have been elucidated in the last decade and mice deficient in the genes involved in the responses have been generated. Initially, the roles of these genes were investigated using viruses that can infect mice. Crossing these knockout mice with PVR-tg mice allowed similar studies to be conducted using poliovirus infections. PVR-tg21 mice were first generated on an ICR background. ICR is not an inbred strain, so the mice were not suitable for certain types of studies, such as the immunological responses to poliovirus infection. A number of knockout mouse strains were maintained on a C57Black/6 (B6) background. Thus, the tg mice were backcrossed with B6 for more than ten generations, and then crossed with knockout mouse strains on a B6 background. Poliovirus preferentially replicates in neurons in the central nervous system (CNS), although the reason for this tissue tropism remains unknown. Thus, PVR-tg mice lacking the interferon (IFN) alpha/beta receptor 1 (*Ifnar1*) gene were highly susceptible to poliovirus infection [38]. Interestingly, poliovirus antigens were detected in the liver, spleen, and pancreas, sites at which viral antigens are seldom found in mice showing a normal IFN response. These results suggested that the innate immune response mediated by type I IFN plays an important role in protecting organs that are not targets of poliovirus. Thus, it was concluded that the IFN response controls the pathogenicity and tissue tropism of poliovirus.

Viral infections are sensed by receptors for pathogen-associated molecular patterns. Abe et al. and Oshiumi et al. investigated the important receptor-signaling pathways that sense poliovirus infection.

They crossed PVR-tg mice with *Rig-I, Mda5, Tlr3, Tlr7, Trif,* and *Myd88*-deficient mice and found that the TLR3-TRIF pathway plays a pivotal role in host innate immune responses [39, 40]. The generation of PVR-tg mice lacking other genes will help us to further understand the roles of genes involved in viral replication and host defense. The present report describes the following methods: (1) mouse neurovirulence testing via intracerebral inoculation; (2) the genotyping of PVR-tg mice; and (3) the detection of poliovirus antigens in PVR-tg mice. These techniques have been used to study the pathogenicity and immune responses to poliovirus in PVR-tg mice.

2 Materials

2.1 Mouse Neurovirulence Test of Poliovirus via Intracerebral Inoculation

1. A needle for intracerebral inoculation (size $27G \times 2$ mm tip, $25G \times 10$ mm neck) (Fig. 1). Alternatively, an intradermal disposable needle (size $27G \times 3$ mm tip, $25G \times 10$ mm neck) can be used.

2. A 0.25 mL glass syringe.

3. Mice: PVR-tg 21 mice on a B6 background (B6-PVR-tg21 mice can be obtained from Tokyo Metropolitan Institute of Medical Science) or an IQI background (IQI-PVR-tg21 mice are sold by CLEA, Japan).

2.2 Genotyping Protocol for the PVR Gene

1. Lysis buffer: 10 mM Tris–HCl (pH 8.3), 50 mM KCl, 1.5 mM $MgCl_2$, 0.45 % (v/v) Nonidet P40, 0.45 % (v/v) Triton X-100, 0.1 % (w/v) gelatin. Sterilize by autoclaving and store the lysis buffer at –20 °C in aliquots of 4 or 8 mL.

2. Proteinase K (recombinant PCR grade): Dissolve the Proteinase K powder in 20 mL of 1 M Tris–HCl (pH 8.3) (final concentration of 5 mg/mL) and store the solution in 100 μL aliquots at –20 °C.

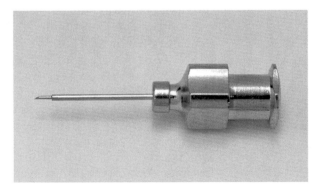

Fig. 1 Needle used for intracerebral inoculation. The needle comprises two parts, i.e., a $27G \times 2$ mm tip and a $25G \times 10$ mm thick neck, which helps to stop the needle at the desired depth

3. Primers:

PVR-4F	5'-ACCCCCCAGAGGTATCCATCTC-3'
MPH-4R	5'-TCATAGTCTGTGGGCTCTGGGTT-3'
IFNAR-51	5'-AGACGAGGCGAAGTGGTTAAAAGT-3'
IFNAR-32	5'-CTGTGTAGAATGGAATAAACGGATCA-3'
neo-N3	5'-GAACACGGCGGCATCAGAGC-3'

4. Thermostable *Taq* DNA polymerase.

5. ×10 PCR buffer: 100 mM Tris–HCl (pH 8.3), 500 mM KCl.

6. Restriction enzyme *Msp*I.

7. ×10 M buffer: 100 mM Tris–HCl (pH 7.5), 500 mM NaCl, 100 mM MgCl$_2$, 10 mM dithiothreitol.

2.3 Detection of Poliovirus Antigens

1. Butterfly catheter (22G).

2. Formalin-fixed, paraffin-embedded tissue.

3. Precoated glass slides.

4. Coplin jars or staining containers.

5. Autoclave.

6. Humidified chamber.

7. Coverslips.

8. Light microscope.

9. Fixative solution: Freshly prepared 4 % paraformaldehyde in 0.1 M phosphate buffer (pH 7.4) (PFA).

10. Decalcifying solution: Ethylenediamine-N,N,N',N'-tetraacetic acid, tetrasodium salt, tetrahydrate (EDTA-4Na) 100 g, citric acid monohydrate 15 g, phosphate-buffered saline (PBS) to make up to 1 L.

11. Wash solution: 10 mM phosphate buffer (pH 7.2–7.3) containing 150 mM NaCl.

12. Antigen retrieval regents: 10 mM citric acid buffer (pH 6.0).

13. Chromogen substrate: DAB (3,3'-diaminobenzidine) substrate solution in 0.1 M Tris–HCl buffer (pH 7.6). (Add 100 μL of H$_2$O$_2$ to 150 mL of DAB solution in Tris–HCl buffer immediately before performing the chromogen substrate reaction.)

14. Antibody diluent: 0.1 % bovine serum albumin in PBS. To preserve the antibodies, add 0.01 % NaN$_2$ to the antibody diluent.

15. Graded ethanol series, i.e., 70, 80, 95, and 100 % ethanol.

16. Xylene.

17. 3 % hydrogen peroxide in methanol.

18. Primary antibody: Rabbit anti-poliovirus serum.

19. Secondary antibody: HRP-labeled polymer conjugated to secondary antibodies.

20. Distilled water.

21. Hematoxylin counterstaining reagent.

3   Methods

3.1   Mouse Neurovirulence Test of Poliovirus via Intracerebral Inoculation

1. Prepare 6- to 7-week-old PVR-tg mice. Ten (≥ 6) mice (the same number of each gender) are recommended for inoculation with each dose (*see* **Note 4.1-1**).

2. Dilute the virus stock in Eagle's Minimum Essential Medium (MEM) and prepare a serial tenfold dilution range from 10^1 to 10^7 plaque-forming units (PFU)/25 μL (*see* **Note 4.1-2**).

3. Anesthetize the mice via the inhalation of 4 % isoflurane. The use of anesthesia apparatus is recommended (*see* **Note 4.1-3**).

4. The mouse is restrained manually on a solid surface. Disinfect the head of the mouse with 70 % ethanol and inoculate 25 μL of virus solution (for each dilution) into the left hemisphere of the brain. The site of injection is approximately halfway between the eye and the ear, and just off the midline (*see* **Note 4.1-4**).

5. Confirm that the mice can walk without any problems after recovering from the anesthetic (*see* **Note 4.1-5**).

6. Keep the mice in an isolator rack and observe daily for 14 or 21 days to monitor paralysis and death.

7. Score the LD_{50} or PD_{50} values according to the formula of Kaerber [41] or Reed and Muench [42] (*see* **Note 4.1-6**).

3.2   Genotyping Protocol for the PVR Gene

A description of the strategy used to establish the *PVR*$^{+/+}$ *Ifnar1*$^{-/-}$ mouse strain is provided as an illustrative example. First, PVR-tg mice are crossed with *Ifnar1*$^{-/-}$ mice [43]. When the resulting F1 mice (*PVR*$^{+/-}$ *Ifnar1*$^{+/-}$) are ≥ 7 weeks old, they are bred with each other to obtain F2 mice. The progeny F2 mice comprise nine different genotypes, including *PVR*$^{+/+}$ *Ifnar1*$^{-/-}$. However, the general PCR protocols for detecting the transgene cannot discriminate between the *PVR*$^{+/-}$ and *PVR*$^{+/+}$ genotypes. Therefore, to select *PVR*$^{+/+}$ mice, it is necessary to amplify the exon 4 region of the human *PVR* gene and the homologous mouse PVR-related gene-2 (*Prr-2*, which is also referred to as *Nectin-2* and *Mph*) simultaneously using a set of PCR primers. The hPVR and mPrr-2 fragments can then be discriminated after *Msp*I digestion because the Prr-2 fragment contains an *Msp*I site. After comparing the band intensities of the hPVR and mPrr-2 fragments, it is possible to identify

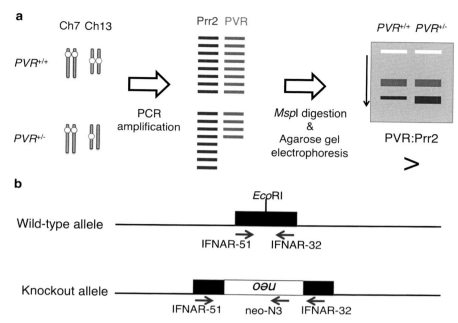

Fig. 2 Method used for identifying the *PVR* genotype (**a**) and the *Ifnar1* genotype (**b**) by PCR. (**a**) The *PVR* transgene (*red circle*) and the mouse *Prr2* gene (*blue circle*) are located on mouse chromosomes 13 and 7, respectively. The DNA ratio of PVR:Prr2 in the genomes of homozygotes is twice that in hemizygotes. The PVR and Prr2 fragments are co-amplified using a common primer. The *red* and *blue bars* represent PVR and Prr2 gene fragments, respectively. The ratio of the amplified fragments reflects the initial ratio. The length of both fragments is 109 bp, but only the mouse Prr2 fragment contains an *Msp*I site and can be cleaved into 87 and 22 bp fragments. The intensity of PVR fragment is higher than that of the Prr2 fragment in the *PVR*^+/+ genotype, whereas the intensity of the two fragments is comparable in the *PVR*^+/− genotype. (**b**) The *Ifnar1* gene is disrupted by inserting the *neo* gene at the *Eco*RI site in the exon 3 region. The primers IFNAR-51 and IFNAR-31 anneal close to the 5′- and 3′- ends of exon 3 of the *Ifnar1* gene, respectively. Neo-N3 anneals to the *neo* gene. The wild-type Ifnar1 fragment is amplified using IFNAR-51 and IFNAR-32 primers (178 bp), and the Ifnar1-neo fusion fragment is amplified using IFNAR-51 and neo-N3 primers (465 bp)

individuals with the *PVR*^+/+ genotype (Fig. 2a). To select the *Ifnar1*^−/− genotype, it is necessary to discriminate between the wild-type allele and the targeted allele using PCR with three different primers (Fig. 2b).

1. Cut a small piece (approximately 1 mm length) of tail from each F2 progeny mouse and place in a 1.5 mL sample tube. At the same time, make nicks in the ears of the mice to distinguish the individuals (*see* **Note 4.2-1**).

2. Add 100 μL of Proteinase K solution in 4 mL of lysis buffer and mix well (*see* **Note 4.2-2**).

3. Add 200 μL of lysis buffer containing Proteinase K to the sampling tube and digest the tail sample at 56 °C for 120 min (*see* **Note 4.2-3**).

4. Inactivate the Proteinase K at 95 °C for 10 min and mix the solution by vortexing.

5. Prepare the reaction tubes. In addition to the samples, the following positive and negative controls are needed.

   (a) Human genomic DNA (positive control for hPVR).

   (b) Mouse (wild-type) genomic DNA (positive control for mPrr-2).

   (c) Lysis buffer (negative control).

   (d) At least three samples of $PVR^{+/+}$ genomic DNA ($PVR^{+/+}$ control), i.e., use genomic DNA from the original $PVR^{+/+}$ mice.

   (e) At least three samples of $PVR^{+/-}$ genomic DNA ($PVR^{+/-}$ control), i.e., use genomic DNA from the F1 mice.

6. Prepare the master mix without template DNA (*see* **Notes 4.2-4** and **5**).

×10 PCR buffer	2.0 μL
dNTP (2 mM each)	2.5 μL
MgCl₂ (25 mM)	2.0 μL
Primer PVR-4F (5 μM)	2.0 μL
Primer MPH-4R (5 μM)	2.0 μL
Taq DNA polymerase (5 units/μL)	0.1 μL
H₂O	12.4 μL (per tube)

7. Add 2 μL of template DNA to the reaction mixture.

8. Amplify the DNA using the following parameters.

Step 1	94 °C for 10 min	
Step 2	94 °C for 30 s	
Step 3	55 °C for 15 s	
Step 4	72 °C for 15 s	(Go back to Step 2 and repeat for 35 cycles)

9. Prepare the *Msp*I enzyme mixture.

*Msp*I (6–20 units/μL)	0.5 μL
× 10 M buffer	0.5 μL
H₂O	4 μL (per tube)

10. Add the enzyme mixture to each tube, mix by pipetting, and incubate at 37 °C for 2 h.

11. Load 5–10 μL of the sample onto a 4 % agarose (3 % NuSieve agarose + 1 % agarose) gel (*see* **Note 4.2-6**).

12. The length of the hPVR fragment is 109 bp. The mouse PRR-2 fragment contains an *Msp*I site and the fragment is cleaved into two fragments: 87 and 22 bp. Compare the intensity of the 109 bp human PVR fragment with that of the 87 bp mPRR-2 fragment (the 22 bp fragment might not be visible) by referring to the $PVR^{+/+}$ and $PVR^{+/-}$ controls. The intensity of the PVR fragment is higher than that of the mouse Prr-2 fragment in the $PVR^{+/+}$ control, whereas the intensities of the two fragments are almost the same in $PVR^{+/-}$ control (Fig. 3a) (*see* **Note 4.2–7** and **8**).

13. Prepare the reaction tubes to genotype the *Ifnar1* locus. In addition to the samples, the following positive and negative controls are required.

    (a) Mouse (wild-type) genomic DNA (positive control for $Ifnar1^{+/+}$).

    (b) Mouse (Ifnar1 KO) genomic DNA (positive control for $Ifnar1^{-/-}$).

    (c) Lysis buffer (negative control).

14. Prepare the master mix without template DNA.

×10 PCR buffer	2.0 μL
dNTP (2 mM each)	2.5 μL
MgCl₂ (25 mM)	2.0 μL
Primer IFNAR-51 (5 μM)	2.0 μL
Primer IFNAR-32 (5 μM)	2.0 μL
Primer neo-N3 (5 μM)	2.0 μL
Taq DNA Polymerase (5 units/μL)	0.1 μL
H₂O	10.4 μL (per tube)

15. Add 2 μL of template DNA to the reaction mixture.

16. Amplify the DNA using the following parameters.

Step 1	94 °C for 5 min	
Step 2	94 °C for 30 s	
Step 3	55 °C for 30 s	
Step 4	72 °C for 60 s	(Go back to Step 2 and repeat for 35 cycles)

17. Load 5–10 μL of the sample onto a 2 % agarose gel.

18. The wild-type and $Ifnar1^{-/-}$ alleles produce bands of 178 bp and 465 bp, respectively. Both bands are detected in the $Ifnar1^{+/-}$ genotype (Fig. 3b).

19. Select males and females with the $PVR^{+/+}$, $Ifnar1^{-/-}$ genotype.

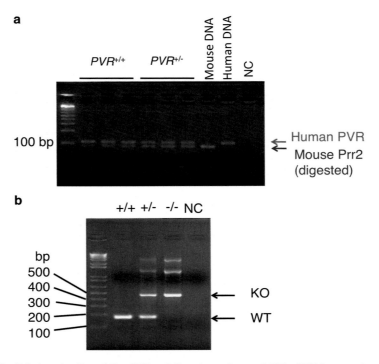

Fig. 3 Determination of the *PVR* and *Ifnar1* genotypes. (**a**) The PVR fragment and the Prr2 fragment (digested with *MspI*) run as bands of 109 bp and 87 bp, respectively. Note that the relative intensity of the PVR fragment to Prr2 fragment is higher in the *PVR+/+* genotype but not in the *PVR+/−* genotype. (**b**) The *Ifnar1+/+*, *Ifnar1+/−*, and *Ifnar1−/−* genotypes yield 178 bp, 178 + 465 bp, and 465 bp fragments, respectively

3.3 Detection of Poliovirus Antigens

Basically, immunohistochemical staining (IHC) and an indirect immunoperoxidase technique are used to detect virus antigens on sections (3 μm thick) of formalin-fixed, paraffin-embedded tissues. At present, polymer-based immunohistochemical systems are standard. These systems are based on a horseradish peroxidase (HRP)-labeled polymer, which is conjugated to secondary antibodies [44]. Commercial kits are available from many companies (*see* **Note 4.3-10**).

1. Inoculate the mice with poliovirus and maintain the infected mice in isolator racks until the day of sampling.

2. Prepare separate syringes filled with 15 mL of PBS and 25 mL of 4 % PFA. Connect the butterfly catheter and two syringes via a three-way stopcock. Purge any air bubbles from the line.

3. Euthanize the infected mice via the inhalation of excess isoflurane.

4. Immediately after breathing stops, open the abdomen and the thoracic cavity, taking care not to cut the major vessels.

5. Cut the right atrium to make a drain using small scissors. Insert the butterfly needle into the left ventricle at the apex toward the aorta and perfuse with 15 mL of PBS first (until the fluid is clear of blood).

6. Adjust the stopcock to perfuse 25 mL of 4 % PFA (*see* **Notes 4.3-1** and **2**).

7. Remove the organs and place them in 4 % PFA solution. Immerse the organs overnight (at least) (*see* **Note 4.3-3**).

8. Remove fat by immersing the organs in 50 % ethanol (three changes, each immersion lasting for 3 days).

9. Decalcify the bones by rocking in the decalcification solution. Change the solution every 3 days (three changes in all) (*see* **Note 4.3-4**).

10. Cut the fixed tissues into appropriate sizes and place in the embedding cassettes (*see* **Note 4.3-5**).

11. Dehydrate the tissues before paraffin embedding by immersing the cassette in the following sequence:
    - 70 % ethanol, two changes, 1 h each, at room temperature (r.t.) with rocking.
    - 80 % ethanol, two changes, 1 h each, at r.t. with rocking.
    - 95 % ethanol, two changes, 1 h each, at r.t. with rocking.
    - 100 % ethanol, three changes, 1 h each, at r.t. with rocking.
    - Xylene, three changes, 1 h each.
    - Paraffin wax, two changes, 1.5 h each at 56 °C.

12. Embed the tissues in paraffin blocks on the heating block.

13. Cool the blocks to allow the paraffin to solidify.

14. Cut off any excess paraffin from the embedded sample.

15. Mount the sample sections on a microtome.

16. Cut 3 μm slices using the microtome.

17. Place the paraffin slices in water at 40–45 °C.

18. Remove the paraffin sections from the water with glass slides and use a brush to position the sections.

19. Dry the sections overnight at 37 °C.

20. Deparaffinize the sections.
    - Wash the slides with three changes of xylene for 5 min each.

21. Rehydrate the sample sections.
    - Wash the slides with three changes of 100 % ethanol for 3 min each.
    - Wash the slides in 95 and 80 % ethanol for 3 min each.

22. Rinse the slides in running (distilled) water for 5 min.

23. Antigen retrieval reaction (*see* **Note 4.3-6**).

    - Heat the slides in retrieval solution (i.e., 10 mM sodium citrate buffer, pH 6.0, at 121 °C for 10 min in an autoclave).

24. After cooling, immerse the glass slides in 3 % hydrogen peroxide in methanol for 30 min at r.t.

25. Wash the slides with three changes of PBS for 5 min each.

26. Block nonspecific binding sites with 10 % normal goat serum in PBS for 5 min at r.t.

27. After draining the blocking solution, add the primary antibody (i.e., rabbit anti-poliovirus serum) at an appropriate dilution (1:500–1:4000) and incubate overnight at 4 °C (*see* **Notes 4.3-7 and 8**).

28. Wash the slides with three changes of PBS for 5 min each.

29. Apply the labeled secondary antibody (i.e., goat anti-rabbit Ig conjugated with dextran polymer on HRP molecules) for 30 min at r.t.

30. Wash the slides with three changes of PBS for 5 min each.

31. React with the chromogen substrate for 5–10 min (i.e., DAB).

32. Stop the color reaction by washing in distilled water.

33. Immerse the slides in aqueous hematoxylin to counterstain.

34. Wash the slides in running (distilled) water for 5 min.

35. Dehydrate the slides using two or three changes of 80, 95, and 100 % ethanol for 1 min each (*see* **Note 4.3–9**).

36. Clear with xylene.

37. Mount the coverslips using mounting medium.

38. Examine the mounted sections and confirm the antigens under a microscope (*see* **Note 4.3–11**).

4   Notes

4.1   Mouse Neurovirulence Test of Poliovirus via Intracerebral Inoculation

1. The sensitivity of mice to poliovirus infection varies with age. In particular, mice aged <5 weeks are highly sensitive to poliovirus. Thus, stable results are obtained if 6- to 7-week-old mice are used. Mice older than 10 weeks are not recommended because the skull is harder and it is difficult to control the depth of the needle.

2. The log LD_{50} or PD_{50} values for virulent strains are about 2–3, whereas those of attenuated strains are >6.

3. If anesthesia apparatus is not available, anesthetize the mice via intraperitoneal injection of 100 mg/kg of ketamine and 2 mg/kg of xylazine in saline.

4. The thick part of the needle helps to prevent it from extending too deeply into the brain, but the injection must be performed as carefully as possible.

5. Mice must be excluded from the dataset if they die on the day of inoculation or on the next day. This is because they might die due to injuries sustained during the injection.

6. If virulent strains are inoculated, all mice showing clinical signs will die. However, some mice will survive with paralysis if attenuated strains are inoculated.

4.2 Genotyping Protocol for the PVR Gene

1. Ten-day-old mice are suitable for sampling. Younger mice are too small to mark the ears by cutting. Mice at 2 weeks-of-age move briskly and are difficult to handle. To avoid cross-contamination, the scissors and forceps should be wiped with absorbent cotton soaked with 70 % ethanol and the residual ethanol must be burned after each operation.

2. Avoid storing Proteinase K at low concentrations. The lysis buffer containing Proteinase K should be prepared immediately before use.

3. Ensure that the small piece of the tail in each tube is soaked in the lysis buffer.

4. The sequences of the primers PVR-4F and MHP-4R do not match perfectly with those of Prr-2 and PVR, respectively; however, they are sufficiently homologous for amplification. Under the conditions used, the primers amplify both genes with almost equal efficiency.

5. This is a competitive PCR, and the theoretical ratio of the two genes does not change before and after amplification. The number of copies of the *PVR* gene in $PVR^{+/+}$ mice is over twice that in $PVR^{+/-}$ mice, whereas the number of copies of the *PRR-2* gene is the same.

6. This assay is delicate. Prepare the gels and run the samples very carefully. It is better to use a wide comb. Use at least three control samples from $PVR^{+/+}$ and $PVR^{+/-}$ mice, and the samples should be loaded on the same gel.

7. After obtaining the male and female $PVR^{+/+}$ candidates, cross them and check that all of the progeny have the same $PVR^{+/+}$ genotype. If all the samples are $PVR^{+/+}$, it can be concluded that they are homozygotes.

8. If the results are unclear, backcross the $PVR^{+/+}$ candidate with wild-type mice to confirm homozygosity.

4.3 Detection of Poliovirus Antigens

1. If perfusion is successful, the liver is bleached soon after perfusion with PBS and muscle contractions are observed soon after beginning perfusion with 4 % PFA. The mouse should be stiff when perfusion is finished.

2. If fluid drips from the mouse's nose, this indicates that the fluid pressure is too high. In this case, reduce the rate of perfusion.

3. During immersion fixation, the volume of 4 % PFA should be at least ten times the weight of the organs.

4. The decalcification step is optional. This step can be omitted if the organs can be excised from the bone.

5. Do not allow the tissues to dry at any time during the staining procedure.

6. Antigen retrieval is often performed with rehydrated sections by heat-mediated retrieval using EDTA buffer (pH 9.0) or citric acid buffer (pH 6.0) as the retrieval solution.

7. The optimal dilutions of antibody used to recognize the poliovirus antigen should be determined based on experiments with positive control tissue samples, such as tissue sections of formalin-fixed, paraffin-embedded cultures or poliovirus-infected animals.

8. The negative controls comprise samples from a non-infected animal or a sequential tissue section cut from each block and incubated with non-immune serum.

9. After reacting with the chromogen substrate, the rehydration and clearing steps should only be used if the chromogen substrate is insoluble in alcohol.

10. Polymer-based systems generally use a two-step IHC staining technique. The labeled polymer is an avidin- or biotin-free system, which avoids false-positive staining due to endogenous avidin-biotin activity. This system is extremely sensitive, and the optimal dilutions of the primary antibody are up to 20 times higher than those used by the traditional peroxidase anti-peroxidase (PAP) technique, as well as being several times higher than those used by the traditional avidin-biotin conjugate (ABC) or labeled avidin-biotin (LSAB) systems [44].

11. Poliovirus antigens can be detected successfully in CNS tissues when mice are fixed on the day of paralysis onset. However, detection becomes difficult at later times because of viral antigen clearance by the host. The detection limit is approximately 10^7 PFU/g. It is difficult to detect the antigens if the viral load is lower.

References

1. Hsiung GD, Black FL, Henderson JR (1964) Susceptibility of primates to viruses in relation to taxonomic classification. In: Buettner-Jaenusch J (ed) Evolutionary and genetic biology of primates, vol 2. Academic, New York, pp 1–23

2. Ida-Hosonuma M, Sasaki Y, Toyoda H, Nomoto A, Gotoh O, Yonekawa H, Koike S (2003) Host range of poliovirus is restricted to simians because of a rapid sequence change of the poliovirus receptor gene during evolution. Arch Virol 148(1):29–44

3. Armstrong C (1941) Cotton rats and white mice in poliomyelitis research. Am J Public Health Nations Health 31(3):228–232

4. Holland JJ (1961) Receptor affinities as major determinants of enterovirus tissue tropisms in humans. Virology 15:312–326

5. Holland JJ, Mc LL, Syverton JT (1959) Mammalian cell-virus relationship. III Poliovirus production by non-primate cells exposed to poliovirus ribonucleic acid. Proc Soc Exp Biol Med 100(4):843–845

6. Holland JJ, Mc LL, Syverton JT (1959) The mammalian cell-virus relationship. IV Infection of naturally insusceptible cells with enterovirus ribonucleic acid. J Exp Med 110(1):65–80

7. Bergelson JM (2010) Receptors. In: Roos RP, Ehrenfeld E, Domingo E (eds) The picornaviruses. ASM Press, Washington, DC, pp 73–86

8. Siddique T, McKinney R, Hung WY, Bartlett RJ, Bruns G, Mohandas TK, Ropers HH, Wilfert C, Roses AD (1988) The poliovirus sensitivity (PVS) gene is on chromosome 19q12 q13.2. Genomics 3(2):156–160

9. Mendelsohn C, Johnson B, Lionetti KA, Nobis P, Wimmer E, Racaniello VR (1986) Transformation of a human poliovirus receptor gene into mouse cells. Proc Natl Acad Sci U S A 83(20):7845–7849

10. Mendelsohn CL, Wimmer E, Racaniello VR (1989) Cellular receptor for poliovirus: molecular cloning, nucleotide sequence, and expression of a new member of the immunoglobulin superfamily. Cell 56(5):855–865

11. Koike S, Horie H, Ise I, Okitsu A, Yoshida M, Iizuka N, Takeuchi K, Takegami T, Nomoto A (1990) The poliovirus receptor protein is produced both as membrane-bound and secreted forms. EMBO J 9(10):3217–3224

12. Koike S, Ise I, Nomoto A (1991) Functional domains of the poliovirus receptor. Proc Natl Acad Sci U S A 88(10):4104–4108

13. Selinka HC, Zibert A, Wimmer E (1991) Poliovirus can enter and infect mammalian cells by way of an intercellular adhesion molecule 1 pathway. Proc Natl Acad Sci U S A 88(9): 3598–3602

14. Ravens I, Seth S, Forster R, Bernhardt G (2003) Characterization and identification of Tage4 as the murine orthologue of human poliovirus receptor/CD155. Biochem Biophys Res Commun 312(4):1364–1371

15. Morrison ME, He YJ, Wien MW, Hogle JM, Racaniello VR (1994) Homolog-scanning mutagenesis reveals poliovirus receptor residues important for virus binding and replication. J Virol 68(4):2578–2588

16. Aoki J, Koike S, Ise I, Sato-Yoshida Y, Nomoto A (1994) Amino acid residues on human poliovirus receptor involved in interaction with poliovirus. J Biol Chem 269(11):8431–8438

17. Ren RB, Costantini F, Gorgacz EJ, Lee JJ, Racaniello VR (1990) Transgenic mice expressing a human poliovirus receptor: a new model for poliomyelitis. Cell 63(2):353–362

18. Koike S, Taya C, Kurata T, Abe S, Ise I, Yonekawa H, Nomoto A (1991) Transgenic mice susceptible to poliovirus. Proc Natl Acad Sci U S A 88(3):951–955

19. Sabin A, Boulger L (1973) History of Sabin attenuated poliovirus oral live vaccine strains. J Biol Stand 1:115–118

20. Nomoto A, Omata T, Toyoda H, Kuge S, Horie H, Kataoka Y, Genba Y, Nakano Y, Imura N (1982) Complete nucleotide sequence of the attenuated poliovirus Sabin 1 strain genome. Proc Natl Acad Sci U S A 79(19): 5793–5797

21. Omata T, Kohara M, Kuge S, Komatsu T, Abe S, Semler BL, Kameda A, Itoh H, Arita M, Wimmer E et al (1986) Genetic analysis of the attenuation phenotype of poliovirus type 1. J Virol 58(2):348–358

22. Kawamura N, Kohara M, Abe S, Komatsu T, Tago K, Arita M, Nomoto A (1989) Determinants in the 5′ noncoding region of poliovirus Sabin 1 RNA that influence the attenuation phenotype. J Virol 63(3):1302–1309

23. Macadam AJ, Pollard SR, Ferguson G, Skuce R, Wood D, Almond JW, Minor PD (1993) Genetic basis of attenuation of the Sabin type 2 vaccine strain of poliovirus in primates. Virology 192(1):18–26

24. Westrop GD, Wareham KA, Evans DM, Dunn G, Minor PD, Magrath DI, Taffs F, Marsden S, Skinner MA, Schild GC et al (1989) Genetic basis of attenuation of the Sabin type 3 oral poliovirus vaccine. J Virol 63(3):1338–1344

25. Haller AA, Stewart SR, Semler BL (1996) Attenuation stem-loop lesions in the 5′ noncoding region of poliovirus RNA: neuronal cell-specific translation defects. J Virol 70(3): 1467–1474

26. Svitkin YV, Cammack N, Minor PD, Almond JW (1990) Translation deficiency of the Sabin type 3 poliovirus genome: association with an attenuating mutation C472 U. Virology 175(1):103–109

27. Horie H, Koike S, Kurata T, Sato-Yoshida Y, Ise I, Ota Y, Abe S, Hioki K, Kato H, Taya C et al (1994) Transgenic mice carrying the human poliovirus receptor: new animal models for study of poliovirus neurovirulence. J Virol 68(2):681–688

28. Abe S, Ota Y, Doi Y, Nomoto A, Nomura T, Chumakov KM, Hashizume S (1995) Studies on neurovirulence in poliovirus-sensitive transgenic

mice and cynomolgus monkeys for the different temperature-sensitive viruses derived from the Sabin type 3 virus. Virology 210(1):160–166

29. Abe S, Ota Y, Koike S, Kurata T, Horie H, Nomura T, Hashizume S, Nomoto A (1995) Neurovirulence test for oral live poliovaccines using poliovirus-sensitive transgenic mice. Virology 206(2):1075–1083. doi:10.1006/viro.1995.1030

30. Dragunsky E, Nomura T, Karpinski K, Furesz J, Wood DJ, Pervikov Y, Abe S, Kurata T, Vanloocke O, Karganova G, Taffs R, Heath A, Ivshina A, Levenbook I (2003) Transgenic mice as an alternative to monkeys for neurovirulence testing of live oral poliovirus vaccine: validation by a WHO collaborative study. Bull World Health Organ 81(4):251–260

31. Koike S, Taya C, Aoki J, Matsuda Y, Ise I, Takeda H, Matsuzaki T, Amanuma H, Yonekawa H, Nomoto A (1994) Characterization of three different transgenic mouse lines that carry human poliovirus receptor gene—influence of the transgene expression on pathogenesis. Arch Virol 139(3-4):351–363

32. WHO (2012) Standard operating procedure neurovirulence test of types 1, 2 OR 3 live attenuated poliomyelitis vaccines (oral) in transgenic mice susceptible to poliovirus, Version 6. World Health Organization, Switzerland

33. Kew O, Morris-Glasgow V, Landaverde M, Burns C, Shaw J, Garib Z, Andre J, Blackman E, Freeman CJ, Jorba J, Sutter R, Tambini G, Venczel L, Pedreira C, Laender F, Shimizu H, Yoneyama T, Miyamura T, van Der Avoort H, Oberste MS, Kilpatrick D, Cochi S, Pallansch M, de Quadros C (2002) Outbreak of poliomyelitis in Hispaniola associated with circulating type 1 vaccine-derived poliovirus. Science 296(5566):356–359. doi:10.1126/science.1068284

34. Jegouic S, Joffret ML, Blanchard C, Riquet FB, Perret C, Pelletier I, Colbere-Garapin F, Rakoto-Andrianarivelo M, Delpeyroux F (2009) Recombination between polioviruses and co-circulating Coxsackie A viruses: role in the emergence of pathogenic vaccine-derived polioviruses. PLoS Pathog 5(5):e1000412

35. Rakoto-Andrianarivelo M, Guillot S, Iber J, Balanant J, Blondel B, Riquet F, Martin J, Kew O, Randriamanalina B, Razafinimpiasa L, Rousset D, Delpeyroux F (2007) Co-circulation and evolution of polioviruses and species C enteroviruses in a district of Madagascar. PLoS Pathog 3(12):e191. doi:10.1371/journal.ppat.0030191

36. Shimizu H, Thorley B, Paladin FJ, Brussen KA, Stambos V, Yuen L, Utama A, Tano Y, Arita M, Yoshida H, Yoneyama T, Benegas A, Roesel S, Pallansch M, Kew O, Miyamura T (2004) Circulation of type 1 vaccine-derived poliovirus in the Philippines in 2001. J Virol 78(24):13512–13521

37. Thorley B, Kelly H, Nishimura Y, Yoon YK, Brussen KA, Roberts J, Shimizu H (2009) Oral poliovirus vaccine type 3 from a patient with transverse myelitis is neurovirulent in a transgenic mouse model. J Clin Virol 44(4):268–271. doi:10.1016/j.jcv.2009.01.014

38. Ida-Hosonuma M, Iwasaki T, Yoshikawa T, Nagata N, Sato Y, Sata T, Yoneyama M, Fujita T, Taya C, Yonekawa H, Koike S (2005) The alpha/beta interferon response controls tissue tropism and pathogenicity of poliovirus. J Virol 79(7):4460–4469

39. Abe Y, Fujii K, Nagata N, Takeuchi O, Akira S, Oshiumi H, Matsumoto M, Seya T, Koike S (2012) The toll-like receptor 3-mediated antiviral response is important for protection against poliovirus infection in poliovirus receptor transgenic mice. J Virol 86(1):185–194. doi:10.1128/JVI.05245-11

40. Oshiumi H, Okamoto M, Fujii K, Kawanishi T, Matsumoto M, Koike S, Seya T (2011) The TLR3/TICAM-1 pathway is mandatory for innate immune responses to poliovirus infection. J Immunol 187(10):5320–5327. doi:10.4049/jimmunol.1101503

41. Kärber G (1931) Beitrag zur kollektiven Behandlung pharmakologischer Reihenversuche. Arch Exp Pathol Pharmakol 162(4):480–483. doi:10.1007/BF01863914

42. Reed LJ, Muench H (1938) A simple method of estimating fifty per cent endpoints. Am J Hyg 27(3):493–497

43. Muller U, Steinhoff U, Reis LF, Hemmi S, Pavlovic J, Zinkernagel RM, Aguet M (1994) Functional role of type I and type II interferons in antiviral defense. Science 264(5167):1918–1921

44. Sabattini E, Bisgaard K, Ascani S, Poggi S, Piccioli M, Ceccarelli C, Pieri F, Fraternali-Orcioni G, Pileri SA (1998) The EnVision++ system: a new immunohistochemical method for diagnostics and research. Critical comparison with the APAAP, ChemMate, CSA, LABC, and SABC techniques. J Clin Pathol 51(7):506–511

Chapter 8

Standardized Methods for Detection of Poliovirus Antibodies

William C. Weldon, M. Steven Oberste, and Mark A. Pallansch

Abstract

Testing for neutralizing antibodies against polioviruses has been an established gold standard for assessing individual protection from disease, population immunity, vaccine efficacy studies, and other vaccine clinical trials. Detecting poliovirus specific IgM and IgA in sera and mucosal specimens has been proposed for evaluating the status of population mucosal immunity. More recently, there has been a renewed interest in using dried blood spot cards as a medium for sample collection to enhance surveillance of poliovirus immunity. Here, we describe the modified poliovirus microneutralization assay, poliovirus capture IgM and IgA ELISA assays, and dried blood spot polio serology procedures for the detection of antibodies against poliovirus serotypes 1, 2, and 3.

Key words Poliovirus, Neutralization, Antibodies, ELISA, IgA, IgM, OPV, IPV, Dried blood spot

1 Introduction

1.1 Poliovirus Microneutralization Assay

The polio microneutralization assay measures neutralizing antibody titers to poliovirus types 1, 2, and 3 using 96-well microtiter plates (it is termed "microneutralization" because the original neutralization assay was performed using larger volumes in culture tubes). The principle of the test is that the anti-poliovirus antibodies in a serum sample will bind to the virus and block infection of susceptible cells. Because poliovirus is cytopathic, virus that is not bound by antibody infects and lyses cells. The amount of neutralizing antibody is quantitated as a titer based on the last serum dilution to protect susceptible cell culture wells from poliovirus infection and cytopathic effect.

The test takes approximately 7 days to complete, from the dilution of sera to staining and reading plates, and data analysis. Each test serum is run in triplicate and diluted from 1:8 to 1:1024; a single 96-well plate contains four test sera (Fig. 1). The three replicates are always tested together in contiguous positions as indicated, and located on the same relative plate number for each

Javier Martín (ed.), *Poliovirus: Methods and Protocols*, Methods in Molecular Biology, vol. 1387,
DOI 10.1007/978-1-4939-3292-4_8, © Springer Science+Business Media New York 2016

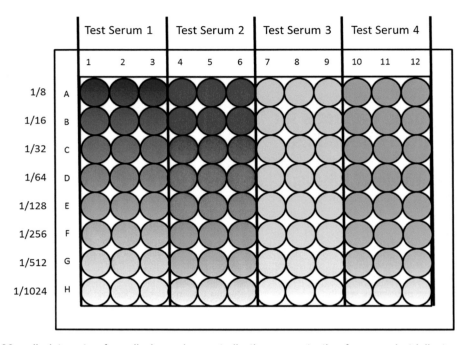

Fig. 1 96-well plate setup for poliovirus microneutralization assay, testing four sera in triplicate against a single virus starting at a 1:8 dilution of sera and twofold dilutions

of the three polio serotypes. This test may be performed manually, automated, or in a combination of the two approaches (Fig. 2). For large studies (>1000 specimens), automation is recommended (*see* **Note 4.4**).

From experience, it was found efficient for work scheduling to design a single run as consisting of up to 92 test sera. Included in each run is a control serum designated In-House Reference Serum (IHRS), which is pooled from serum samples with high neutralizing antibody titers to each Sabin poliovirus. The IHRS is tested in multiple replicates, on multiple plates, in multiple positions in each run (at least four in each run), to provide a measure of assay variability within and between a run. If more than seven sera are being tested, the samples must be randomized using a balanced block randomization scheme. Control plates are generated for each run and consist of three back titration plates with no antibody added (one for each Sabin virus) and a cell control plate (no virus or antibody added, to assess cell viability). At the end of each run, control plates are checked for accurate dilution of each Sabin poliovirus (back titration plates) or cell monolayer confluency (cell control plate).

1.2 Dried Blood Spot Poliovirus Serology

The collection, storage, transport and processing of serum presents a challenge in resource-poor settings and when surveying hard-to-reach or vulnerable populations. Dried blood spots (DBS) are commonly collected for a variety of clinical and public health

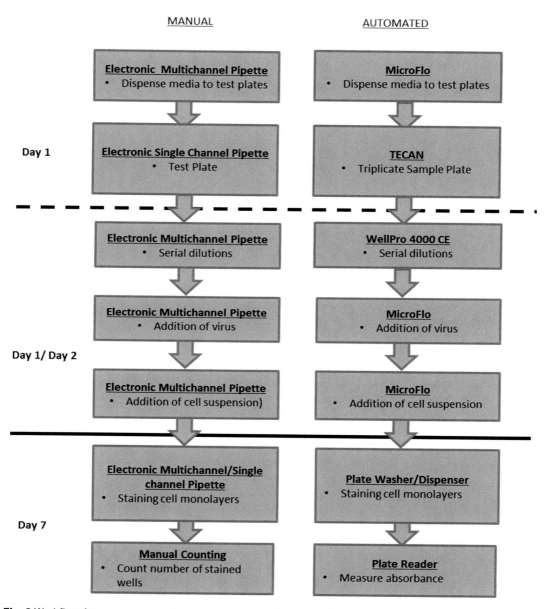

Fig. 2 Workflow for manual and automated poliovirus microneutralization assay

studies, so they are often available for other studies as well. As a result, DBS have been identified as a viable alternative to serum for the detection of a wide range of infections and immune responses. A 6 mm punch from a DBS card contains approximately 6 μl of sera. To achieve a 1:8 serum dilution this assay uses two 6 mm punches to test for polio neutralizing antibodies against Sabin 1, 2, and 3, starting at the standard 1:8 dilution.

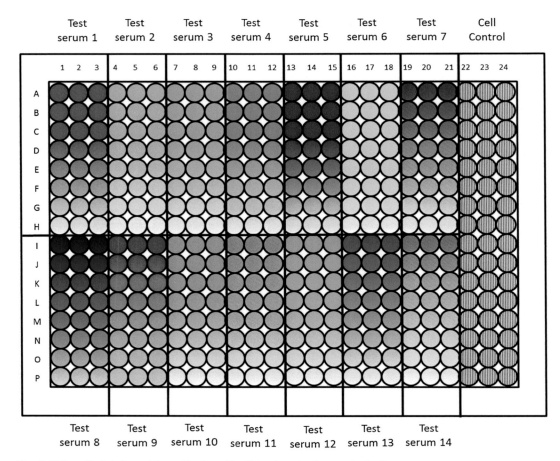

Fig. 3 384-well plate layout for poliovirus dried blood spot microneutralization assay

This procedure is a modification of the SOP for poliovirus serology in 96-well tissue culture plates. The method has been adapted to 384-well plates to allow use of smaller volumes of test sera that are eluted from DBS. Each test serum is run in triplicate and diluted from 1:8 to 1:1024; a single 384-well plate contains 14 test specimens plus cell controls (Fig. 3). The test takes approximately 7 days to complete, from the elution of dried blood spot punches to adding luminescent reagent and reading plates (Fig. 4).

For a single run, up to 252 sera can be tested against each of the three poliovirus serotypes, requiring fifty-four 384-well plates. New control plates are generated for each run and consist of one back titration containing each of three Sabin viruses. At the end of each run, control plates are checked for accurate dilution of each Sabin poliovirus (back titration plate).

If more than 14 sera are being tested, the samples must be randomized using a balanced block randomization scheme with integrated controls (*see* **Note 4.2**). Included in each run is a con-

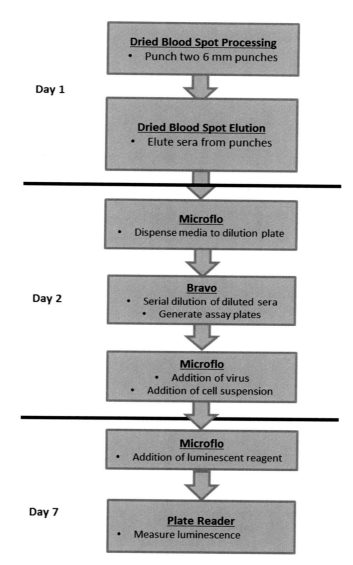

Fig. 4 Workflow for poliovirus dried blood spot microneutralization assay

trol serum designated In-House Reference Serum (IHRS), which is pooled from serum samples with high neutralizing antibody titers to each Sabin poliovirus (*see* **Note 4.1**). Due to the size of the wells and accuracy required for liquid handling on the scale that is necessary for 384-well plates, it is <u>not</u> recommended that this procedure be performed manually.

1.3   Poliovirus IgA/ IgM Capture ELISA

The use of the inactivated polio vaccine (IPV) and oral polio vaccine (OPV) has been effective at preventing acute paralytic poliomyelitis and in containing the spread of wild poliovirus almost to the point of eradication. The induction of neutralizing antibody to each serotype protects against disease and can be detected using a modified microneutralization assay. However, the detection of anti-poliovirus antibody in an ELISA format may represent a more rapid approach to the characterization of a vaccine-induced antibody response and early detection of exposure to live virus, which will be more important in the final stages of eradication and in the immediate post-eradication era.

Immunoglobulin M (IgM) exists as a membrane-bound monomer or a secreted pentamer and is the first antibody produced in response to antigen. Thus, the detection of poliovirus- and enterovirus-specific serum IgM has been proposed as an early detector of exposure to live virus or vaccine. In infants, the presence of antigen-specific IgM in cord blood indicates exposure to the antigen because maternal IgM is unable to cross the placental barrier.

Immunoglobulin A (IgA) is present in both serum and mucosal secretions, such as saliva, stool, and breast milk, as monomers or dimers, and functions to neutralize pathogens and toxins at mucosal surfaces. The mucosal immune response plays a significant role in the response to poliovirus infection and is an important mediator of protection induced by the oral polio vaccine. Poliovirus-specific IgA has been shown to be present in saliva and serum of OPV recipients and individuals exposed to wild poliovirus.

Poliovirus-specific IgM and IgA in biological fluids can be detected with a μ-chain or α-chain capture ELISA or a sandwich ELISA, by using serotype-specific monoclonal antibodies to detect antigen bound by captured IgA or IgM. However, the published capture ELISA protocols use reagents generated in-house and are not easily accessible to other public health and surveillance laboratories worldwide.

2 Materials and Equipment

Most of these items may be substituted with equivalent items from other manufacturers/suppliers but alternative materials must be validated against an appropriate standard. The materials below have been validated for consistent performance in this assay.

2.1 Polio Microneutralization Assay

2.1.1 Consumables and Small Equipment

- T-150 tissue culture flasks (Corning, #430823).
- 96-well tissue culture, clear, sterile plates (Corning, #3997).
- Low evaporation lids (Corning, #3931).
- Plastic wrap (e.g., Saran Wrap).
- Deep-well 96-well microplate, 2.2 ml capacity. (VWR, #40002-014).
- Mat lids for 2.0 ml microplates (VWR, #400002-018).
- Sterile pipettes; 10, 25 ml (Falcon, #357551, #357325).

- Single-channel pipettes:
  - Manual: LTS 20, 200, 1000 (Rainin).
  - Electronic: 10-300, 50-1000 (Biohit).
- 12-channel pipettes.
  - Electronic: 10-300, 50-1200 (Biohit).
- Pipette tips.
  - Rainin: RT-L200F, RT-L1000F, RT-L10F.
  - Biohit: 350, 1000, 1200 µl.

2.1.2 Cells and Media

- HEp-2(C)cells (ATCC # CCL23).
- MEM + 10 % fetal bovine serum (FBS) (*see* **Note 4.6**).
- MEM + 2 % FBS (*see* **Note 4.6**).

2.1.3 Antigens and Control Sera

- In-House Reference Sera (developed in-house) (*see* **Note 4.1**).
- Sabin virus stocks grown in HEp-2(C)cells (*see* **Note 4.3**).

2.1.4 Staining

- Crystal violet stain (0.05 % crystal violet, 0.5 % Tween 20, 24 % ethanol, in H_2O).

To prepare crystal violet stock solution:	
2 g	Crystal violet (Sigma, C-3386)
1000 ml	95 % ethanol

Mix together overnight, with stirring, until dissolved; may be stored up to 1 year at room temperature

To prepare working dilution crystal violet solution:	
250 ml	Crystal violet stock solution
5 ml	Tween 20 (Fisher Scientific, #BP337-100)
745 ml	Deionized H_2O

2.1.5 Other Equipment

- CO_2 water-jacketed incubator (Thermo Fisher, Model 3110 or equivalent).

2.2 Dried Blood Spot Poliovirus Serology

2.2.1 Consumables and Small Equipment

- 150 cm^2 tissue culture flasks (Corning, #430823).
- 384-well, white, opaque cell culture plates w/lids (PerkinElmer, #3596).
- 384-well, v-bottom plates (Matrix, Technologies; Fisher Scientific, Cat no. 50-823-850).
- Deep-well 96-well microplate, 0.5 ml capacity (VWR, #40002-022).
- 96-well, filter-bottom plates (PALL, Cat no. 8079).

- MicroClime Environmental lid (Labcyte, Cat no. LL-0301-IP).
- Plastic wrap (e.g., Saran Wrap).
- Sterile pipettes; 10, 25 ml (Falcon, #357551, #357325).
- Single-channel pipettes:
  - Manual: LTS 20, 200, 1000 (Rainin).
  - Electronic: 10-300, 50-1000 (Biohit).
- 12-channel pipettes.
  - Electronic: 10-300, 50-1200 (Biohit).
- Pipette tips.
  - Rainin: RT-L200F, RT-L1000F, RT-L10F.
  - Biohit: 350, 1000, 1200 μl.
  - Viaflow: 125 μl, 384 tips; filtered, sterile (#4425).

2.2.2 Cells and Media

- HEp-2(c)cells (ATCC # CCL23).
- MEM + 10 % FBS (*see* **Note 4.6**).
- MEM + 2 % FBS (*see* **Note 4.6**).

2.2.3 Antigens and Control Sera

- In House Reference Sera (generated in-house).
- Sabin virus stocks grown in HEp-2(c) cells.

2.2.4 Other Equipment

- CO_2 water-jacketed incubator (Thermo Fisher, Model 3110 or equivalent).
- ATPlite (PerkinElmer).
- Eppendorf refrigerated centrifuge (Model 5810R or equivalent with rotor capable of accommodating 96-well plate).
- MicroFlo (BioTek), or equivalent.
- BioStack2 (microplate Stacker) (BioTek), or equivalent.
- Automatic cell counter (Bio-Rad, TC-20), or equivalent.
- Victor X4 Multimode Reader (PerkinElmer), or equivalent.
- Bravo Liquid Handling System (Agilent) with 384-channel disposable tip manifold, or equivalent.
- Wallac DBS Puncher (PerkinElmer), or equivalent or manual puncher.
- 8-channel spanning-head pipette, Viaflow Voyager (Integra), or equivalent.

2.3 Poliovirus IgA/IgM Capture ELISA

2.3.1 Antibodies

1. **Capture antibodies:**
   - Affinity-purified antibody, anti-human IgA (α) (Kirkegaard & Perry Laboratories, Inc.; catalog no. 01-10-01).
   - Affinity-purified antibody, anti-human IgM (μ) (Kirkegaard & Perry Laboratories, Inc.; catalog no. 01-10-03).

2. **Monoclonal antibodies:**

Each monoclonal antibody was screened for specificity to their respective antigen by direct ELISA.

- Anti-poliovirus type-1 IPV (Antibody Shop ; HYB 295-17-02).
- Anti-poliovirus type-2 IPV (Antibody Shop ; HYB 294-06).
- Anti-poliovirus type-3, IPV (Antibody Shop ; HYB 300-06).

Antibody Shop monoclonal antibodies are supplied in 200 μl volumes. To store the monoclonal antibodies properly, add 200 μl of glycerol and store the monoclonal antibodies at −20 °C.

- Anti-poliovirus type-1, Sabin (Millipore ; MAB8560).
- Anti-poliovirus type-2, Sabin, (Millipore ; MAB8562).
- Anti-poliovirus type-3, Sabin, (Millipore ; MAB8564).

3. **Detector antibody:**

- Goat anti-mouse IgG (H+L), human-serum-adsorbed, horseradish-peroxidase-labeled, (Kirkegaard & Perry Laboratories, Inc., catalog no. 074-1806).

2.3.2 Equipment and Supplies

- Immulon 2HB, 96-well polystyrene, flat-bottom, high-binding. Thermo no. 3455 (Thermo Fisher cat. no. 14-245-61).
- Class II biosafety cabinet (BSC).
- Incubator (Precision) (37 °C).
- Multichannel Pipette (Gilson 50–200 μl).
- Pipettes: (Gilson LTS 2, 10, 20, 200, 1000 μl).
- Deep well dilution racks (VWR).
- 15 and 50 ml centrifuge tubes (Falcon).
- 2, 5, 10, and 25 ml serological pipettes (Falcon).
- Reagent reservoirs (Costar).
- Balance (Ohaus, Adventurer).
- Stir plate (Thermolyne).
- ELISA washer (BioTek), inside BSC.
- ELISA plate reader (Victor X4, or equivalent).

2.3.3 Buffers

- **Carbonate buffer:** 0.05 M, pH 9.6, plus 0.02 % sodium azide.
- **Wash buffer:** PBS: phosphate-buffered saline (0.01 M) pH 7.2 with Tween 20 (0.05 %).
- **Blocking and dilution buffer:** P-G-T, PBS (0.01 M) pH 7.2, 0.5 % gelatin, 0.15 % Tween 20: PBS (0.01 M, pH 7.2), Difco Gelatin, Fisher cat. no. DF0143-17-9, polyoxyethylene-sorbitan monolaurate (Tween 20), Sigma cat. no. P-1379.

2.3.4 Antigens

- IPV1: RIVM inactivated poliovirus type-1, Pu96-1285-907.
- IPV2: RIVM inactivated poliovirus type-2, Pu97-273-907.
- IPV3: RIVM inactivated poliovirus type-3, Pu05-3454-907.
- Sabin 1: NIBSC 01/528, RD2, 2/10/2009. ($5.1 \log CCID_{50}$).
- Sabin 2: NIBSC 01/530, RD2, 2/10/2009. ($5.1 \log CCID_{50}$).
- Sabin 3: NIBSC 01/532, RD2, 2/10/2009. ($5.3 \log CCID_{50}$).

3 Protocol

3.1 Polio Microneutralization Assay

3.1.1 In the Week Preceding the Test Runs

1. Assign sera randomly to each run using a balanced block randomization scheme (*see* **Note 4.2**).

2. Use the list generated by the randomization scheme to label plates and organize sera to be tested. Each test serum is run in triplicate, so four sera may be run on each plate (Fig. 1). Each plate is duplicated two more times, yielding three plates, one for each poliovirus serotype. There will also be a back-titration plate for each serotype, and one cell control plate per run. In a typical 96 sera run, there are 76 plates numbered as follows: PV1 virus back titration: 1; PV2 virus back titration: 2; PV3 virus back titration: 3; Cell control: 4; Sera against PV1: 5–28; Sera against PV2: 29–52; and Sera against PV3: 53–76.

3. *See* Table 1 for an example of a randomized sample list. In this example, serum sample number 0000000009 will be in run number 420, position 4 on plates 5 (PV1), 28 (PV2), and 53 (PV3). The in-house reference serum (IHRS) is in position 2 on plates 7 (PV1), 30 (PV2), and 55 (PV3).

4. For each run, the IHRS is tested an average of 4–6 times, depending on the number of samples being randomized. The IHRS sera are randomized with the test sera and are not in the same plate or position for every run.

5. Prepare MEM+ 2 % FBS and MEM + 10 % FBS according to **Note 4.6**.

3.1.2 Prior to Each Test Run

1. 24–48 h before each run, seed T-150 flasks with 30 ml of HEp-2(C) cells at 5×10^5 cells/ml.

    (a) Approximately 3–4 flasks are needed for one run of 96 sera.

2. Incubate flasks at 37 °C, 5 % CO_2, in a humidified atmosphere for 24–48 h to ensure that cell monolayers are confluent the day assay runs are started.

3.1.3 Optional: 1 Day before Test Run

The following steps can be done the day of the run or the day before the addition of the virus. Steps 6 and 7 can be performed using automation (see Subheading "Data Collection" in Note 4.4)

Table 1
Example of randomized sample list for poliovirus microneutralization assay

RUN[a]	PV1 Plate[b]	PV2 Plate[b]	PV3 Plate[b]	Position[c]	Sample ID[d]
420	5	28	53	1	0000000001
420	5	28	53	2	0000000002
420	5	28	53	3	0000000006
420	5	28	53	4	0000000009
420	6	29	54	1	0000000011
420	6	29	54	2	0000000030
420	6	29	54	3	0000000025
420	6	29	54	4	0000000041
420	7	30	55	1	0000000013
420	7	30	55	2	IHRS
420	7	30	55	3	0000000038
420	7	30	55	4	0000000051

[a]Each run of 1–96 sera
[b]PV1, polio type 1; PV2, polio type 2; PV3, polio type 3
[c]See Fig. 1
[d]Unique specimen ID and IHRS (in-house reference serum)

1. Manually aliquot 100 μl of each serum sample into deep-well polypropylene microplate, sealed with a mat lid to prevent contaminations, and heat-inactivated at 56 °C in a water bath for 30 min.

2. Store at 4 °C until ready to transfer samples to assay plates (no more than 24 h).

3. Prepare IHRS for testing (see **Note 4.1**).

4. Use a multichannel pipette to add 300 μl MEM + 2 % FBS to 100 μl heat-inactivated test serum aliquots (for a final dilution of 1:4)

5. For a full run (i.e., 96 sera), label two stacks of 12 microplates for each serotype (Fig. 5).

   (a) Use a lidded microplate for the top of each stack, with the low evaporation lid; the 11 remaining microplates should be lidless.

   (b) Cell control and back titration plates can be set up in lidded plates.

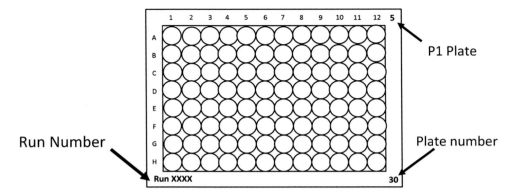

Fig. 5 Suggested labeling scheme for assay plates. Each plate should indicate the run number in the *bottom left* corner, the plate number in the *bottom right* corner, and the P1 plate number in the *upper left* corner as a reference point

6. Use an electronic multichannel pipette to dispense 25 μl MEM + 2 % FBS to the test plates.

   (a) Add 25 μl MEM + 2 % FBS to each well of the back titration plates (PV1, PV2, and PV3).

   (b) Add 50 μl MEM + 2 % FBS to each well of the cell control plate.

7. Use an electronic single-channel pipette to transfer 25 μl of each test serum (from **step 4**) to each test plate in triplicate (Fig. 6a).

8. Cover the top plate of each stack of plates and wrap in plastic wrap.

9. Plates can be stored overnight at 4 °C.

3.1.4 Day of Run

Steps 1, 4, and 9 can be performed using automation (see Subheading "Day of Run" in Note 4.4). These steps should be performed under a biological safety cabinet.

1. Using a multichannel pipette, make serial twofold dilutions from row A to row H. (serum dilution will range from 1:8 to 1:1024) (Fig. 7).

   (a) Discard 25 μl from row H to make final volume for all wells 25 μl.

2. Dilute each virus in MEM + 2 % FBS to contain 100 $CCID_{50}/25$ μl.

   (a) *See* Table 2 for example dilution scheme.

   (b) Prepare sufficient virus challenge suspension for the number of sera to be tested; each plate requires approximately 2.5 ml of diluted challenge virus.

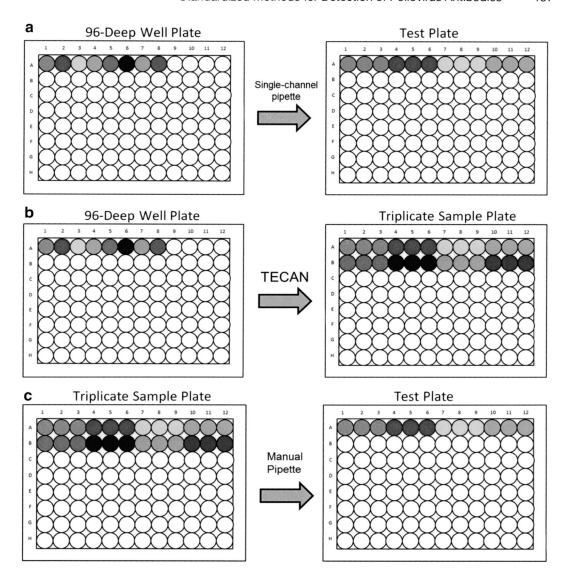

Fig. 6 For each run, 100 μl of sample is transferred from sample plate to dilution plates in triplicate (enough for starting three 25 μl sample dilutions). For simplicity, only the first eight samples are shown in the deep 96-well plate. (a) Using a single-channel pipette, transfer each sample from the 96-deep well plate to a test plate in triplicate. (b) The TECAN configuration will allow for triplicates in consecutive order. This step can transfer 96 samples from one 96-deep well plate to a 96-well triplicate sample plate in 22 min. (c) Using a 12-channel manual pipette, each row of from the triplicate sample plate is transferred to a test plate in triplicate for each serotype

3. Prepare the back titrations of each poliovirus serotype in MEM + 2 % FBS. Titrate each virus from 100 $CCID_{50}$ a further 3 tenfold steps (Table 2).

4. Use an electronic, multichannel pipette to add 25 µl of 100 $CCID_{50}$ of relevant poliovirus antigen to all wells in the diluted test plates.

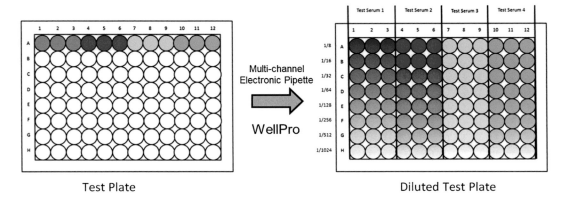

Fig. 7 Each test serum is serially diluted (twofold) from 1:8 to 1:1024 in triplicate. This is repeated for each test plate designated for poliovirus serotypes 1, 2, or 3 and can be done using a multichannel electronic pipette or automation

Table 2
Example virus dilution for 100 TCID$_{50}$ and back titration plate for Sabin type 1, 2, and 3

Sabin 1 (NIBSC 01/528)						100 TCID$_{50}$	10 TCID$_{50}$	1 TCID$_{50}$	0.1 TCID$_{50}$
$100\ TCID_{50} = 10^{-5.28}$									
Virus	100 µl	100 µl	100 µl	100 µl	800 µl	7 ml	100 µl	100 µl	100 µl
Medium	38 µl	900 µl	900 µl	900 µl	7.2 ml	63 ml	900 µl	900 µl	900 µl
Sabin 2 (NIBSC 01/530)									
$100\ TCID_{50} = 10^{-5.20}$									
Virus	100 µl	100 µl	100 µl	100 µl	800 µl	7 ml	100 µl	100 µl	100 µl
Medium	82 µl	900 µl	900 µl	900 µl	7.2 ml	63 ml	900 µl	900 µl	900 µl
Sabin 3 (NIBSC 01/532)									
$100\ TCID_{50} = 10^{-4.68}$									
Virus	100 µl	100 µl	100 µl		800 µl	7 ml	100 µl	100 µl	100 µl
Medium	82 µl	900 µl	900 µl		7.2 ml	63 ml	900 µl	900 µl	900 µl
						Working stock	Back titration plate		

5. Prepare back titration plate (Fig. 8) for each Sabin strain using dilutions prepared in **step 2** (Table 2).

   (a) Add 25 µl of 100 TCID$_{50}$ of virus to rows A and B (i.e., 24 wells/dilution)

   (b) Add 25 µl of the next 3 tenfold dilutions to rows C and D, E and F, and G and H, respectively (*see* Table 2 for dilutions).

   (c) ***Change pipette tips between each dilution.***

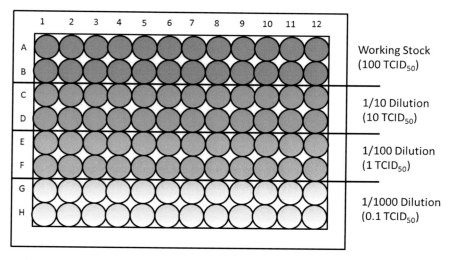

Fig. 8 Layout for back titration plate for the poliovirus microneutralization assay. A back titration plate is made for each virus tested in the assay

6. Wrap all plates in plastic wrap and incubate for 3 h at 35 °C and 5 % CO_2.

7. During serum-virus incubation, wash HEp-2(C) monolayer cell cultures (from 150 cm² flasks), trypsinize, and count cells using automatic cell counter (Bio-Rad) or a hemocytometer.

8. Prepare a HEp-2(C) cell suspension in MEM + 10 % FBS to contain 3×10^5 cells/ml. Prepare a sufficient volume of cells: each plate requires approximately 2.5 ml of cell suspension, and every run requires 3–4 confluent 150 cm² flasks. Store cells in glass bottle at 4 °C until ready to use.

9. Use a repeating, multichannel pipette to add 25 µl of prepared cell suspension to each well of every plate.

10. Wrap all plates in plastic wrap, in stacks of 12–13 plates. To prevent spills and cross-contamination, _**avoid abrupt handling of plates.**_

11. Carefully transfer plates to incubator for 5 days incubation at 35 °C and 5 % CO_2.

3.1.5 Plate Washing and Staining

*The steps in Subheading 3.1.5 can be performed using automation (see Subheading "**Plate Washing and Staining**" in Note 4.4). These steps should be performed in a biological safety cabinet.*

1. After 5 days incubation, aspirate/discard media with multichannel pipette or a vacuum into freshly made 0.5 % sodium hypochlorite solution.

2. Using a repeating, multichannel pipette, add 50 µl crystal violet stain (0.05 %) to all plates.

3. Incubate for a minimum of 40 min at room temperature.

4. Aspirate/discard stain with multichannel pipette, fill each well with tap water (approximately 250–300 µl), and discard.

5. Repeat washing **step 3** more times.

6. Allow plates to dry at room temperature for at least 2 h under a biological safety cabinet.

7. Once dried, plates can be stored outside of a biological safety cabinet at room temperature until results are calculated and reported.

3.1.6 Data Collection

*The steps in Subheading 3.1.6 can be performed using automation (see Subheading "**Data Collection**" in Note 4.4).*

1. For each triplicate test serum, count the total number of wells positive for neutralization (i.e., purple wells).

2. To calculate a neutralization titer:

$$\text{Titer} = \left(\# \text{positive wells} / \# \text{replicates} \right) + 2.5$$

3. To calculate reciprocal titer:

$$\text{Reciprocal titer} = 1 : 2^{\text{titer}}$$

4. For the neutralization titers, the upper limit of detection is 10.5 and the lower limit is 2.5, which is considered negative.

3.1.7 Cross-Checking Stained Plates

To insure accuracy, each plate is cross checked to verify correct order of plates and compare plate staining pattern to electronic data file to verify titer (Fig. 9).

1. Due to biological and technical (i.e., pipetting errors) variations in the assay, there is a likelihood that some wells will have virus not neutralized by antibody despite neutralization at lower serum dilutions.

2. Within one dilution of the endpoint this can be stochastic because of small amounts of antibody. Further from the endpoint it can be either obvious or not related to technical problems and/or errors. Another possibility is that virus is omitted from a well or row of wells and then incorrectly appear as neutralization.

3. To account for this, stained plates should be checked for these situations and the data adjusted. In the examples below, refer to Fig. 9.

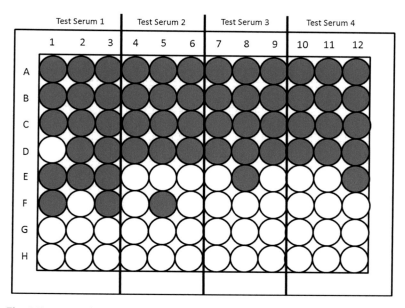

Fig. 9 Example of poliovirus microneutralization assay plate stained with crystal violet for cross-checking. Due to variations in the assay (biological and technical), there is a likelihood that the stained plates will have some wells negative for neutralization despite neutralization at lower serum dilutions

Example 1

(a) For Test Serum 1, the reader software will automatically read this as a titer of 7.83 (1:227).

(b) By cross-checking the plate, we will count D1 as positive for neutralization because the 1:128 and 1:256 dilutions (E1 and F1) are positive for neutralization (i.e., purple).

(c) Using Formula 1, the titer is adjusted to 8.17 (1:288).

Example 2

(a) For Test Serum 2, the reader software will automatically read this as a titer of 6.83 (1:114).

(b) By cross-checking the plate, we will count F5 as negative for neutralization because the 1:64 dilutions (D4, D5, and D6) are negative for viral growth.

3.1.8 Quality Control Analysis for Polio Microneutralization Assay

In-House Reference Serum

1. As the data is collected for each run, the neutralization titers for the in-house reference serum should be collected and summarized for quality control analysis.

2. The median IHRS neutralization titer for each run should be within a ± 1.0 $\log_2$ range from the established titer for a stock of IHRS.

3. Within a run, the standard deviation for the IHRS neutralization titer should not exceed ± 0.5 $\log_2$.

4. If there is any deviation from the ranges indicated above, the microneutralization run should be repeated. These IHRS titers should be monitored over time.

Back Titration Plates

1. As the data is collected for each run, the data for the back titration plates for each poliovirus tested should be collected and summarized for quality control analysis.

2. Calculate the titer for each poliovirus using the following formula:

$$Log\text{CCID}_{50} = S - 0.5, \text{ where}$$

S = sum of proportion of positive wells.

3. The expected titer calculated from each back titration plate is 2.00 ($\log_{10}$), corresponding to 100 CCID_{50}.

4. For each run, the titer for each poliovirus tested should be between 1.5 and 2.5 corresponding to 32 and 320 CCID_{50}, respectively.

5. If there is any deviation from this range, the CCID_{50} for the virus stocks should be repeated, the dilution for the microneutralization assay recalculated, and the microneutralization run should be repeated. These virus titers should also be monitored over time.

3.2 Dried Blood Spot Poliovirus Serology

1. Assign sera randomly to each run using a balanced block randomization scheme (*see* **Note 4.2**).

3.2.1 In the Week Preceding the Test Runs

2. Each test serum is run in triplicate, so 14 sera may be tested on each plate (Fig. 3). Each plate is duplicated two more times for the other two poliovirus serotypes. For each run, a single back-titration plate is required to monitor the quality for each run.

3. In Table 3, serum sample number 0000000009 will be in run number 420, position 4 on plates 2 (PV1), 20 (PV2), and 38 (PV3). The in house reference serum (IHRS) is in position 10 on plates 2 (PV1), 20 (PV2), and 38 (PV3).

4. For each run, the IHRS is tested 5–10 times, depending on the number of samples being randomized.

5. Prepare MEM+ 2 % FBS and MEM + 10 % FBS according to **Note 4.6**.

3.2.2 Dried Blood Spot Punch Collection and Elution

1. Pretreat 96-well filter-bottom plate with 200 μl MEM + 2 % FBS to wet filter.

2. Centrifuge at 1500 × g for 10 min. Discard flow-through.

3. Using PerkinElmer DBS Processor or manual punch, remove three 6 mm punch from dried blood spots into 96-well filter bottom plate (Fig. 10).

4. Add 68 μl MEM +2 % FBS to all wells of filter plate.

Table 3
Example of randomized sample list for poliovirus microneutralization assay

RUN[a]	PV1 Plate[b]	PV2 Plate[b]	PV3 Plate[b]	Position[c]	Study ID[d]
420	2	20	38	1	0000000001
420	2	20	38	2	0000000002
420	2	20	38	3	0000000006
420	2	20	38	4	0000000009
420	2	20	38	5	0000000011
420	2	20	38	6	0000000030
420	2	20	38	7	0000000025
420	2	20	38	8	0000000041
420	2	20	38	9	0000000013
420	2	20	38	10	IHRS
420	2	20	38	11	0000000038
420	2	20	38	12	0000000051
420	2	20	38	13	0000000098
420	2	20	38	14	0000000120
420	3	21	39	1	0000000087
420	3	21	39	2	0000000067
420	3	21	39	3	0000000199

[a]Each run of 1–270 sera
[b]PV1,polio type 1; PV2, polio type 2; PV3, polio type 3
[c]See Fig. 1
[d]Unique specimen ID

5. For wells designated for IHRS, add 12.5 µl of diluted IHRS following guidelines for preparation of the IHRS (see **Note 4.1**).

6. Replace plate cover and wrap in plastic wrap.

7. Elute DBS punches by incubation at 4 °C overnight.

8. Remove DBS filter plate from refrigerator.

9. Set elution plate on 96-deep well 0.5 ml plate. Centrifuge at $1500 \times g$ for 10 min.

10. Replace lid on V-bottom plate and discard filter plate.

    (a) DBS elution plate contains 50 µl of 1:4 dilution of each serum from DBS punch.

11. Store at 4 °C until ready to begin serology run (not to exceed 24 h).

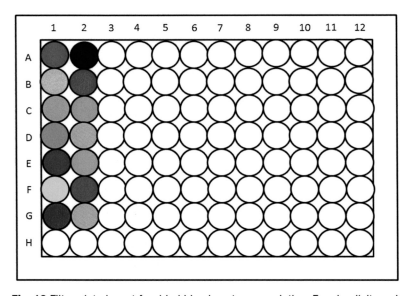

Fig. 10 Filter plate layout for dried blood spot serum elution. For simplicity, only the first 14 specimens are shown

3.2.3 Day of Serology Run

1. Use the Microflo to add 30 μl MEM + 2 % FBS to all rows of a 384-well V-bottom plate.

2. Using the Viaflow Voyager, transfer 30 μl of the first seven diluted sera to row A of dilution plate in triplicate (Fig. 11).

   (a) Repeat for the second set of seven sera to row I.

   (b) This will generate the 1:8 dilution.

3. Use the Bravo to perform twofold serial dilutions in rows A-H and rows I-P to generate a dilution plate (Fig. 12). For a single run, there will be 18 dilution plates.

4. Use the Bravo to transfer 3 μl of each well from a dilution plate to three assay plates.

   (a) Each assay plate corresponds to the serotype being tested.

5. Dilute each poliovirus serotype in MEM + 2 % FBS to contain 100 $CCID_{50}/3$ μl.

   (a) Prepare sufficient virus challenge suspension for the number of sera to be tested. A single plate requires approximately 1.5 ml of diluted challenge virus.

   (b) It is recommended to prepare an additional 1.5 ml of virus to account for priming the Microflo prior to diespensing.

6. Prepare the back titrations of each poliovirus serotype in MEM + % FBS. Dilute each virus from 100 $CCID_{50}$ a further 3 tenfold steps. These dilutions will be used in **step 9**.

7. Use the MicroFlo to add 3 μl of 100 $TCID_{50}$ of relevant poliovirus to all wells in the test plates. Use a different sterile dispensing cartridge for each virus.

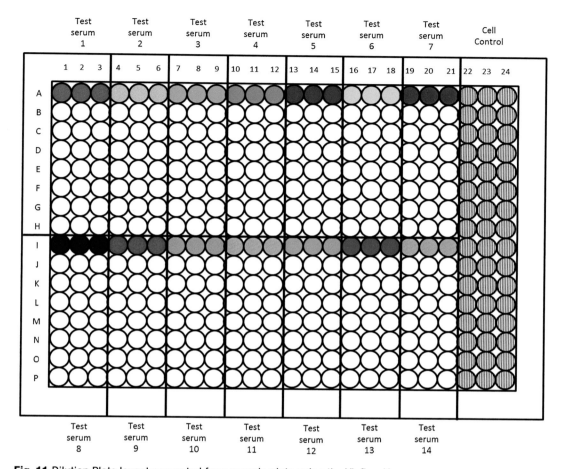

Fig. 11 Dilution Plate layout generated from sample plate using the Viaflow Voyager

(a) Quick spin all plates on a table top plate centrifuge.

8. Prepare back titration plate (Fig. 13) for each Sabin strain using dilutions prepared in **step 5**.

9. Use the Microflo to add 3 μl of MEM + 2 % FBS to all wells of a 384-well flat bottom white opaque plate.

   (a) Add 3 μl of 100 $CCID_{50}$ of Sabin 1, 2 , and 3 to rows A, F, and K (i.e., 24 wells/dilution).

   (b) Add 3 μl of the next 3 tenfold dilutions to rows B, C, and D for Sabin 1; G, H, I for Sabin 2; L, M, and N for Sabin 3.

   (c) Rows E, J, O, and P are reserved for cell only controls.

   (d) Add an additional 3 μl of MEM + 2 % FBS to cell control wells.

10. Replace the top plate with a MicroClime environmental lid that has been moistened with sterile water.

11. Wrap all plates in plastic wrap (ten plates per stack) and incubate for 3 h at 35 °C and 5 % CO_2.

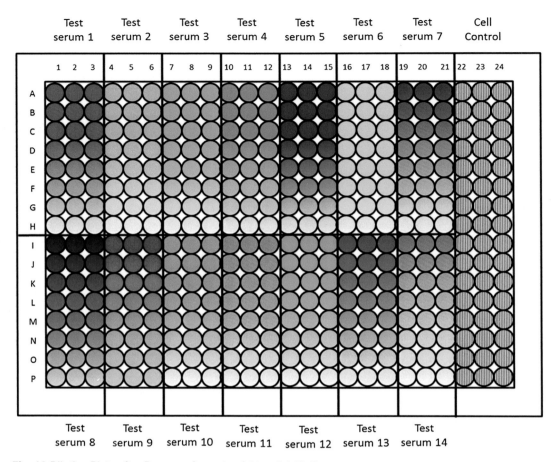

Fig. 12 Dilution Plate after Bravo performs twofold serial dilution

12. During antigen-serum incubation, prepare HEp-2(c) cell suspension in MEM + 10 % FBS to contain 7.5×10^4 cells/ml.

   (a) Prepare a sufficient volume of cells; each plate requires approximately 9 ml of cell suspension, and every 20 plates require one to two 150 cm² flasks of confluent cells.

   (b) Store cells in glass bottle at 4 °C until ready to use.

13. Use the MicroFlo to add 20 μl of prepared cell suspension to each well of every plate (Fig. 5).

14. Wrap all plates in plastic wrap and tap gently to mix. Incubate for 5 days at 35 °C and 5 % CO_2.

3.2.4 Detection of Viral CPE

1. After 5 days incubation, remove ATPlite Kit from refrigerator and allow Mammalian Cell lysis buffer, substrate, and substrate buffer to reach room temperature.

2. Use the Microflo to add 13 μl of Mammalian Cell Lysis buffer to all plates for each serotype.

3. Incubate for at least 10 min at room temperature.

4. For the Sabin 1 test plates (18 plates for one run), use the Microflo to add 13 μl of Substrate solution.

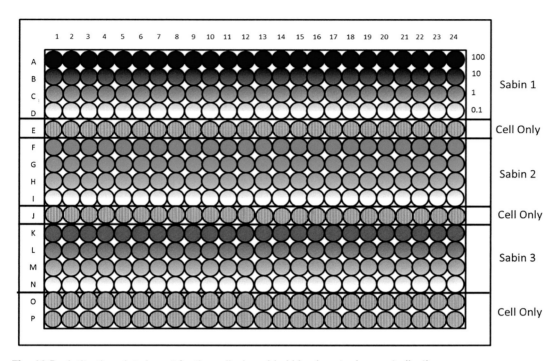

Fig. 13 Back titration plate layout for the poliovirus dried blood spot microneutralization assay

5. Incubate for 10 min at room temperature.

6. Read plates with Victor X4 Multimode plate reader using the luminescent function.

7. As the Sabin 1 plates are read, repeat **steps 5** and **6** for Sabin 2 and 3.

3.2.5 Data Collection

1. For each plate, a cell control is included in columns 22–24 for cutoff calculations.

2. The cutoff for positive/negative wells for neutralization is determined by calculating **80 %** of the average luminescence signal for the cell control wells on each plate.

(a) Wells on test plate below cutoff are considered negative for neutralization (assigned a value of 0).

(b) Wells on test plate above the cutoff are considered positive for neutralization (assigned a value of 1).

3. Titers can be calculated in Excel using the following formulas:

Formula 1

$$\text{Titer} = \left(\#\text{positive wells} / \#\text{replicates}\right) + 2.5$$

(a) To calculate reciprocal titer:

Formula 2

$$\text{Reciprocal titer} = 1:2^{\text{titer}}$$

3.3 Polio IgA/IgM Capture ELISA

3.3.1 Preparation of Capture and Detector Antibodies

Always check stocks of capture and detector antibodies. The following should be performed the day before microplates are coated with the capture antibodies if no stock is available at −20 °C.

1. Add 500 μl of sterile water to the lyophilized pellet for the capture and/or detector antibodies and rotate until dissolved.

2. Add 500 μl of glycerol and thoroughly mix by pipetting.

3. Allow to reconstitute at −20 °C for 18–24 h.

3.3.2 Coating Capture Antibodies

1. Coat the capture antibody onto the solid phase of a microplate.

   For IgA:

   (a) Using a multichannel pipette, coat labeled 96-well microplates with 50 μl/well of a 1:500 dilution of goat anti-human IgA in Carbonate Buffer (0.1 μg/well).

   For IgM:

   (b) Using a multichannel pipette, coat labeled 96-well microplates with 50 μl/well of a 1:500 dilution of goat anti-human IgM in Carbonate Buffer (0.1 μg/well).

2. Lay the microplates flat in a moist chamber (i.e., not stacked) and incubate for 60 min at 37 °C.

3. Use the BioTek EL406 to wash the microplates (*see* **Note 4.5**).

3.3.3 Blocking Buffer

1. Using a multichannel pipette, add 200 μl of P-G-T/well.

2. Lay the microplates flat in a moist chamber (i.e., not stacked) and incubate for 60 min at 37 °C.

3. Use the BioTek EL406 to wash the microplates (*see* **Note 4.5**).

3.3.4 Addition of Serum Samples

1. Using a multichannel or single-channel pipette, add 50 μl of a 1:200 dilution of the positive-control serum and 1:200 dilution of the test serum in P-G-T in duplicate down each pair of columns. For the negative control, add 50 μl of P-G-T only (Fig. 14).

2. Incubate the microplates for 60 min, 37 °C (*plates flat in moist chamber, not stacked*).

3. Aspirate serum, fill wells with wash buffer, and allow plates to soak for 6 min.

4. Use the BioTek EL406 to wash the microplates (*see* **Note 4.5**).

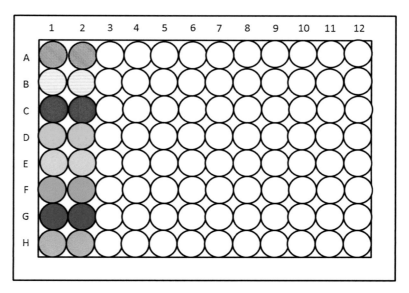

Fig. 14 Plate layout for the IgA/IgM ELISA. Wells A1-A2 and B1-B2 are designated for the positive and negative control, respectively. For simplicity, only the first 6 test specimens are indicated in C1-C2, D1-D2, E1-E2, F1-F2, G1-G2, and H1-H2. A single 96-well plate can accommodate up to 46 test specimens in duplicate

3.3.5 Addition of Type-Specific Antigen

The following steps should be performed in a biological safety cabinet if the live Sabin viruses are being used for the ELISA antigen.

1. Using a multichannel pipette and working in BSC, add 50 µl of a 1:50 dilution of each antigen in P-G-T to the appropriate plates and wells. (Antigen in columns 1–12, rows A–H).

2. Incubate the microplates overnight at room temperature (*plates flat in moist chamber, not stacked*).

3. Aspirate antigen, fill wells with wash buffer, and allow plates to soak for 6 min.

4. Use the BioTek EL406 to wash the microplates (*see* **Note 4.5**).

3.3.6 Addition of Type-Specific Monoclonal Antibodies

1. Using a multichannel pipette, add 50 µl of a 1:40,000 dilution of each monoclonal antibody in P-G-T to the appropriate plates and wells.

2. Incubate the microplates for 60 min, 37 °C (*plates flat in moist chamber, not stacked*).

3. Aspirate monoclonal antibodies, fill wells with wash buffer, and allow plates to soak for 6 min.

4. Use the BioTek EL406 to wash the microplates (*see* **Note 4.5**).

3.3.7 Addition of Horseradish Peroxidase Labeled Conjugate

1. Using a multichannel pipette, add 50 µl of a 1:4000 dilution of goat anti-mouse IgG (H+L) horseradish peroxidase labeled conjugate in P-G-T. (HRP conjugate in columns 1–12, rows A–H).

2. Incubate the microtiter plates for 60 min, 37 °C (*plates flat in moist chamber, not stacked*).

3. Aspirate conjugate, fill wells with wash buffer, and allow plates to soak for 6 min.

4. Use the BioTek EL406 to wash the microplates (*see* **Note 4.5**).

3.3.8 Addition of Substrate and Stop Solutions

1. Warm SureBlue Reserve™ Substrate to room temperature. Using a multichannel pipette, add 50 µl of substrate into each well.

2. Incubate the microplates for 15 min undisturbed at room temperature. **Do not stack plates**.

3. Warm TMB BlueStop™ Solution to room temperature. Using a multichannel pipette, add 50 µl of stop solution to each well and incubate for 10 min at room temperature.

4. Read absorbance of each plate within 10 min after addition of the BlueStop™ Solution using a standard ELISA reader with a 620 nm filter.

3.3.9 Qualitative Analysis

For qualitative analysis, negative control absorbance values from all plates corresponding to an immunoglobulin type (IgM or IgA) and antigen serotype (OPV1, OPV2, OPV3, IPV1, IPV2, IPV3) should be averaged. The procedure below uses SAS to determine the cutoff, however multiple statistical software packages are appropriate.

1. To determine the cutoff value for qualitative analysis, Perform the procedure outlined above without adding test sera (i.e., an entire plate of "blank" wells) for each polio antigen for IgM and IgA.

    (a) Ideally, "true negatives" would be used, but these are difficult to identify given the very high levels of polio immunity worldwide.

2. Use unrounded absorbance values for each isotype and antigen combination.

3. Fit the data to the Johnson family of distributions using JMP (SAS statistical package, SAS Institute).

4. Plot the cumulative density function of the Johnson distributions to show fit of data and determine 95 % confidence values (i.e., the proportion of the population with a value less than $x = 95$ %).

5. Repeat analysis 1000 times to get an average of fits for the 95 %.

6. Apply the 95 % confidence value cutoff to each isotype and antigen serotype combination. Any absorbance value above the 95 % confidence value is considered to be "positive."

4   Notes

4.1   In-House Reference Serum for Neutralization Assays

1. The in-house reference serum should be established by measuring polio neutralizing antibody titers in a population of immunized subjects.

2. Multiple sera with high neutralization titers ($\geq 7.5 \log_2$) should be pooled to generate a high-volume control.

3. Make 100 μl aliquots of the pooled sera to be stored at $-20\,^{\circ}C$ for future use. Once thawed, IHRS aliquots must be stored at $4\,^{\circ}C$ for no more than 3–4 days.

4. To prepare the IHRS for use in the serology assay, adjust the initial dilution such that the endpoint titer is reached on the 3–5 dilution on an assay plate.

5. Make the initial dilution of the IHRS and transfer to the wells that are designated for the IHRS in the randomized list (*see* Table 1).

6. When generating a new stock of IHRS, test the new stock with the old stock in parallel in the microneutralization assay (at least three times) to calibrate the new stock.

4.2   Randomization for Neutralization Assays

1. For each study, the randomization process requires a specimen number, study site, study arm, patient number, and unique ID for each specimen.

2. Each run needs to have 4–6 in-house reference sera.

3. Based on the total number of runs and the number of divisions per plate (4 divisions for 96-well plate or 14 divisions for 384-well plate), the program will also determine how many plates will be required.

4. Determine the number of unique subjects and unique study groups exist.

5. A study group is defined as a unique combination of study site and study arm for multi-site or multi-arm studies.

6. Each study group must be evenly represented in each run, so that the subjects within each study group are assigned randomly and proportionally amongst all of the runs.

7. After each subject identifier is assigned a run number, all sera associated with those subject identifiers are assigned to the same run, since comparison of sera from a single individual is usually the most important comparison (e.g., seroconversion or trends in titer).

8. Finally, plate numbers and plate positions are randomly assigned to the sera within each run.

9. Each randomized list for a study should be inspected to verify that all samples for testing are included and that the appropriate number of IHRS has been included.

10. The randomization is best performed using Visual Basic for Excel or MATLAB (available from the authors).

4.3 Poliovirus Stock Preparation and Titration for the Microneutralization Assay

4.3.1 Materials and Equipment

- T-150 tissue culture flasks (Corning, #430823).
- Single-channel pipettes:
  - Manual: LTS 20, 200, 1000 (Rainin).
  - Electronic: 10-300, 50-1000 (Biohit).
- Pipette tips.
  - Rainin: RT-L200F, RT-L1000F, RT-L10F.
  - Biohit: 350, 1000, 1200 µl.
- HEp-2(C)cells (ATCC # CCL23).
- Cell culture media.
  - Eagle's Minimum Essential Media (EMEM)(Gibco, #11095-072).
  - Penicillin/streptomycin (Gibco, #15140).
  - Fetal Bovine Serum—Optima (Atlanta Biologicals, #S12450).
  - 0.05 % Trypsin-EDTA (Gibco, #25300).
- Poliovirus stock for expansion.
- Cryovials (Wheaton, #), or equivalent.
- 96-well tissue culture, clear, sterile plates (Corning, #3997).
- Inverted microscope.
- CO_2 water-jacketed incubator (Thermo Fisher, Model 3110 or equivalent).

4.3.2 Poliovirus Stock Preparation for the Microneutralization Assay

1. Each serotype of stock poliovirus should be prepared separately.

2. For a single poliovirus, prepare confluent monolayers of Hep-2(C) cells in MEM + 10 % FBS in 150 cm² culture flasks at 37 °C. Prepare an additional flask to serve as an uninfected cell culture.

3. For all flasks, decant the media and wash with 10 ml of serum-free MEM.

4. Add 3 ml of MEM + 2 % FBS to each flask.

5. For the infected flask, add 3.0×10^5 $TCID_{50}$ in 100 µl of serum-free MEM.

6. Incubate the infected and uninfected flasks at 35 °C for 60 min.

7. Add 12 ml of MEM + 2 % FBS to each flask and incubate at 35 °C, 5 % CO_2 until cytopathic effect is observed (approximately 24–48 h post-infection).

8. To harvest virus, freeze the infected flask at –70 °C and thaw at room temperature. Repeat this freeze-thaw two more cycles.

9. Collect the contents of the infected flask and transfer to 15 ml conical centrifuge tube.

10. Clarify the harvested virus to remove cellular debris by centrifugation at $3000 \times g$ for 5 min at 4 °C.

11. Aliquot the supernatant at 100 µl per cryovial and store at –70 °C for titration.

4.3.3 Poliovirus Stock Titration for the Microneutralization Assay

1. For each stock of poliovirus, thaw an aliquot and make tenfold serial dilutions ranging from 1:10 to $1:10^9$ in MEM + 2 % FBS.

2. Prepare a HEp-2(c) cell suspension in MEM + 10 % FBS to contain 3×10^5 cells/ml.

3. On a sterile 96-well cell culture plate, add 100 µl of each dilution to columns 1–10 for rows A/B ($1:10^6$), C/D ($1:10^7$), E/F ($1:10^8$), and G/H ($1:10^9$) (Fig. 15).

4. For the cell control, add 100 µl of MEM + 2 % FBS to columns 11 and 12 (Fig. 15).

5. Add 100 µl of HEp-2(C) cell suspension to all wells.

6. Incubate for 5 days at 35 °C, 5 % CO_2.

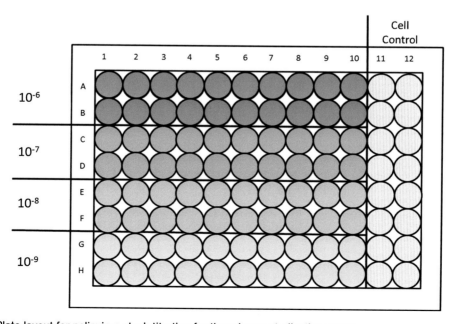

Fig. 15 Plate layout for poliovirus stock titration for the microneutralization assay

7. Count the number of wells positive for cytopathic effect for each virus dilution using an inverted microscope.

8. Calculate the virus titer using the Kärber formula:

$$LogCCID_{50} = L - d(S - 0.5), \text{ where}$$

L = log of lowest dilution used in the test
d = difference between log dilution steps
S = sum of proportion of positive wells.

9. Use the titer, to calculate the dilution factor needed to obtain 100 $CCID_{50}$ in 25 μl.

4.4 Materials and Equipment for Automation of the Polio Microneutralization Assay

Most of these items may be substituted with equivalent items from other manufacturers/suppliers but alternative materials must be validated against an appropriate standard. The materials below have been validated for consistent performance in this assay.

4.4.1 Additional Consumables for Automation

- Tecan Freedom EVO conductive tips, Black, non-filtered (Phenix Research; #TRX-HTR200BK).

- 200 μl sterile tips for use with ProGroup WellPro(Phenix Research; #TRX-WP200BRS).

4.4.2 Equipment and Automation

- CO_2 water-jacketed incubator (Thermo Fisher Model 3110 or equivalent).

- WellPro 4000 CE (ProGroup).

- EL406 plate washer and dispenser (BioTek).

- MicroFlo dispenser (BioTek).

- Tecan Evo100 or equivalent (Tecan).

- TC20 Automatic Cell Counter (Bio-Rad).

- Victor X4 Multimode plate reader (PerkinElmer) or equivalent.

4.4.3 Protocol

Optional: 1 Day Before Test Run

1. Use the MicroFlo reagent dispenser to add 25 μl MEM 2 % FBS to the test plates and the back titration plates for PV1, PV2 and PV3.

   (a) For the cell control plate, a total of 50 μl MEM + 2 % FBS should be added to each well.

2. Use the Tecan Evo100 to make the triplicate sample plates (Fig. 4a, b) by transferring 100 μl of the test serum in a deep-well plate (from **step 1**) in triplicate to a triplicate sample plate.

   (a) Using electronic multichannel pipette, transfer 25 μL of test serum from triplicate sample plate (generated in **step 3**) to row A of three test plates (one for each serotype) (Fig. 4c).

<table>
<tr><td>Day of Run</td><td>

1. Use the WellPro to perform twofold serial dilutions (Fig. 5).

    (a) The WellPro will transfer 25 μl of the diluted serum in row A to row B, then from row B to row C, etc.

    (b) For row H, 25 μl will be aspirated and discarded with the tips, to maintain the correct total volume.

    (c) For each transfer, the program will mix the diluted serum a total of four times.

    (d) Using this program will produce serum dilutions ranging from 1:8 to 1:1024.

    (e) This step generates the test plates, ready for addition of virus and cells.

2. Use the MicroFlo to add 25 μl of 100 $CCID_{50}$ of relevant poliovirus to all wells in the diluted serum test plates. Use a different sterile dispensing cartridge for each virus.

3. Use the MicroFlo to add 25 μl of prepared cell suspension to each well of every plate.

</td></tr>
</table>

Plate Washing and Staining

Due to the chances of aerosolization of virus in media from the microneutralization plates, this step should be done in a biological safety cabinet.

(a) After 5 days incubation, use the washer to aspirate and add 50 μl of a 0.05 % crystal violet solution to all wells (Fig. 8).

(b) Incubate for a minimum of 40 min at room temperature.

(c) Using the washer, aspirate the crystal violet stain then fill all wells with 250 μl of tap water and aspirate.

(d) This process will be repeated three more times to completely remove any excess crystal violet stain in the test plates prior to reading.

4.4.4 Data Collection

1. Read the crystal violet-stained microplates with an ELISA reader at a 595 nm wavelength.

2. To determine the absorbance cutoff, determine 80 % of the average absorbance value for the cell control plate.

3. Absorbance data from plate reader can be processed using macros written in Visual Basic for Microsoft Excel.

4.5 Polio IgA/IgM Capture ELISA Wash Procedure

1. Fill wash reservoir with the prepared wash buffer.

2. The program will aspirate the 96-well plate and dispense 250 μl of wash buffer to each well.

4.5.1 Procedure

3. Soak the microplates for 6 min.

4. Repeat this program three more times to complete the washing procedure.

5. Tap plates on a paper towel to remove residual liquid.

4.6 Cell Culture Media Preparation

4.6.1 Materials and Equipment

- Cell culture media
  - Eagle's Minimum Essential Media (EMEM)(Gibco, #11095-072).
  - Penicillin/streptomycin (Gibco, #15140).
  - Fetal Bovine Serum—Optima (Atlanta Biologicals, #S12450).
  - 0.05 % Trypsin-EDTA (Gibco, #25300).
- Nalgene 500 ml, 0.20 μm filter (Nalgene, #450-0020).

4.6.2 Procedure

1. FBS must be inactivated at 56 °C for 30 min and filtered with a Nalgene 0.2 μm filter prior to media preparation.
2. Prepare MEM+ 2 % FBS and MEM + 10 % FBS:
   (a) Add 1 ml of streptomycin/penicillin to 1000 ml of EMEM (0.1 % final concentration).
   (b) Add 20 ml (2 %) or 100 ml (10 %) of fetal bovine serum to 1000 ml bottle of EMEM.
   (c) For serum free MEM, add no fetal bovine serum.
   (d) MEM + 10 % FBS is used for maintaining HEp-2(C) cells and generating the cell suspensions.
   (e) MEM + 2 % FBS is used for filling the assay plates, diluting sera, and diluting virus.

Chapter 9

Molecular Properties of Poliovirus Isolates: Nucleotide Sequence Analysis, Typing by PCR and Real-Time RT-PCR

Cara C. Burns, David R. Kilpatrick, Jane C. Iber, Qi Chen, and Olen M. Kew

Abstract

Virologic surveillance is essential to the success of the World Health Organization initiative to eradicate poliomyelitis. Molecular methods have been used to detect polioviruses in tissue culture isolates derived from stool samples obtained through surveillance for acute flaccid paralysis. This chapter describes the use of realtime PCR assays to identify and serotype polioviruses. In particular, a degenerate, inosine-containing, panpoliovirus (panPV) PCR primer set is used to distinguish polioviruses from NPEVs. The high degree of nucleotide sequence diversity among polioviruses presents a challenge to the systematic design of nucleic acid-based reagents. To accommodate the wide variability and rapid evolution of poliovirus genomes, degenerate codon positions on the template were matched to mixed-base or deoxyinosine residues on both the primers and the TaqMan™ probes. Additional assays distinguish between Sabin vaccine strains and non-Sabin strains. This chapter also describes the use of generic poliovirus specific primers, along with degenerate and inosine-containing primers, for routine VP1 sequencing of poliovirus isolates. These primers, along with nondegenerate serotype-specific Sabin primers, can also be used to sequence individual polioviruses in mixtures.

Key words Poliovirus, Poliomyelitis, Sequencing, Sanger, Molecular, Polymerase chain reaction, Molecular serotyping

1 Introduction

1.1 Early Poliovirus Diagnostics

Virologic surveillance is essential to the success of the World Health Organization initiative to eradicate poliomyelitis [1, 2]. Expanded surveillance for cases of acute flaccid paralysis has been complemented by the development of improved methods for poliovirus identification. Two independent approaches were employed for routine identification of polioviruses: (1) antigenic characterizations using cross-adsorbed antisera [3] or (2) molecular characterizations using genotype-specific probes ([4, 5] Fig. 1) or PCR primers [6, 7]. Serologic methods can generally differentiate vaccine-related isolates from wild polioviruses [3] but have a limited capacity to differentiate among wild poliovirus genotypes. One of the earliest molecular approaches to poliovirus diagnostics

Javier Martín (ed.), *Poliovirus: Methods and Protocols*, Methods in Molecular Biology, vol. 1387,
DOI 10.1007/978-1-4939-3292-4_9, © Springer Science+Business Media New York 2016

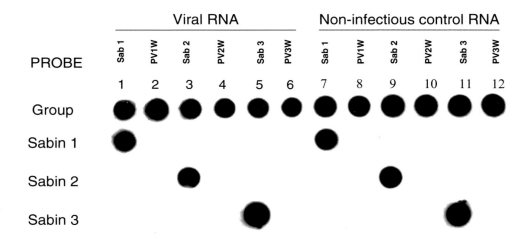

1 Sabin 1 viral RNA
2 PV1W viral RNA
3 Sabin 2 viral RNA
4 PV2W viral RNA
5 Sabin 3 viral RNA
6 PV3W viral RNA
7 Sabin 1 control RNA
8 PV1W control RNA
9 Sabin 2 control RNA
10 PV2W control RNA
11 Sabin 3 control RNA
12 PV3W control RNA

Fig. 1 Dot blot hybridization of poliovirus RNA

was the use of oligonucleotide RNA fingerprinting [8]. RNA fingerprinting was once widely used for identifying RNA virus isolates by allowing the recognition of viral genetic relationships using a two-dimensional pictorial form of the ribonuclease T1 digested RNA. As diagnostic technologies improved, additional molecular methods were developed which include the use of either probe hybridization or RT-PCR in order to directly identify Sabin and/ or some wild genotype viruses. Both of these methods give identifications that are in excellent agreement with results obtained by the more detailed analyses of genomic sequencing [5–7, 9]. In addition, restriction fragment length polymorphism assays of PCR-amplification products were also developed [10, 11].

Wild polioviruses could be identified indirectly by their non-reactivity with the Sabin-strain-specific molecular reagents. Direct identification of wild polioviruses was only possible with wild genotype-specific probes [4] and PCR primer sets [7]. At that time, the current catalog of wild-genotype-specific molecular reagents did not cover all of the many different poliovirus geno-types still in circulation worldwide [9]. Consequently, identifica-tion of wild polioviruses through the exclusive use of Sabin-strain-specific molecular reagents is dependent upon the accurate typing of virus isolates. The standard methods for poliovi-rus typing with neutralizing antibodies were comparatively time-consuming and laborious [12]. Poliovirus neutralizing antibodies are type specific [12], and no well-characterized poliovirus-specific group antigen has been described [12–14]. Molecular reagents, in the form of nucleic acid probes [4, 5] and PCR primers [7, 15, 16] targeted to highly conserved nucleotide intervals within the 59 untranslated region, have been developed for the reliable detection of nearly all members of the enterovirus group (including poliovi-ruses). Other reagents were described which permitted the detec-tion of some, but not all, polioviruses by PCR [17, 18]. An alternate method to screen out nonpolio enteroviruses (NPEVs) is to isolate virus in recombinant murine cells expressing the human gene for the poliovirus receptor [19]. This approach also permit-ted the detection of underlying polioviruses in virus mixtures, and avoided the laborious virus neutralization and recultivation steps of standard methods [12]. To overcome these diagnostic limitations, we developed a degenerate, inosine-containing, panpoliovirus (panPV) PCR primer set that could be used to distinguish poliovi-ruses from NPEVs.

2 Materials and Equipment

2.1 Realtime PCR Kit Components

The kit is supplied in one box containing six vials of primers and probes in Buffer A (Serotype 1, Serotype 2, Serotype 3, Pan-Poliovirus, Pan-Enterovirus, Sabin Multiplex), two vials of Buffer B (to which DTT and enzymes should be added prior to the first use) and one vial of DTT. The box also contains the appropriate positive controls for each primer set, a tube of water and one copy of this package insert. Additional required reagent and enzymes, not supplied with the kit, are Protector RNase inhibitor, Transcriptor Reverse Transcriptase, and Taq DNA polymerase from Roche Applied Science. The listed products were used in the development and evaluation of this kit and do not constitute a specific product endorsement. Enzyme availability from manufac-turers may vary with each laboratory. Therefore, it is the responsibility of each laboratory to find appropriate substitutes when necessary.

2.2 10× RT-PCR Buffer

RT-PCR buffer is made at a 10× concentration so that when diluted to 1× in a normal PCR reaction, the final concentration of the buffer in reaction mixtures is 67 mM Tris (pH 8.0), 17 mM

NH_4SO_4, 6 µM EDTA, 2 mM $MgCl_2$, and 1 mM dithiothreitol (DTT). Below is the protocol for making 10 ml of 10× RT-PCR buffer. To a 15 ml conical tube, add the following:

0.2 ml $MgCl_2$ (1 M).

6.7 ml Tris, pH 8.8 (1 M).

1.7 ml NH_4SO_4 (1 M).

1.2 µl 0.5 M EDTA

1.4 ml PCR grade H_2O (nuclease-free).

Mix well and aliquot 1.0 ml of the buffer into 1.5 ml microcentrifuge tubes. Freeze the aliquots at –20 °C. Add 5.0 µl of 1 M dithiothreitol (DTT) to each 1.0 ml aliquot before using for the first time. The buffer + DTT will be stable for at least 1 month after the addition of DTT (re-freeze the buffer after each use).

2.3 Oligonucleotides

Synthetic oligodeoxynucleotides were prepared, purified, and analyzed as described previously (Yang et al. [6]). Primer polarities are indicated by A (antisense or antigenome polarity) or S (sense or genome polarity). *See* Table 2 for a list of poliovirus Reverse Transcriptase (RT) PCR and sequence primers

PCR thermocycler.

Certified RNase- and DNase-free 1.5 ml and 200 µl microcentrifuge tubes.

Sterile nuclease-free H_2O.

Ethanol (96–100 %).

QIAamp Viral RNA Mini Kit (*Qiagen, Inc; Cat. # 52906*); 250 extractions (smaller kit available).

Sharps container.

4 mm disposable biopsy punch (VWR 21909-140).

Cutting Mat (VWR WB100020).

Bleach (for 10 % solution).

Optional: RNase AWAY individual wipes (VWR 89025-862).

Tissue wipes (VWR 82003-824).

70 °C heat block.

Optional: suitable vacuum manifold dedicated to RNA extraction ONLY.

Inoculating loop (1 µl) (VWR 90001-096) *or*

Inoculating needle (VWR 90001-104) *or*

Disposable tweezers (Enviro Safety Products 77233).

TE-1 buffer: 10 mM TRIS–HCl, 0.1 mM EDTA pH 7.6.

Glycogen, Nuclease Free (Ambion, AM9510).

1 M Dithiothreitol (DTT).

RNA process buffer (200 µl/sample): 191.6 µl TE-1 buffer, 0.4 µl 1 M DTT, 8.0 µl 5 mg/ml nuclease-free glycogen.

10× PCR Buffer.

dNTP mix (concentration of 10 mM each dNTP, *Roche #11814362001*).

RNase inhibitor (40 U/µl, *Roche #3335402001*).

Reverse transcriptase (20 U/µl, *Roche #3531287001*).

Taq polymerase (5 U/µl, *Roche #11596594001*).

Gel electrophoresis System (gel tray, sample combs, buffer chamber, power supply leads, and power supply).

Agarose gel powder (*GeneMate Gene Pure LE Agarose, BioExpress #E-3120-500*).

Electrophoresis buffer (Tris-borate-EDTA = TBE).

6× load buffer with bromophenol blue.

GelStar, GelRed, or any fluorescent dye that intercalates with DNA.

Molecular weight gel marker (such as 100 bp ladder, *Roche #11721933001*).

Ethanol (96–100 %).

QIAquick PCR Purification Kit (*Qiagen, Inc; Cat. # 28106*; 250 extractions (smaller kit available)).

QIAquick Gel Extraction Kit (Qiagen, Inc; Cat. # 28706; 250 extractions).

3 M sodium acetate, pH 5.0.

ABI Automated DNA sequencer (models 3130, 3130XL).

BigDye® Terminator™ v1.1 or v3.1 Cycle Sequencing kit.

MicroAmp Optical 96-well Reaction Plates (*Applied Biosystems #N801-0560*).

Digital Vortex-Genie 2 or any vortex with shock absorbing feet, microplate adapter for automated capillary sequencer plates and elastic bands.

Swinging bucket centrifuge (with microplate adapters).

Multichannel pipettor (20–200 µl).

Troughs (for holding BigDye® Terminator™ reaction mix).

MicroAmp clear adhesive film (*Applied Biosystems #4306311*) or PCR cap strips of 8 count (*ABgene #AB-0602*).

Adhesive film applicator (*Applied Biosystems #4333183*).

Wide bore pipette tips (for pipetting BigDye® Terminator™ solution).

BigDye® Terminator™ purification kit (*Applied Biosystems #4376487, processes ~2000 10 µl reactions, smaller sizes available*).

Swinging Bucket Centrifuge with microplate carriers.

Centri-Sep 8-well prehydrated strips (*Princeton Separations, Inc #CS-912, box of 12*)

OR

Centri-Sep 96-well prehydrated plates (*Princeton Separations, Inc #CS-963, 25 plates*).

Multichannel pipettor (2–20 μl).

Collection plate (deep-well reservoir for effluent from prehydrated columns/plates).

PCR tube/plate rack.

Hi-Di formamide (*Applied Biosystems, #4311320C*).

POP6 or POP7 (depending on the model of sequencer; see Applied Biosystems sequence products for order numbers and details).

10× Running Buffer (Applied Biosystems part number depends on quantity ordered).

3   Methods

3.1   Use of Deoxyinosine

The key to developing an assay specific for any group of viruses is to identify a structural feature unique to that group. The high degree of nucleotide sequence diversity among polioviruses presents a challenge to the systematic design of nucleic acid-based reagents. Genomic sequences that encode intervals of strong amino acid conservation can still be highly degenerate. To accommodate this wide variability, degenerate codon positions on the template were matched by mixed-base or deoxyinosine residues on the primers [20]. In the case of polioviruses, virion surface determinants are unsuitable targets because they are type specific [12, 21]. Internal capsid antigens and nonstructural protein antigens also appear to be unsuitable because they tend to be shared among enteroviruses [13, 14]. However, polioviruses bind to a cell receptor that is distinct from those used by other enteroviruses [19]. The canyon structure on the poliovirion surface that is postulated to bind the cell receptor is primarily formed from conserved intervals of VP1 and VP2. Genetic studies have suggested that VP1 residue Met-132, at the end of the "TYSRFDM" amino acid interval, whose codons are targeted by panPV PCR-2, interacts with the cell receptor [22]. In contrast, the structural role of the highly conserved "NNGHALN" amino acid sequence, whose codons are targeted by panPV PCR-1, is unknown. Many different synonymous codon combinations could potentially occur within the primer-binding sequences (432 for panPV PCR-1 and 512 for panPV PCR-2, assuming usage only of the observed codons). If all degenerate codon combinations are permitted within both target sequences,

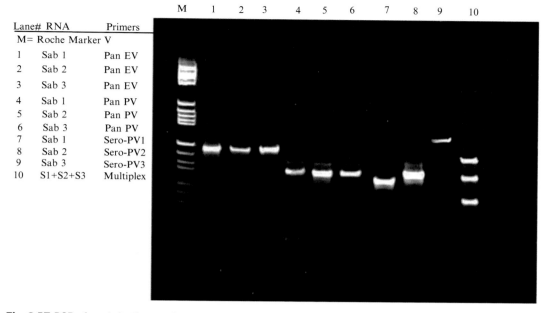

Lane#	RNA	Primers
M= Roche Marker V		
1	Sab 1	Pan EV
2	Sab 2	Pan EV
3	Sab 3	Pan EV
4	Sab 1	Pan PV
5	Sab 2	Pan PV
6	Sab 3	Pan PV
7	Sab 1	Sero-PV1
8	Sab 2	Sero-PV2
9	Sab 3	Sero-PV3
10	S1+S2+S3	Multiplex

Fig. 2 RT-PCR of noninfectious poliovirus RNAs with polio diagnostic primers

these panPV PCR primers would have to match 200,000 different sequence combinations. All polioviruses tested with this degenerate panPV primer set have been successfully identified..

3.2 Molecular Serotyping

The approach taken for the development of poliovirus serotype-specific PCR primers followed that taken earlier to develop the panPV group primers ([23]; Fig. 2). We first identified amino acid sequences that were characteristic for each serotype. We then prepared sets of candidate primers for testing against a large collection of wild poliovirus isolates representing all known contemporary genotypes. The primers showing the best specificities and sensitivities were used in routine characterizations of recent wild isolates from many different countries. When a template was inefficiently amplified in our PCR assays, its target sequences were determined, and the design of the primers was further optimized. As with the poliovirus group-specific primers, the target nucleotide sequences were highly degenerate. These new primers fill the gap between group-specific PCR primers that recognize all polioviruses and genotype-specific primers that recognize Sabin vaccine strain-related isolates or particular wild poliovirus genotypes. The typing of polioviruses by PCR offered several advantages over typing by the standard serologic methods [3, 24]. First, the primers are chemically defined reagents having uniform and predictable properties. Second, the primers can be prepared in effectively inexhaustible quantities. Third, the PCR typing assays are rapid, highly specific, and readily standardized. Finally, the exceptional sensitivity of PCR permitted the detection of very low amounts of underlying polioviruses in mixtures of poliovirus or NPEV serotypes.

3.3   Real-Time RT-PCR

In order to increase the sensitivity and timeliness of the poliovirus group and serotype assays within the WHO Global Polio Laboratory Network (GPLN), these assays were adapted to the Real-time TaqMan™ probe platform. To accommodate the wide variability and rapid evolution of poliovirus genomes, degenerate codon positions on the template were matched to mixed-base or deoxyinosine residues on both the primers and the TaqMan™ probes [25]. Designing the degenerate TaqMan™ probes (10–20 pmol probe for each assay) was especially challenging because of the need to use longer sequences to obtain good hybrid stabilities while simultaneously compensating for the high level of degeneracy of sequences between primer binding sites. Although hybrid stabilities can be estimated by physicochemical calculations [26], development of the optimal primer and probe sets was a highly empirical process because variation within the target sequences was not predictable. All of the TaqMan™ probes contain numerous mixed-base and inosine-containing residues to compensate for the high levels of variability in capsid region target sequences within and across poliovirus serotypes. We also developed a novel Real-time run method which takes advantage of the lower anneal temperatures required by the presence of inosine residues which reduce the T_m of the oligo binding. These lower anneal/extension temperatures were combined with higher anneal/extension temperatures (which adds to the increased amplicon yields), all within the same amplification run. We have found that this single tube dual-stage amplification run method (15 cycles of 95–44–60 °C, followed immediately with 40 cycles of 95–47–65 °C) increases the sensitivities of both the inosine- and non-inosine-containing assays (Fig. 3). These group and serotype intratypic differentiation (ITD) Real-time assays are routinely used within the GPLN.

3.3.1   Real-Time RT-PCR Reactions

1. Fill out PCR worksheet with name, date, primers, samples, and sample order, as well as thermocycler and program identifiers.

   (a) Name wells using thermocycler software for samples and controls (positive and reagent).

   (b) One positive control: noninfectious control RNA supplied with Polio rRT-PCR kit.

   (c) One reagent control: Buffer A + B with no template.

2. Thaw virus isolates and PCR reagents at room temperature.

3. Making Buffer B + enzyme mix: The first time a vial of Buffer B 1 mL is used, add **2.8 μl 1 M DTT, 27.6 μl 40 U/μl RNase inhibitor, 18.0 μl 20 U/μl RT (or 14.4 μl 25 U/μl RT), and 54.8 μl 5 U/μl Taq polymerase** (CAUTION, do not use error correcting Taq polymerases like Pfu and Pwo; they will not work with inosine primers) and mix. The enzyme mix should be stable for 6 months at 4 °C. Once the enzymes have been added, mark "+E" on the cap with an indelible

Amplification Plot

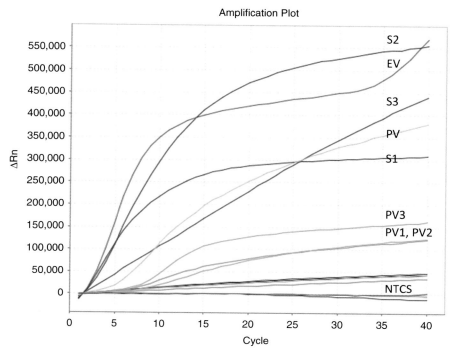

Fig. 3 Specificities of poliovirus TaqMan rRT-PCR primers and probes, demonstrated by using Sabin strain RNA templates (S1, Sabin 1; S2, Sabin 2; S3, Sabin 3; NTC, no-template control). Individual assays used the following fluorophores: FAM = panEV, panPV, seroPV1, seroPV2, and seroPV3. A multiplex Sabin assay used a combination of primers/probes for the following assays; CY5 = Sab 1, FAM = Sab 2, and ROX = Sab 3. The no-template control was used to manually set the zero baseline fluorescence emissions. The efficiency of each assay was ~90 % or greater, based on tenfold dilutions of control RNAs (10 ng to 1 pg). Cycle threshold values of 30 or more were observed to approach the sensitivity limits of the real-time detection system; therefore, cycle threshold values of < or =30, were considered positive detections of the target template

marker. For long term storage (>6 months), aliquot and freeze Buffer B + E at −20 °C.

4. Making reaction solution: For each primer set, mix 19 μl Buffer A (vortex to resuspend probe before use) and 5 μl Buffer B + E; dispense 24 μl reaction solution into each well. For testing large sample numbers, create a master mix of Buffers A + B (i.e., 8 samples × 19 μl Buffer A = 152 μl; 8 × 5 μl Buffer B = 40 μl), and dispense 24 μl of the A + B master mix per reaction well. (We recommend using the first well on your 8-well strip to make the A + B master mix since some commercial eppendorf tubes may bind the probe.)

5. Sample preparation: Take 50 μl virus cell culture and place into a tube and spin it (benchtop microcentrifuge at 2000 × g or full speed (2000 × g) of Tube-Strip PicoFuge) at room temperature for 2 min. (**Once the samples have been spun, they can be stored at −20 °C and reused if needed. You need to respin the sample after being stored at −20 °C.**)

6. Take 1.0 μl of cell culture supernatant (or 1 μl of Control RNA) for each sample and add into the appropriate reaction strip/plate well. rRT-PCR does **_NOT_** require a 95 °C heat step. One microliter of extracted RNA can be used, but it's not generally required.

7. Place strips in real-time thermocycler and cycle as shown below. If using a thermocycler with a rapid ramp speed, program the ramp from 44 to 60 °C for 45 s (note for the ABI 7500, you can use 25 % ramp speed between the anneal and extension temperatures for all assays). Thermocyclers with regular ramp speeds can use the default ramp time; Stratagene Mx3000P and similar machines do not have adjustable ramp capabilities. An additional intermediate step between the lower and higher temperature in the PCR cycle compensates for the inability to adjust the ramp time between anneal and extension:

   (a) RT reaction, 42 °C, 45 min.

   (b) Inactivate RT, 95 °C, 3 min.

   (c) PCR cycles (**all primer sets**):

   Using a Stratagene MX3000P: 95 °C for 24 s, 44 °C for 24 s, 52 °C for 30 s, 60 °C for 24 s for **7 cycles**, followed by Stage 2: 95 °C for 24 s, 47 °C for 24 s, 57 °C for 10 s, 65 °C for 24 s for **40 cycles.**

   Using an ABI 7500: Stage 1: 95 °C for 24 s, 44 °C for 30 s, then a 25 % ramp speed to 60 °C for 24 s, for **15 cycles**, followed by Stage 2: 95 °C for 24 s, 47 °C for 30 s, then a 25 % ramp speed to 65 °C for 24 s for **40 cycles. The end point fluorescent data is collected at the end of the Stage 2 anneal step**.

   (d) Select the appropriate dye filter to correspond with the assay being used.

8. The dual-stage run method for the VDPV assay is: Stage 1: 95 °C for 24 s, 44 °C for 30 s, 60 °C for 24 s for **5 cycles**, followed by Stage 2: 95 °C for 24 s, 50 °C for 30 s, then a 25 % ramp speed to 65 °C for 24 s, for **40 cycles**. The end point fluorescent data is collected at the end of the Stage 2 anneal step. VDPVs were identified by VP1 sequencing (Burns et al. [34]).

3.4   VDPV
Dual-Stage Assay

In addition to the ITD Real-time assays, specific assays have been developed to screen for the presence of vaccine-derived polioviruses [27]. Vaccine-related viruses excreted by patients with vaccine-associated paralytic poliomyelitis (VAPP) show only limited sequence divergence from the parental OPV strains. However, more highly divergent circulating vaccine-derived polioviruses (cVDPVs), indicative of prolonged replication or circulation, can arise in areas with suboptimal vaccine coverage ([28]). New cVDPV outbreaks continue to emerge as vaccine coverage declines in key high-risk countries and population immunity gaps widen, particularly to type 2. In order to develop VDPV screening assays, the primary amino acid targets for each serotype were identified by sequence analysis as those amino acids which are most likely to change in VDPVs, relative to the normal Sabin virus sequences [29–34]. Once an isolate appears non-Sabin like by screening with these new VDPV assays, they are further analyzed by sequencing the VP1 gene to confirm whether the virus is indeed a VDPV. The reaction set up for the VDPV assays is similar to that described in Subheading 3.3.1, with a modified run method. These Real-time assays also use a single tube dual-stage run method (5 cycles of 95–44–60 °C, followed immediately with 40 cycles of 95–50–65 °C [27]; Fig. 4, Table 1). These VDPV assays have several advantages including (1) sharply reducing the workload to sequence vaccine-related isolates to screen for cVDPVs; (2) being more sensitive (at least for detecting S2 VDPVs) than the ELISA assay in detecting early genetic changes associated with VDPV emergence [34] and (3) yielding accurate results on serotype mixtures without

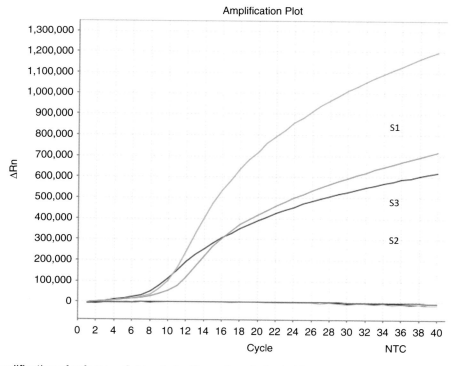

Fig. 4 Amplification of reference Sabin strain (S, serotype indicated by number) sequences in the rRT-PCR VDPV assays. Assays used the FAM fluorophore for targeting the VP1 gene. The no-template controls were used to manually set the zero baseline fluorescence emissions. The efficiency of each assay was >90 %, based on tenfold dilutions of control RNA (10 ng to 1 pg). Cycle threshold values of 30 or more were observed to approach the sensitivity limits of the real-time detection system; therefore, cycle threshold values of <30 were considered positive detections of the target template

Table 1
VDPV primers/probes

Primer/Probe	Sequence (5′ → 3′)a	Positiona
S1 VDPV-S	CATGCGTGGCCATTATA	2753–2769
S1 VDPV-A	TAAATTCCATATCAAATCTA	2902–2883
S1 VDPV-Probe	FAM-CACCAAGAATAAGGATAAGC-BHQ-1	2789–2809
S2 VDPV-S	GACATGGAGTTCACTTTTG	2890–2908
S2 VDPV-A	CTCCGGGTGGTATATAC	2989–2973
S2 VDPV-Probe	FAM-CATTGATGCAAATAAC-BHQ-1	2925–2940
S3 VDPV-S	CATTTACATGAAACCCAAAC	3276–3295
S3 VDPV-A	TGGTCAAACCTTTCTCAGA	3400–3382
S3 VDPV-Probe	FAM-TAGGAACAACTTGGAC-BHQ-1	3360–3374

aPosition relative to the positions reported by Toyoda et al. *S* sense, *A* antisense

the need for neutralizing one or more of the serotypes in the mixture, as now required when using the ELISA assay.

4   Sequencing

Once poliovirus isolates of program importance have been identified through diagnostic testing, additional vital information is obtained through sequence analysis of the viruses [28]. This information determines genetic relationships to other wild poliovirus isolates or VDPVs to demonstrate chains of transmission, sources of imported viruses causing outbreaks, and estimates of the duration of virus circulation [35]. To achieve this, VP1 sequences are determined for all wild polioviruses and potential VDPVs isolated worldwide. Systematic design of nucleic acid-based poliovirus diagnostic reagents targeting the capsid region (or any coding region) is especially challenging because nucleotide sequence diversity among poliovirus isolates is exceptionally high [20, 23] and nucleotide intervals encoding conserved amino acid sequences are typically highly degenerate. To accommodate the wide variability and rapid evolution of poliovirus genomes [35], degenerate codon positions on the template were matched to mixed-base or deoxy-inosine residues on both PCR primers. A particular challenge is to sequence the individual components of isolates containing poliovirus mixtures. The standard method to resolve the mixtures has been by blocking the growth of individual components with type-specific neutralizing antibodies followed by limit-dilution culture or plaque-purification, procedures that are both time-consuming and laborious. Generic poliovirus specific primers, along with degenerate and inosine-containing primers are used for routine VP1 sequencing of poliovirus isolates (Table 2). These primers, along with nondegenerate serotype-specific Sabin primers, can also be used to sequence individual polioviruses in mixtures. By using various combinations of these primers, the VP1 regions of >7000

Table 2
Poliovirus VP1 RT-PCR and sequencing primers

Primer	Sequence (5′ → 3′)[a]	Position[b]
Y7	GGGTTTGTGTCAGCCTGTAATGA	2419–2441
Y7R[c]	GGTTTTGTGTCAGCITGYAAYGA	2419–2441
PV1,2S[c]	TGCGIGAYACIACICAYAT	2459–2477
246S-S1	CGAGATACCACACATATAGA	2461–2480
247S-S2	CGAGATACAACACACATTAG	2461–2480
248S-S3	CGAGACACCACTCACATTTC	2461–2480
255S-S1	GGGTTAGGTCAGATGCTTGAAAGCATG	2500–2526
256S-S2	GGAATTGGTGACATGATTGAAGGG	2500–2523
257S-S3	GGTATTGAAGATTTGATTTC	2500–2519
P1-SENSE[c]	GGICARATGYTIGARAGIATG	2506–2526
P2-SENSE[c]	CTIGGIGAYATIMTIGARGG	2503–2522
P3-SENSE[c]	ACIGARGTIGCICARKGYGC	2515–2535
AFRO P3W 6S[c]	CCIAARCCRCAIAAYRGHC	2531–2549
AFRO P1W 2S[c]	CRGTICAAACYAGRCAYGTYA	2658–2678
SOAS P1W 2S	ACRGGRGCYACRAACCCNTT	2624–2643
SOAS P3W 6S	GTYRTACARCGRCGYAGYAGRA	2666–2687
AFRO P3W 7A[c]	GAYTCIATKGTIGAYTCBGT	2705–2686
AFRO P1W 1A	GMRAAYARYTTRTCYTTRGA	2795–2776
SOAS P1W 1A	ACTGARAAYARYTTRTCYYTKGA	2801–2779
SOAS P3W 5A[c]	TCYTTRTAIGTRATGCGCCAAG	2812–2791
249A-S1	CACTGTAAATAGCTTATCC	2820–2802
250A-S2	AACCGAAAACAATCTGCTG	2820–2802
251A-S3	CATGGCAAATAGTTTCTGT	2820–2802
PanPV4S[c]	ACITAYAARGAYACIGTICA	2830–2849
PanPV2S[c]	CITAITCIMGITTYGAYATG	2876–2895
PanPV1A[c]	TTIAIIGCRTGICCRTTRTT	2954–2935
SeroP3A[c]	CCCCIAIPTGRTCRTTIKPRTC	3176–3157

(continued)

Table 2
(continued)

Primer	Sequence (5′ → 3′)[a]	Position[b]
252A-S1	ATATGTGGTCAGATCCTTGGTG	3407–3387
253A-S2	ATAAGTCGTTAATCCCTTTTCT	3407–3387
254A-S3	ATATGTGGTTAATCCTTTCTCA	3407–3387
PV8A[c]	GCYTTRTTITGITGICCRAA	3431–3412
PV10A[c]	GTRTAIACIGCYTTRTTYTG	3440–3421
Q8	AAGAGGTCTCTRTTCCACAT	3527–3508

[a]IUB ambiguity codes; I inosine, P, pyrimidine base analog (T + C)
[b]Position relative to the positions reported by Toyoda et al.
[c]Inosine-containing primers, use degenerate PCR conditions

wild polioviruses, representing all known genotypes were amplified and sequenced, including all combinations of serotype mixtures (including ternary mixtures), whether wild or vaccine-related [36]. The VP1 regions of separate components of homotypic mixtures (e.g, Sabin 1-related poliovirus + type 1 wild poliovirus) can also be sequenced. These mixtures are difficult to resolve using type-specific polyclonal antisera and may also be difficult to resolve using highly specific neutralizing monoclonal antibodies in some situations because of the rapid antigenic reversion of vaccine-related poliovirus isolates [3].

The complete VP1 region can be amplified as a single DNA amplicon and then sequenced using four primers (sense and antisense). An alternative method is to amplify and sequence two overlapping fragments of the VP1 region into a single contig representing a consensus sequence; however, the preferred method is to amplify the complete VP1 region.

4.1 Poliovirus Sequencing

Poliovirus sequencing can be broken down into nine basic steps as follows:

- RNA extraction from tissue cell culture isolates of poliovirus or from an FTA card spotted with poliovirus isolate.

- Reverse transcription of the RNA and PCR amplification of the resulting cDNA using the primers Y7R and Q8 (covering the entire VP1 region of the poliovirus genome).

- Visualizing the DNA amplicon on an agarose gel.

- Purifying the DNA amplicon to remove unincorporated nucleotides and any nonspecific PCR products (optional step).

- Performing the sequence PCR protocol for automated capillary DNA sequencing using the primers Y7, PV1A, PV4S, and

Q8 (covering the VP1 region in both sense and antisense directions).

- Removal of unincorporated BigDye terminators from the sequence PCR reactions.

- Analyzing the reactions on the automated capillary sequencer.

- Viewing and editing the raw sequence data.

- Construction of contigs and consensus poliovirus VP1 sequences.

A flow chart outlining these basic steps can be seen in Fig. 5.

4.1.1 RNA Extraction from Cell Culture Isolates

1. Freeze-thaw the tissue cultures twice to release intracellular virus. Collect the supernatant after the second freeze-thaw and centrifuge at $13,000 \times g$ for 1 min to pellet cell debris. Remove the supernatant to a sterile, RNase-free microcentrifuge tube.

2. Follow the QIAamp Viral RNA Mini Spin Protocol or the QIAamp Viral RNA Mini Vacuum Protocol.

4.1.2 RNA Extraction from FTA Cards Spotted with Poliovirus Isolates

1. Prepare RNA processing buffer for the FTA cards to be extracted; 200 μl will be needed per card processed. Aliquot into 1.7 ml nuclease-free microcentrifuge tubes and place on ice.

2. Clean the cutting mat with 10 % bleach followed by high quality DI water and 70 % ethanol OR wipe the mat with an RNAse AWAY wipe. Dry the mat with a clean tissue wipe.

3. Take the FTA card with air dried poliovirus isolate and place the area to be punched on the cutting mat.

4. Using a 4 mm disposable biopsy punch, cut out 6 punches per card and place the punches into a microfuge tube with 200 μl of RNA processing buffer. All six punches can be made sequentially and then removed from the biopsy punch with any of the following: a sterile disposable inoculating loop (with the end removed), a sterile disposable inoculating needle, or sterile disposable forceps.

5. Dispose of the biopsy punch in a sharps container. Use a new biopsy punch for each FTA card processed, as well as new inoculating loop/needle or forceps.

6. If more than one FTA card is processed, always place each card on a different spot on the cutting mat to avoid any chance of cross contamination between specimens. Repeat **steps 4** and **5**.

7. Vortex the tubes containing the punches in RNA processing buffer. Incubate the tubes at 70 °C for 5 min, vortexing at the halfway point of the incubation period. Open the tubes during the first 2 min of the incubation period to release any pressure inside the tube.

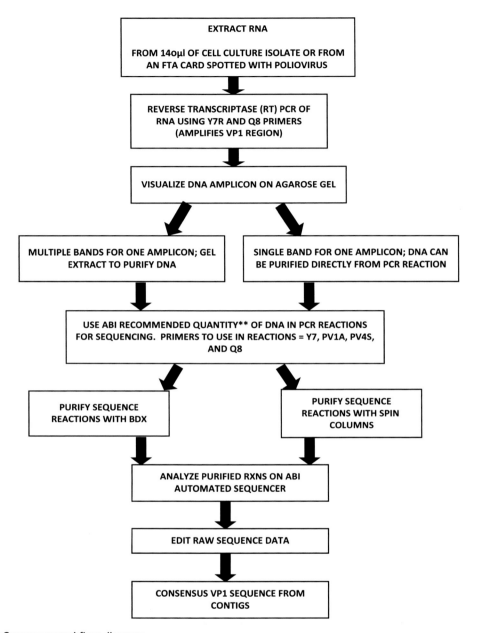

Fig. 5 Sequence workflow diagram

8. After incubation is complete, spin the tubes for 1 min at full speed in a microcentrifuge.

9. Remove the supernatant after centrifugation and place into an appropriate number of labeled microfuge tubes containing 620 μl of AVL buffer (QIAmp Viral RNA mini kit) plus the appropriate amount of carrier RNA according to the QIAmp protocol.

10. Mix the tubes containing the AVL buffer and supernatant by pulse vortexing for 15 s. Briefly spin the tubes in a microcentrifuge to remove any liquid from the cap.

11. Add 620 µl of absolute ethanol to each tube with FTA supernatant plus AVL buffer. Mix by pulse vortexing for 15 s and briefly spin the tubes in a microcentrifuge as in **step 10**.

12. At this point, follow the QIAamp Viral RNA Mini Spin Protocol or the QIAamp Viral RNA Mini Vacuum Protocol at the stage of adding the supernatant, AVL buffer, ethanol mixture to the QIAmp columns.

13. Store the extracted RNA at –70 °C (preferred) or at –20 °C (~1–2 year storage).

4.1.3 Degenerate RT-PCR for Sequence Amplicons

1. Log into a laboratory notebook a record of the RNA samples to be used in the PCR run. If RNA was extracted the same day of the PCR amplification, keep the tubes containing the RNA on ice throughout the procedure. If frozen RNA is used, thaw the samples on ice and keep them on ice for the remainder of the procedure. DO NOT THAW AT ROOM TEMPERATURE.

2. Remove reagents from the freezer and thaw on ice or place into a tube rack (dedicated for PCR reactions) at 4 °C until the reagents are thawed; then place the reagents on ice.

3. Take out the required number of PCR tubes needed for the PCR run (the number of RNA samples to be processed plus a positive and a negative RNA control sample). Label the tubes appropriately for cross reference to the lab notebook log of the RNA samples to be processed. Place the tubes on ice.

4. Prepare a master mix for the PCR reactions to give a final volume of 47 µl/PCR tube. Below is an example for making a master mix for eight reactions plus two controls.

Reagent	Amt/PCR tube	×10 Rxns
PCR water (DNase–RNase free)	37 µl	370 µl
10× PCR buffer	5 µl	50 µl
dNTPs (10 mM each)	2 µl	20 µl
RNase inhibitor (40 U/µl)	0.25 µl	2.5 µl
Reverse transcriptase (20 U/µl)	0.25 µl	2.5 µl
Y7R primer (40 pmol/µl)	1 µl[a]	10 µl
Q8 primer (10 pmol/µl)	1 µl[b]	10 µl
Taq polymerase (5 U/µl)	0.5 µl	5 µl
Total	47 µl	470 µl

[a]Concentrated stock of 40 pmol/ul contains 10 pmol/µl of each primer species because of degeneracy.
[b]Concentrated stock of 10 pmol/µl contains 5 pmol/µl of each primer species because of degeneracy.

5. Combine the master mix by gently pipetting up and down, avoiding formation of air bubbles; centrifuge the tube briefly. Aliquot 47 µl of master mix into each PCR reaction tube kept on ice. Keeping the reaction tubes on ice, aliquot 3 µl of the sample RNA into each tube and mix by gently pipetting the reaction up and down several times. When you have finished aliquoting all the RNA samples to be tested, store the poliovirus RNA at –20 °C.

6. Place the PCR reactions into the thermocycler and cycle as shown below:

   (a) RT reaction, 42 °C, 45 min.

   (b) Inactivate RT, 94 °C, 3 min.

   (c) PCR cycle (all primers): 94 °C for 30 s , 42 °C for 30 s, ramp at 0.4 °C/s to 60 °C, 60 °C for 2 min for 35 cycles.

   (d) Immediately follow the cycle program with a final extension time of 5 min at 60 °C.

7. At the completion of the PCR run, reaction products may be immediately examined by agarose gel or may be frozen at –20 °C for later evaluation.

4.1.4 Visualizing DNA Amplicons

1. Assemble the gel tray into a casting assembly and place a sample comb with the desired number of sample wells into the gel tray.

2. Prepare a 1 % agarose gel by weighing out an appropriate amount of agarose powder (e.g., 1.0 g for 100 ml gel) and mix with 1× TBE buffer in a glass flask. Heat the gel mixture in a microwave oven until melted (the agarose is completely dissolved), cool to approximately 55 °C, and add fluorescent dye directly to the gel according to the dye manufacturer guidelines.

3. Allow the gel to cool slightly, pour into the gel tray and allow it to solidify at room temperature or at 4 °C.

4. Place the solidified gel into the electrophoretic buffer chamber. Pour enough 1× TBE buffer into the chamber to cover the gel; remove the sample comb at this point and flush out the sample wells with 1× TBE buffer.

5. Take out enough PCR tubes to account for the number of reactions you want to run on the gel. Aliquot 1 µl of 6× load dye into each tube and then add 5 µl of each PCR reaction. Place the remainder of the PCR reaction either on ice or at 4 °C.

6. Load the PCR reaction plus load dye (total 6 µl) into the wells made by the sample comb. In the first and last lanes of the gel, load the molecular weight marker (with load dye) to

help determine the size of the DNA amplicon of the PCR reaction.

7. Close the gel chamber, attach the leads to a power supply and apply current (~70 V for 100 ml gel). **DNA will migrate towards the positive electrode, which is colored red.** Continue to run the gel at constant voltage/current.

8. The distance the DNA has migrated through the gel can be judged by visually monitoring the migration of the bromophenol blue dye in the load buffer (bromophenol blue migrates at about the same rate of double stranded DNA fragments of ~300 bp). Also note that interference from the blue dye may be noticed.

9. After sufficient migration has occurred, turn off the power supply and remove the gel tray (wear disposable gloves while handling the gel tray), Place the gel tray on a UV transilluminator with a plexiglass cover and check for PCR amplicons of ~1.1 kb size (*WARNING: wear proper PPE (personal protection equipment) including protective eye gear and a UV blocking face shield to minimize exposure to UV light*). Photograph the gel to document the results.

10. There should be only one band per PCR reaction (band size is ~1.1 kb). If this is true, then the remaining product stored at 4 °C may be column-purified for use in DNA sequencing reactions (*see* Subheading 4.1.5). If there are nonspecific bands in addition to the correct 1.1 kb fragment, the VP1 fragment can be extracted from the gel (*see* Subheading 4.1.6).

4.1.5 Direct Purification of PCR DNA Amplicon

1. Remove the correct number of microcentrifuge tubes corresponding to the PCR amplicons to be purified and label the tubes accordingly.

2. Using the QIAquick PCR Purification Kit, add 5 volumes of PB buffer (with indicator dye added) to each microcentrifuge tube for each volume of PCR sample (e.g., a 50 μl PCR reaction would require 250 μl of PB buffer).

3. Add the PCR reaction to the corresponding microcentrifuge tube with PB buffer.

4. Mix by vortexing. Check the color of the mixture; it should be dark yellow. If the color is orange or violet, add 10 μl of 3 M sodium acetate, pH 5.0 (color should now be yellow).

5. Follow the QIAquick PCR purification microcentrifuge protocol or the QIAquick PCR purification vacuum manifold protocol. Elute the DNA from the column using either the elution buffer provided in the kit or with nuclease-free water.

6. The purified DNA is now ready for sequencing. The DNA may be quantified by running a sample adjacent to standards containing known quantities of the same-sized DNA fragment or by spectrophotometric analysis for nucleic acids.

4.1.6   Gel Extraction
of PCR DNA Amplicon

1. Make a 1 % agarose gel with an appropriate gel stain as outlined in **step 3** of this manual. After the gel has solidified, place the gel into the electrophoresis chamber of the gel box and add 1× TBE buffer to cover the gel; remove the sample comb and flush the wells with 1× TBE buffer.

2. Mix 9 μl of 6× load buffer (*see* **step 3**) with the remaining 45 μl of PCR product and load into the sample wells of the gel. REMEMBER: In the first and last lanes of the gel, load the molecular weight marker (with load dye) to help determine the size of the DNA amplicon of the PCR reaction.

3. Close the gel chamber, attach the leads to a power supply and apply current (~70 V for 100 ml gel). Keep the voltage/current constant.

4. The distance the DNA has migrated through the gel can be judged by visually monitoring the migration of the bromophenol blue dye in the load buffer (bromophenol blue migrates at about the same rate of double stranded DNA fragments of ~300 bp).

5. After sufficient migration has occurred (i.e., the bromophenol blue dye has migrated ~ halfway through the gel), turn off the power supply and remove the gel tray. Place the gel tray on a UV transilluminator with a plexiglass cover and check for PCR amplicons of ~1.1 kb size (*WARNING: wear proper PPE (personal protection equipment) including protective eye gear and a UV blocking face shield to minimize exposure to UV light*). Photograph the gel to document the results.

6. Wearing proper laboratory PPE, including protective eyewear and a protective UV blocking face shield, use a surgical scalpel to excise the 1.1 kb bands from the gel. Place each gel piece into a separate labeled microcentrifuge tube.

7. Weigh the gel slice in the tube. Use the QIAquick Gel Extraction Kit to extract the DNA from the gel slice. Add 3 volumes of QG buffer from the kit to 1 volume of gel (100 mg of gel ~100 μl).

8. Incubate the gel slice + QG buffer at 50 °C for at least 10 min, vortexing every 2 min to aid in gel dissolution. If the gel slice is not completely dissolved after 10 min, the incubation time can be extended for a few minutes. THE GEL SLICE MUST BE COMPLETELY SOLUBILIZED.

9. Check the color of the gel solution—it should be the same color (yellow) as the as the QG buffer. If the color is orange or violet, add 10 μl of 3 M sodium acetate, pH 5.0.

10. Add the sample to a labeled QIAquick spin column and follow the QIAquick PCR purification microcentrifuge protocol or the QIAquick PCR purification vacuum manifold protocol. Please allow the PE wash buffer to remain on the column for a 5 min incubation time prior to removal of the wash.

11. Elute the DNA from the column using either the elution buffer provided in the kit or with nuclease-free water.

12. The purified DNA is now ready for sequencing. The DNA may be quantified by running a sample adjacent to standards containing known quantities of the same-sized DNA fragment or by spectrophotometric analysis for nucleic acids.

4.1.7 Degenerate Sequence PCR Reactions

1. Log into a laboratory notebook a record of the DNA samples to be used in the PCR run. If the DNA amplicon was purified the same day of the sequence PCR amplification, keep the tubes containing the DNA on ice throughout the procedure. If frozen DNA is used, thaw the samples on ice and keep them on ice for the remainder of the procedure. DO NOT THAW AT ROOM TEMPERATURE.

2. Take all reagents for the reaction (Ready Reaction Premix, primers, and BigDye Sequencing dilution buffer) out of the freezer and thaw on ice or place into a PCR reaction dedicated tube rack at 4 °C until the reagents are thawed; then place the reagents on ice.

3. Take out the required number of PCR tubes needed for the PCR run (the number of DNA samples × 4). Label the tubes appropriately for cross reference to the lab notebook log of the DNA sample/primer to be processed. Place the tubes on ice.

4. For each DNA template there will be a total of four reactions, one for each of the following primers: Y7, PanPV1A, PanPV2S, and Q8. Oligonucleotide composition and priming position on the poliovirus genome for each of these primers are listed in Table 2. Use of four primers ensures coverage of the entire VP1 region in both the sense and antisense directions (*see* Fig. 2). The amount of DNA template used in the PCR sequence reaction will depend on the method used to purify the sequence reaction post thermocycling. The PCR reaction (10.0 µl) is prepared as follows: 2.0 µl of Ready Reaction premix (v1.1 or v3.1), 1.0 µl 5× Big Dye Sequence buffer, 3.2 pmol primer (dilute from stock as needed to give a final added volume of 1.0 µl), 20–40 ng DNA template (sephadex column purification) OR 5–10 ng DNA template (BDXterminator purification), nuclease-free water added to a final volume of 10.0 µl.

5. Mix the PCR reactions by pipetting up and down. If necessary, spin the tubes briefly to remove any bubbles.

6. Place the PCR tubes into the PCR thermal cycler and run the following PCR program (all primers): 94 °C 20 s, 42 °C 15 s, ramp at 0.4 °C/s to 60 °C, 60 °C for 4 min for 25 cycles.

7. At the end of the PCR sequencing program, remove the tubes and proceed to Subheading 4.1.8 (Purifying Sequence Reactions with BigDye® XTerminator™) or Subheading 4.1.9 (Purifying Sequence Reactions with Spin Columns).

4.1.8 Purifying Sequence Reactions with BigDye® XTerminator™

1. Remove the PCR sequence reactions from the PCR thermocycler.

2. Follow the BigDye® XTerminator™ purification protocol as outlined in the premix pipetting section of the process. Use a trough to hold the final premix solution (SAM + XTerminator™) and a multichannel pipettor to aliquot 55 µl to each PCR reaction. When pipetting from the trough, keep the premix agitated by rocking the trough back and forth lengthwise to create a wave motion.

3. After the premix is aliquoted as indicated by the BDXterminator protocol, seal the plate using the adhesive film applicator. An alternative method of sealing the plate is to cap the plate wells with PCR strip caps (rows of 8).

4. Place the sealed plate on the Digital Vortex-Genie 2 and vortex for 30 min at $1800 \times g$ at room temperature.

5. After vortexing is complete, spin the plate in a swinging bucket centrifuge for 2 min at $1000 \times g$ at room temperature.

6. The samples may now be analyzed on the DNA sequencer (Subheading 4.1.10) OR you can store the sealed plate for up to 48 h at room temperature or up to 10 days at 4 °C.

4.1.9 Purifying Sequence Reactions with Spin Columns

1. Place a 96-well sequence plate onto a PCR plate rack.

2. Aliquot 15 µl of HiDi Formamide into the wells of the plate for the number of sequencing reactions that you have.

3. Remove the number of strips that you will need to purify your sequence reactions or take out a Centri-Sep plate.

4. Using a swinging bucket centrifuge, pre-spin the columns or Centri-Sep plates into the collection plate according to the manufacturer's specifications (2 min at $750 \times g$ for the column strips, $1500 \times g$ for the Centri-Sep plate; please *see* the centrifugation notes at the end of this section) to remove the interstitial fluid present in the columns/plate.

5. After centrifugation, place the columns into the formamide-filled wells of the 96-well sequence plate OR place the Centri-Sep plate above the formamide filled 96-well plate.

6. Using the multichanneled pipettor, pipette the 10 µl sequencing reaction into the center of each well of the Centri-Sep strip/plate, making sure not to touch the top of the gel in the well.

7. Spin the loaded Centri-Sep column/plate according to manufacturer's instructions. The purified fluorescently tagged DNA will spin through the column directly into formamide in the 96-well sequence plate.

8. Remove the Centri-Sep columns/ plate from the top of the 96-well sequence plate. The samples (now in the 96-well sequence plate) are ready to be analyzed on the DNA sequencer (Subheading 4.1.10).

4.1.10 Analyze Purified Sequence Reactions on the Automated DNA Sequencer

1. Take the 96-well sequence plate containing the purified sequence reactions and cover it with a plate septa. Place the plate into the plate base /retainer assembly and put it into the automated sequencer according to manufacturer specifications.

2. Prepare 1× Running buffer (dilute 1 part 10× running buffer to 9 parts nuclease-free water (PCR grade)). Add the buffer to the two buffer chambers in the sequencing instrument according to ABI instructions.

3. Prepare a sample sheet according to the user's guide for the automated DNA sequencer, following a logical naming convention. (Please note that an electronic sample sheet template may be imported.) Make sure that the proper instrument/run protocols are chosen based on the version of BigDye used (v1.1 or v3.1), the length of the capillary in the machine, and the purification method used. (*NOTE: if BigDye XTerminator was used to purify the post sequencing PCR reactions, make sure that the proper run modules are available on the sequencer. These run protocols are freely distributed by Applied Biosystems on their website.*)

4. Begin the sequence run. After the run is complete, download the sequence data from the data collection software run files and edit the raw sequence data as described in Subheading 4.2—Viewing and Editing Raw Sequence Data.

4.2 Viewing and Editing Raw Sequence Data

4.2.1 Software

You will need a software package capable of analyzing trace files, constructing contigs and editing base pairs. Examples are the Sequencher analysis package distributed by Gene Codes Corporation Ann Arbor, Michigan; Lasergene distributed by DNASTAR Inc. Madison, Wisconsin; and Bionumerics (Applied Maths, Inc., Austin, Texas). All subsequent sequence chromatograms and contig representation presented in this section are generated through Sequencher Ver. 4.8

4.2.2 Sequence Quality and Trimming

1. The trace data files are imported into the Sequencher software, and the chromatograms of each file are examined for quality. Peaks present in the chromatogram should have a good signal to noise ratio (distinct with minimal background) and base-calling should be clean and concise.

2. Trimming—Frequently, the beginning (close to the primer read) of the sequence data will have ambiguous base calls (N's) and/or will have a great deal of background noise. Generally, the first 15–25 base pairs of the sequence displayed in the chromatogram will need to be trimmed. The tail end (3′ end) of the sequence will also need to be trimmed; read lengths of the sequence will vary based on the following criteria: (1) quality of the sequence read and (2) the primer used for the sequence. As a general rule, more than three ambiguous bases and/or a deterioration of sequence quality (i.e., loss of peak resolution and base calling accuracy) signals the need to trim the end portion of the sequence read.

4.2.3 Troubleshooting Sequence Quality Issues

Noisy background and/or trailing peak width early in the sequence read could be due to a number of factors including

1. Errors generated during sequence run—Check the error log and rerun the sample.

2. Polymer bubbles—Check the sequencer to make sure no bubbles are present in the line from the polymer supply or in the capillary array.

3. High number of sequence runs on the capillary—ABI sequence capillaries are guaranteed for 100 sequence runs only. Capillaries should be changed between 100 and 150 runs. Please note that the capillaries can be run more than 150 runs, but this is not recommended because sequence quality may suffer.

4. Impure DNA amplicon—check the starting DNA material on a 1 % agarose gel to ensure that only the 1.1 kb Y7/Q8 amplicon is present. If other bands can be seen, extract the 1.1 kb band from the gel and gel purify the fragment using a kit such as the QiaQuik gel purification system from Qiagen. Re-sequence the purified DNA.

5. Mixtures of serotypes/genotypes present in the DNA amplicon—Check the original PCR result from the diagnostic PCR to ensure that no mixture is present such as a wild–sabin mix or a wild mixture of different serotypes (e.g., a PV1/PV3 wild poliovirus mixture).

See Fig. 6 for examples of unacceptable sequence quality.

4.3   Construction of Contigs and Consensus of Poliovirus VP1 Sequences

1. After sequence quality is ascertained, the sequences corresponding to the Y7, PV1A, PV4S, and Q8 primers for each DNA amplicon analyzed can be assembled into a single contig. Minimum contig assembly parameters should be at least 85 % base pair match with a 20 base pair overlap between primer sequences (*see* Fig. 7).

4.3.1 Contig Assembly

2. Good sequence overlap between the sense and the antisense primers over the VP1 region is critical to ascertain agreement

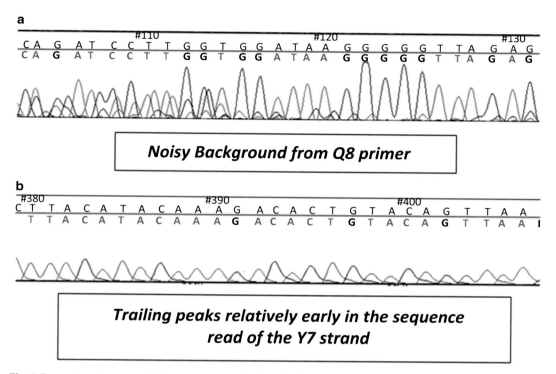

Fig. 6 Examples of unacceptable sequence quality. Panel a has high background (a low signal to noise ratio) in a portion of the chromatogram, and Panel b shows trailing peaks in the Y7 primer sequence

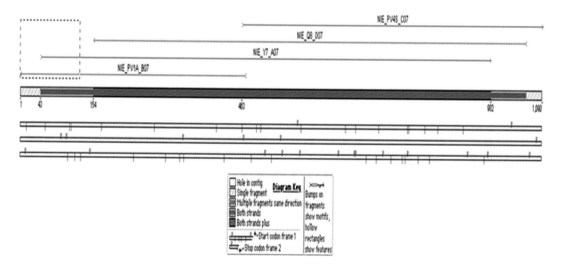

Fig. 7 Sequence contig formed using Sequencher software (GeneCodes) demonstrating good overlap with both sense primers (Y7 and PV4S shown in *green*) and antisense primers (PV1A and Q8 shown in *red*) in the VP1 gene region of poliovirus. Primers extend into the VP3 and 2A gene regions to ensure complete coverage of VP1

between the sequence strands; each primer provides sequence from a different region of VP1 and strengthens the validity of the base pair calls. THERE MUST BE COMPLETE COVERAGE OF THE VP1 REGION WITH BOTH THE SENSE AND ANTISENSE STRANDS OF THE CONTIG. There should be no region of single-stranded sequence or absence of a sense or antisense strand.

4.3.2 Base Pair Agreement and Editing

Contigs should be checked for ambiguous base pair calls and gaps between the assembled sequences. The software package used should have the ability to check and to edit any ambiguous base call(s) in the contig

The chromatograms for bases in the contig should be examined to get a clear consensus for the proper base pair calls in the VP1 region (Fig. 8). From the three aligned chromatograms in Fig. 8, the ambiguous base in the PV1A sequence is a 'T' in agreement with the other two sequence primer strands. The consensus base pair call would be 'T'. Notice that the location of the base in question differs in each of the three primer sequence strands; contig assembly effectively aids in simultaneous editing base calls in all sequence strands for a specific Y7/Q8 DNA amplicon.

After all ambiguous base pairs have been resolved, the contig should be renamed with a descriptive nomenclature to help identify the specimen sequenced. An example would be a unique identifier such as a Lab Id that is used only once and not rolled over from year to year.

A consensus sequence may be generated from the edited contig. The consensus sequence should be trimmed to the VP1 region (906 base pairs for PV1, 903 base pairs for PV2, and 900 base pairs for PV3) (Fig. 9).

4.4 Sequencing Poliovirus Mixtures

When sequencing poliovirus specimens, mixtures of different serotypes and genotypes are often encountered and can present challenges in obtaining good sequence data. The degree of difficulty in resolving individual polio serotypes/genotypes depends on the type of mixture present in the specimen. Mixture types can be classified as follows, from the easiest type of mixture to the most difficult:

1. Sabin–Sabin mixtures—The specimen contains combinations of the three Sabin strains (Sabin, Sabin2, Sabin3). These are the easiest mixtures to resolve due to the availability of Sabin specific primers that can be used in the initial RT-PCR reactions to produce Sabin-specific DNA amplicons and/or used in the sequencing PCR reaction.

2. Wild–Wild mixtures—The specimen contains a combination of two different serotypes (e.g., PV1 and PV3) but no Sabin or Sabin derived polioviruses (VDPVs). Serotype specific primers

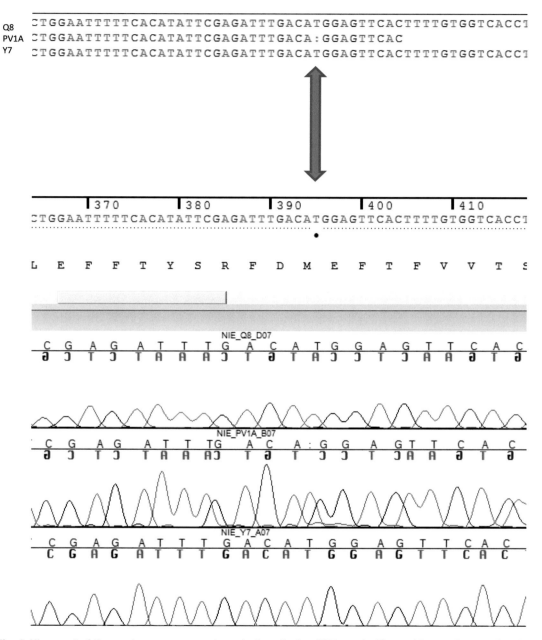

Fig. 8 Alignment of three primer sequences in a single poliovirus VP1 contig. The ambiguous base call in the PV1A sequence is a 'T' comparing the base calls in the same position in the Y7 and Q8 primer sequences

in the initial RT-PCR reaction produce specific DNA amplicons, and the primers can also be used in sequencing PCR reactions to resolve these mixtures.

3. Wild–VDPV–Sabin mixtures—The specimen contains a combination of wild and at least one Sabin or Sabin derived strain

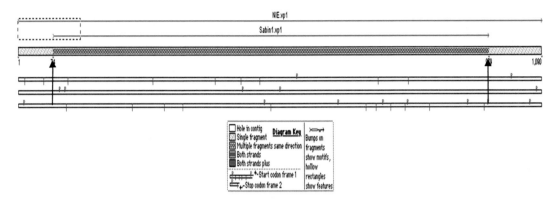

Fig. 9 Alignment of a poliovirus serotype 1 VP1 sequence (generated from the consensus sequence of a VP1 contig) with the reference Sabin 1 strain sequence for appropriate trim length (906 bp). The consensus sequence would be trimmed at the *arrow* points

of different serotypes (e.g., PV1 wild–PV2 VDPV–Sabin3). Combinations of Sabin specific, serotype-specific and genotype-specific primers are used in both RT-PCR reactions (to produce specific DNA amplicons) and in the PCR sequence reactions to resolve these mixtures.

4. Homotypic mixtures—The specimen contains a combination of wild, Sabin or Sabin-derived strains of the same serotype. This is the most difficult mixture to resolve, as serotype-specific primers will target both wild and Sabin/VDPVs in the mix. Strategies for resolution of these mixtures involve production of several DNA amplicons using both serotype- and genotype-specific primers and also sequencing the amplicons with different combinations of these primers (*see* Table 2).

The genotype- and serotype-specific primers used in resolving poliovirus mixtures are described in Table 2.

The frequency of each type of mixture encountered is dependent on the specimen type. Poliovirus mixtures from AFP specimens are predominantly Sabin–Sabin mixtures. Environmental specimens, which are composite samples from sewage originating from many households (as opposed to one single patient), can have all four mixture types, particularly in polio endemic regions. Strategies for resolving the four mixture types are described in detail in the next sections.

4.4.1 Sabin–Sabin Poliovirus Mixtures

The following primer pairs may be used in the RT-PCR reaction (same PCR conditions as used for the Y7R/Q8 amplification) to produce Sabin specific DNA amplicons for the VP1 region:

1. 246S/Q8—Produces Sabin 1 DNA amplicon. Primers amplify the end of VP3 and the beginning of 2A to give ~1.1 kb PCR product.

2. 247S/Q8—Produces Sabin 2 DNA amplicon. Primers amplify the end of VP3 and the beginning of 2A to give ~1.1 kb PCR product.

3. 248S/Q8—Produces Sabin 3 DNA amplicon. Primers amplify the end of VP3 and the beginning of 2A to give ~1.1 kb PCR product.

The Q8 primer is used as the antisense primer with the Sabin-specific sense primer to produce the full length Sabin VP1 sequences. A combination of Sabin-specific and generic primers can be used to sequence the purified DNA amplicon. Figure 10 gives an overview of the strategies and primers used to resolve Sabin–Sabin mixtures.

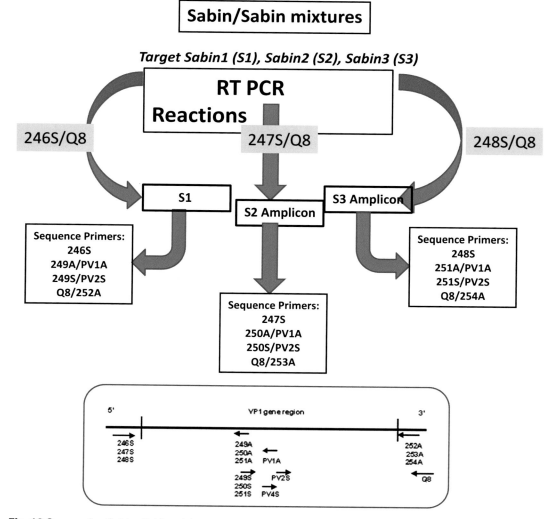

Fig. 10 Sequencing Sabin–Sabin mixtures

The best strategy to resolve wild–wild mixtures is to have serotype- and wild genotype-specific primers in the RT-PCR reaction to produce single serotype DNA amplicons. The following primers are used to resolve each serotype component in the mixture:

1. PV1 wild component—generate 2 DNA amplicons.

   (a) P1 Sense/Q8—Produces a serotype 1-specific DNA amplicon (~1 kb PCR product).

   (b) Y7R/(AFRO P1W 1A or SOAS P1W 1A)—Produces a serotype 1/wild genotype-specific DNA amplicon (~400 bp PCR product). The choice of genotype-specific primer is dependent upon the suspected wild genotype of the sample.

2. PV3 wild component—generate 2 DNA amplicons.

   (a) P3 Sense/Q8—Produces a serotype 3-specific DNA amplicon (~1 kb PCR product).

   (b) Y7R/(AFRO P3W 7A or SOAS P3W 5A)—Produces a serotype 3/wild genotype-specific DNA amplicon (~300 bp PCR product). The choice of genotype-specific primer is dependent upon the suspected wild genotype in the sample.

The P1 and P3 Sense primers amplify the beginning of the VP1 region; therefore, in order to obtain a full-length VP1 sequence on the 5′ end, a second, smaller amplicon is generated with Y7R (amplifies the end of VP3) and a genotype/serotype-specific primer. The purified DNA amplicons are sequenced according to serotype with a combination of serotype/genotype-specific primers (P1 Sense, P3 Sense, AFRO P1W1A/SOAS P1W 1A, AFRO P3W 7A/SOAS P3W 5A, 2952S, and 2922S) and generic primers. Figure 11 gives an overview of the strategies and primers used to resolve wild–wild mixtures.

Wild–Sabin mixtures that also have a VDPV component are rarely encountered in specimens from AFP cases but are more frequent in environmental specimens from polio-endemic areas. The Sabin components of the mixture can be resolved by using Sabin-specific primers, and the wild component is resolved using a combination of serotype-specific and serotype/genotype-specific primers. An example of the strategy to resolve a poliovirus mixture that was determined by rRT-PCR to be PV1 wild, PV2 discordant, and PV3 discordant is given below:

1. PV1 wild component—generate 2 DNA amplicons.

   (a) P1 Sense/Q8—Produces a serotype 1 specific DNA amplicon (~1 kb PCR product).

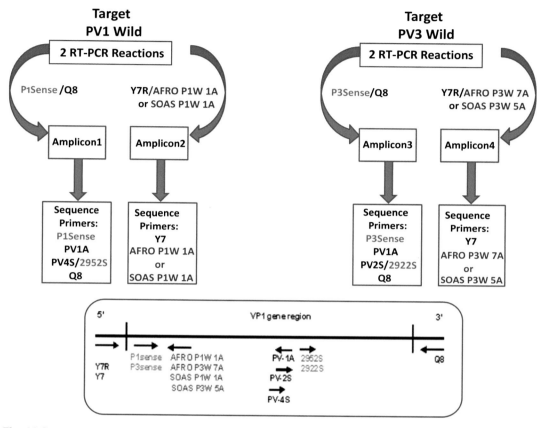

Fig. 11 Sequencing wild–wild mixtures. Primers are color coded as follows; primers in *red* are serotype specific, primers in *blue* are serotype/genotype specific, primers in *black* are generic

(b) Y7R/(AFRO P1W 1A or SOAS P1W 1A)—Produces a serotype 1/wild genotype-specific DNA amplicon (~400 bp size PCR product). The choice of genotype-specific primer is dependent upon the suspected wild genotype in the sample.

2. PV2 Sabin component—generate a PV2 Sabin-specific amplicon using 247S/Q8.

3. PV3 Sabin component—generate a PV3 Sabin-specific amplicon using 248S/Q8.

Overall, four DNA amplicons will be produced. The P1 DNA amplicons are sequenced with a combination of serotype 1, wild genotype-specific, and generic primers to get a full-length VP1

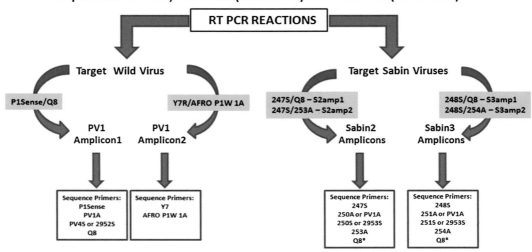

Fig. 12 Sequencing components of wild and Sabins (discordant in VDPV assay) from a poliovirus specimen. *The Q8 sequence primer listed for the Sabin amplicons will only sequence the amplicons S2amp1 and S3amp1

PV1 wild sequence. The PV2 and PV3 amplicons are sequenced with a combination of Sabin-specific and generic primers to achieve full length Sabin VP1 sequences for each serotype. It may be noted that if Sabin-specific amplicons cannot be achieved using the Q8 primer on the 3′ end, Sabin-specific primers can be used at this position (e.g., 247S/253A produce an ~940 bp amplicon). However, a full length VP1 sequence may not be obtained with complete sequence coverage in the both the sense and antisense directions. Figure 12 gives an overview of the strategies and primers used to resolve wild–VDPV–Sabin mixtures.

4.4.4 Homotypic Poliovirus Mixtures

Homotypic poliovirus mixtures refer to specimens that contain combinations of wild, VDP, and Sabin genotypes that share the same serotype (e.g., PV1 wild and PV1 Sabin). These mixtures are by far the most difficult to resolve of all the mixture types. Serotype-specific primers may not be informative because the viruses are of the same serotype. Genotype-specific primers are helpful, but some cross reactivity can occur. Depending on the complexity of the mixture (titer of each component, presence of multiple Sabins of the same serotype) only partial VP1 sequences may be obtained. Virus plaque purifications of the mixtures can be done to separate the viruses, but this process is time consuming and not done routinely.

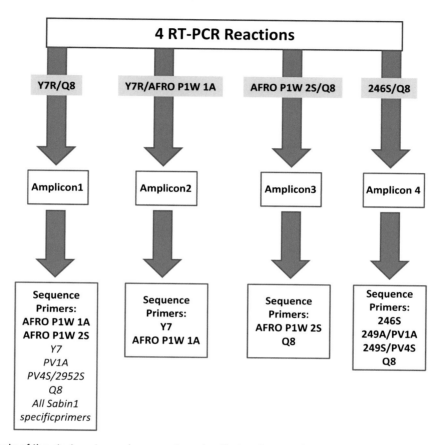

Fig. 13 Example of the strategy to resolve a serotype 1 poliovirus homotypic mixture

Figure 13 gives on overview of a strategy to resolve a PV1 wild, PV1 Sabin poliovirus mixture. Overall, at least four RT-PCR reactions with combinations of generic, serotype-, and genotype-specific primers need to be done to target the wild and Sabin components of the mixture. The resulting DNA amplicons are sequenced with generic and specific (serotype- and genotype-) primers to produce VP1 sequences for each genotype.

4.5   Notes

1. Check all kit reagents for expiration dates.

2. All reagents and disposable plastic ware used in the procedure should be certified nuclease-free.

3. Check Buffer AVL solution for precipitate. If necessary, redissolve precipitate by warming at 80 °C for no more than 5 min.

4. Make all solutions according to kit instructions and write the date of first use on each solution bottle.

5. Perform RNA extractions on a positive poliovirus control and negative cell control, preferably from the same cell line as the isolates being tested.

6. This protocol assumes that the user has a thorough knowledge of how to handle RNA and the hazard to RNA by RNases. All reagents and disposable plastic ware used in the procedure should be certified RNase-free.

7. Store the RNA at −70 °C (preferable) or at −20 °C (1–2 year storage limit). Most viral RNAs are stable for many years (≥4 years) at −70 °C.

8. Always wear protective eyewear and a protective UV blocking face shield when observing DNA on a transilluminator to prevent damage to the eyes from UV light.

9. Several commercially available kits can be used for DNA purification and extraction from gels; any of these kits may be used as long as the purified product is suitable to be used in downstream sequence reactions.

10. Thermocycled sequence reactions may be stored dark and frozen at −20 °C for no more than 24 h prior to purification and subsequent analysis on the automated sequencer.

11. Purified sequence reactions may be stored according to the purification method used:

    (a) Centri-Sep columns/plates—the purified reactions should be dried in a vacuum centrifuge (no heat or low heat) and stored dark and sealed (adhesive film) at −20 °C for up to 10 days. Do not store at temperatures below −20 °C.

    (b) BigDye® Terminator™—reactions cleaned with BigDye XTerminator in plates sealed with adhesive film can be stored as follows: room temperature (20–25 °C) for no more than 48 h, 4 °C up to 10 days, or −20 °C for 1 month.

12. The process described for BigDye XTerminator purification of sequence reactions assumes that the user will be using an automated DNA capillary sequencer (ABI 3130 or 3730 instruments) equipped with BigDye XTerminator run modules to analyze the fluorescently tagged PCR product.

13. The process described for purifying sequence reactions with spin columns mentions Centri-Sep prehydrated 8-well strips (for small numbers of samples) and Centri-Sep prehydrated 96-well plates (for processing large numbers of samples) but any gel purification method specifically formulated for sequence reaction purification could work just as well.

14. If the RCF (relative centrifugal force) or g force of a centrifuge rotor is unknown (i.e., the centrifuge only measures RPM

speeds), it can be calculated by the following formula: $RCF = r \times 11.18 \times (rpm \div 1000)^2$ where RCF is relative centrifugal force in "g," r is the rotor radius in centimeters, and rpm is the rotor speed (revolutions per minute).

15. To assist in trimming the VP1 sequence to the proper length, the contig assembly parameters in the Sequencher software can be adjusted to a 75 % sequence match (Sequencher ver 4.8) and the consensus sequence assembled with the reference strain VP1 sequences for Sabin1 (AY184219), Sabin2 (AY184220), or Sabin3 (AY184221) into a single contig. The consensus sequence may be trimmed according to the sequence length of the reference strain (Fig. 9).

References

1. Hull H et al (1994) Paralytic poliomyelitis: seasoned strategies, disappearing disease. Lancet 343(8909):1331–1337

2. Organization WH (2008) Progress towards interrupting wild poliovirus transmission worldwide, January 2007-April 2008. Wkly Epidemiol Rec 83(19):170–177

3. van der Avoort HG et al (1995) Comparative study of five methods for intratypic differentiation of polioviruses. J Clin Microbiol 33(10): 2562–2566

4. da Silva EE et al (1991) Oligonucleotide probes for the specific detection of the wild poliovirus types 1 and 3 endemic to Brazil. Intervirology 32(3):149–159

5. De L et al (1995) Identification of vaccine-related polioviruses by hybridization with specific RNA probes. J Clin Microbiol 33(3):562–571

6. Yang CF et al (1991) Detection and identification of vaccine-related polioviruses by the polymerase chain reaction. Virus Res 20(2):159–179

7. Yang CF et al (1992) Genotype-specific in vitro amplification of sequences of the wild type 3 polioviruses from Mexico and Guatemala. Virus Res 24(3):277–296

8. Kew BK et al (1984) Applications of oligonucleotide fingerprinting to the identification of viruses. Methods Virol 8:41–84

9. Kew OM et al (1995) Molecular epidemiology of polioviruses. Semin Virol 6(6):401–414

10. Balanant J et al (1991) The natural genome variability of poliovirus analyzed by a restriction fragment length polymorphism assay. Virology 184(2):645–654

11. Schweiger B et al (1994) Differentiation of vaccine and wild-type polioviruses using polymerase chain reaction and restriction enzyme analysis. Arch Virol 134(1–2):39–50

12. Melnick J (1990) Poliomyelitis. In: Warren K, Mahmoud A (eds) Tropical and geographical medicines, 2nd edn. McGraw-Hill, New York, pp 558–576

13. Mertens T, Pika U, Eggers HJ (1983) Cross antigenicity among enteroviruses as revealed by immunoblot technique. Virology 129(2): 431–442

14. Emini E et al (1985) Antigenic conservation and divergence between the viral-specific proteins of poliovirus type 1 and various picornaviruses. Virology 140(1):13–20

15. Hyypia T, Auvinen P, Maaronen M (1989) Polymerase chain reaction for human picornaviruses. J Gen Virol 70:3261–3268

16. Olive DM et al (1990) Detection and differentiation of picornaviruses in clinical samples following genomic amplification. J Gen Virol 71(9):2141–2147

17. Abraham R et al (1993) Shedding of virulent poliovirus revertants during immunization with oral poliovirus vaccine after prior immunization with inactivated polio vaccine. J Infect Dis 168(5):1105–1109

18. Egger D et al (1995) Reverse transcription multiplex PCR for differentiation between polio- and enteroviruses from clinical and environmental samples. J Clin Microbiol 33(6): 1442–1447

19. Hovi T, Stenvik M (1994) Selective isolation of poliovirus in recombinant murine cell line expressing the human poliovirus receptor gene. J Clin Microbiol 32(5):1366–1368

20. Kilpatrick DR et al (1996) Group-specific identification of polioviruses by PCR using primers containing mixed-base or deoxyinosine residue at positions of codon degeneracy. J Clin Microbiol 34(12):2990–2996

21. Minor PD et al (1986) Antigenic and molecular evolution of the vaccine strain of type 3 poliovirus during the period of excretion by a primary vaccinee. J Gen Virol 67(4):693–706

22. Colson E, Racaniello V (1994) Soluble receptor-resistant poliovirus mutants identify surface and internal capsid residues that control interaction with the cell receptor. EMBO J 13(24):5855–5862

23. Kilpatrick DR et al (1998) Serotype-specific identification of polioviruses by PCR using primers containing mixed-base or deoxyinosine residues at positions of codon degeneracy. J Clin Microbiol 36(2):352–357

24. Melnick J (1996) Enteroviruses: polioviruses, coxsackieviruses, echoviruses, and newer enteroviruses. In: Fields BN, Knipe DM, Howley PM et al (eds) Fields virology, 3rd edn. Lippincott-Raven, Philadelphia, pp 655–712

25. Kilpatrick DR et al (2009) Rapid group-, serotype-, and vaccine strain-specific identification of poliovirus isolates by real-time reverse transcription-PCR using degenerate primers and probes containing deoxyinosine residues. J Clin Microbiol 47(6):1939–1941

26. Rychlik W, Spencer WJ, Rhoads RE (1990) Optimization of the annealing temperature for DNA amplification in vitro. Nucleic Acids Res 18(21):6409–6412

27. Kilpatrick DR et al (2014) Identification of vaccine-derived polioviruses using dual-stage real-time RT-PCR. J Virol Methods 197:25–28

28. Kew OM et al (2005) Vaccine-derived polioviruses and the endgame strategy for global polio eradication. Annu Rev Microbiol 59:587–635

29. Jegouic S et al (2009) Recombination between polioviruses and co-circulating Coxsackie A viruses: role in the emergence of pathogenic vaccine-derived polioviruses. PLoS Pathog 5(5):e1000412

30. Kew O et al (2002) Outbreak of poliomyelitis in Hispaniola associated with circulating type 1 vaccine-derived poliovirus. Science 296(5566):356–359

31. Martin J et al (2000) Evolution of the Sabin strain of type 3 poliovirus in an immunodeficient patient during the entire 637-day period of virus excretion. J Virol 74(7):3001–3010

32. Shimizu H et al (2004) Circulation of type 1 vaccine-derived poliovirus in the Philippines in 2001. J Virol 78(24):13512–13521

33. Yakovenko ML et al (2006) Antigenic evolution of vaccine-derived polioviruses: changes in individual epitopes and relative stability of the overall immunological properties. J Virol 80(6):2641–2653

34. Burns C et al (2013) Multiple independent emergences of type 2 vaccine-derived polioviruses during a large outbreak in Northern Nigeria. J Virol 87(9):4907–4922

35. Jorba J et al (2008) Calibration of multiple poliovirus molecular clocks covering an extended evolutionary range. J Virol 82(9):4429–4440

36. Kilpatrick DR et al (2011) Poliovirus serotype-specific VP1 sequencing primers. J Virol Methods 174(1–2):128–130

Chapter 10

Isolation and Characterization of Vaccine-Derived Polioviruses, Relevance for the Global Polio Eradication Initiative

Wenbo Xu and Yong Zhang

Abstract

Stool specimens were collected from children with acute flaccid paralysis (AFP) and their contacts, and viral isolation was performed according to standard procedures. If the specimens tested positive for poliovirus, then intratypic differentiation (ITD) methods were performed on the viral isolates to determine whether the poliovirus isolates were wild or of vaccine origin, these include a poliovirus diagnostic ITD real-time PCR method and a vaccine-derived poliovirus (VDPV) screening real-time PCR method.

Viral RNA was extracted from the poliovirus isolates by using the QIAamp Mini Viral RNA Extraction Kit (Qiagen) and was used for RT-PCR amplification by the standard method. The entire *VP1* region of the poliovirus isolates was amplified by RT-PCR with primers that flanked the *VP1*-coding region. After purification of the PCR products by the QIAquick Gel Extraction Kit (Qiagen), the amplicons were bidirectionally sequenced with the ABI PRISM 3130 Genetic Analyzer (Applied Biosystems). A neurovirulence test of polioviruses isolates was carried out using PVR-Tg21 mice that expressed the human poliovirus receptor (CD155). And the temperature sensitivities of polioviruses isolates were assayed on monolayer RD cells in 24-well plates as described.

Key words Vaccine-derived polioviruses, Viral isolation, Intratypic differentiation, VP1 region sequencing, Neurovirulence test, Temperature sensitivities test

1 Introduction

Polio eradication was declared as a programmatic emergency for global public health by World Health Organization (WHO). The live, attenuated oral polio vaccine (OPV), which was successfully used for controlling and preventing the circulation of wild polioviruses in the WHO program for the global eradication of poliomyelitis, also frequently undergoes such genetic changes throughout their genomes while replicating in human guts because

Javier Martín (ed.), *Poliovirus: Methods and Protocols*, Methods in Molecular Biology, vol. 1387, DOI 10.1007/978-1-4939-3292-4_10, © Springer Science+Business Media New York 2016

of their inherent genetic instability [1, 2]. It appears likely that the only source of poliovirus infection worldwide will be from OPV in post eradication era, and VDPV surveillance has already been implemented through careful analysis of polioviruses isolated from AFP patients (including their contacts) in supporting the Global Polio Eradication Initiative.

The genetic instability of OPV strains due to a RNA-dependent RNA polymerase error and recombination also appear to underlie the occurrence of poliomyelitis outbreaks associated with circulating VDPVs (cVDPVs) [3, 4], which exhibit ≤99 % (for type I and type III) or ≤99.5 % (for type II) *VP1* region sequence homology to the OPV strains. Two genetic characteristics—nucleotide mutations and recombination—seem to underlie the occurrence of poliomyelitis outbreaks associated with cVDPVs [5, 6].

To date, there have been several outbreaks of circulating VDPVs worldwide, for example, in Egypt [7], Hispaniola (Haiti and the Dominican Republic) [8], The Philippines [9], Madagascar [10, 11], China [12, 13], Indonesia [14], and Nigeria [15, 16]. Some phenotypic properties of VDPVs resemble those of wild polioviruses rather than those of vaccine-related polioviruses; these include properties such as the capacity for sustained person-to-person transmission, higher neurovirulence, reversion or recombination of critical attenuating sites, "non-vaccine-like" antigenic properties, the existence of replicates at a higher temperature. Hence, VDPVs may greatly hinder the polio eradication initiative taken worldwide.

2   Materials

Prepare all solutions using deionized water to attain a sensitivity of 18 MΩ cm at 25 °C, and analytical grade reagents. Prepare and store all reagents at room temperature (unless indicated otherwise).

2.1   Isolation of Viruses in Continuous Cell Lines

1. Tube cultures of L20B and RD cells.
2. Maintenance medium.
3. 1 and 5 ml plastic disposable pipettes.

2.2   Poliovirus Diagnostic Intratypic Differentiations Real-Time PCR

1. 1.5 ml microcentrifuge tubes.
2. 8-well PCR strips, 0.2 ml per well, and optical caps to specific to the real-time PCR machine.
3. Microcentrifuge with 8-well strip adaptor.
4. Vortex mixer.
5. Real-time PCR machine.

6. CDC Poliovirus Diagnostic rRT-PCR Kit.

7. Enzymes: Protector RNase Inhibitor 40 U/μl, Taq DNA Polymerase 5 U/μl, and Transcriptor RT 20 U/μl (or AMV RT 25 U/μl).

2.3 VDPV Screening Real-Time PCR

1. 1.5 ml microcentrifuge tubes.

2. 8-well PCR strips, 0.2 ml per well, and optical caps to specific to the real-time PCR machine.

3. Microcentrifuge with 8-well strip adaptor.

4. Vortex mixer.

5. Real-time PCR machine.

6. CDC VDPV Screening rRT-PCR Kit.

7. Enzymes: Protector RNase Inhibitor 40 U/μl, Taq DNA Polymerase 5 U/μl, and Transcriptor RT 20 U/μl (or AMV RT 25 U/μl).

2.4 Extraction of Virus RNA Using QIAamp Viral RNA Mini Column

1. QIAGEN QIAamp® Viral RNA Mini Kit.

2. Ethanol (96–100 %).

3. Refrigerated centrifuge.

2.5 VP1 Region Amplification Using RT-PCR

1. 0.2 ml thin-walled PCR tube.

2. Reverse transcriptase, RNasin, and Taq polymerase.

3. Microcentrifuge.

4. PCR thermal cycler.

2.6 Agarose Gel Electrophoresis of PCR Products

1. 1.7 % agarose gel.

2. 1× TAE buffer.

3. Ethidium bromide (5 mg/μl).

4. Microwave.

5. Electrophoresis tank.

6. UV-transilluminator.

2.7 Purification of PCR Products Using QIAGEN QIAquick Gel Extraction Kit

1. QIAGEN QIAquick Gel Extraction Kit.

2. 1.7 % agarose gel.

3. Isopropanol.

4. Refrigerated centrifuge.

2.8 Sequencing Reactions

1. Bigdye Terminator v3.1 Sequencing Kit (Applied Biosystems).

2. BigDye Terminator v3.1 5× Sequencing Buffer (Applied Biosystems).

3. PCR thermal cycler.

2.9 Purification of Sequence Product	1. Sephadex G-50 Fine.
	2. Centri-Sep Spin column (Applied Biosystems).
	3. Refrigerated centrifuge.
2.10 Assay for Temperature Sensitivity (One-Step Growth Curve)	1. Maintenance medium.
	2. Health RD cell.
	3. Sterile 96-well microplate with lid.
	4. Sterile 5 ml dilution tube.
	5. 1 and 5 ml plastic disposable pipettes.
	6. 36 and 39 °C CO_2 incubator.
	7. Vortex mixer.
2.11 Neurovirulence Testing in PVR-Tg21 Mice	1. PVR-Tg21 transgenic mice.

3 Methods

3.1 Viral Isolation in Two Continuous Cell Lines

1. Maintain two continuous cell lines, RD (human rhabdomyosarcoma cell) and L20B (murine cell line expressing the human poliovirus receptor) cell lines.

2. Microscopically examine recently monolayered cultures to be sure that the cells are healthy and at least 75 % confluent.

3. Remove the growth medium and replace with 1 ml maintenance medium.

4. Label two tubes of RD and two of L20B for each specimen to be inoculated (specimen number, date, passage number).

5. Label one tube of each cell type as a negative control.

6. Inoculate each cell tube with 0.2 ml of stool specimen extract and incubate in the stationary sloped (5°) position at 36 °C.

7. Examine cultures daily, using a standard or inverted microscope, for the appearance of cytopathogenic effects (CPE).

8. Record all observations of inoculated and control cultures, recording CPE as 1+ to 4+ to indicate the percentage of cells affected (*see* **Note 1**), toxicity (*see* **Note 2**), degeneration or contamination (*see* **Note 3**).

9. If no CPE appears after at least 5 days of observation, perform a blind passage (*see* **Note 4**) in the same cell line and continue examination for a further 5 days. Cultures should be examined for a total of at least 10 days before being judged as negative and discarded.

10. If characteristic enterovirus CPE appears at any stage after inoculation (rounded, refractile cells detaching from the surface of the tube), record observations and allow CPE to develop until at least 75 % of the cells are affected (>3+ CPE).

11. At this stage, perform a second passage in the other cell line (*see* **Note 5**).

3.2 Poliovirus Diagnostic Intratypic differentiations Real-Time PCR

1. Fill out PCR worksheet with name, date, primers, samples, and sample order, as well as thermocycler and program identifiers (*see* **Note 6**).

2. Thaw virus isolates and PCR reagents at room temperature.

3. Making Buffer B + enzyme mix: a vial of Buffer B 1 ml, add 2.8 μl 1 M DTT, 27.6 μl 40 U/μl RNase inhibitor, 18.0 μl 20 U/μl Transcriptor RT, and 54.8 μl 5 U/μl Taq polymerase (*see* **Note 7**) and mix.

4. Making PCR reaction solution: For each primer set, mix 19 μl Buffer A (vortex to resuspend probe before use) and 5 μl Buffer B + E; dispense 24 μl reaction solution into each well. For testing large sample numbers, create a master mix of Buffers A + B (*see* table), and dispense 24 μl of the A + B master mix per reaction well.

5. Sample preparations: Take 50 μl virus cell culture and place into a tube and centrifuge (benchtop microcentrifuge at $2500 \times g$) at room temperature for 2 min.

6. Take 0.5 μl of cell culture supernatant (or 1 μl of Control RNA) for each sample and add into the appropriate reaction strip/plate well. rRT-PCR does NOT require a 95 °C heat step.

7. Place strips in real-time thermocycler and run cycle as shown below. If using a thermocycler with a rapid ramp speed, program the ramp from 44 to 60 °C for 45 s (*see* **Note 8**).

    (a) Reverse transcript reaction, 42 °C, 45 min.

    (b) Inactivate RT, 95 °C, 3 min.

    (c) PCR cycles (all primers sets): 95 °C for 24 s, 44 °C for 30 s, then a 25 % ramp speed to 60 °C for 24 s, for 40 cycles, and the endpoint fluorescent data is collected at the end of the anneal step.

    (d) Select the appropriate dye filter to correspond with the assay being used.

8. These Ct values are calculated automatically by the software of the real-time PCR machine, and the results are interpreted by looking for a Ct value. The Ct value cutoff is 30, with values less than 30 as positive and values more than 30 as negative (*see* **Note 9**).

3.3 VDPV Screening Real-Time PCR

1. Fill out PCR worksheet with name, date, primers, samples, and sample order, as well as thermocycler and program identifiers.

2. Thaw virus isolates and PCR reagents at room temperature.

3. Making Buffer B + enzyme mix: a vial of Buffer B 1 ml, add 2.8 µl 1 M DTT, 27.6 µl 40 U/µl RNase inhibitor, 18.0 µl 20 U/µl Transcriptor RT, and 54.8 µl 5 U/µl Taq polymerase (*see* **Note 7**) and mix.

4. Making reaction solution: For each primer set, mix 19 µl Buffer A (vortex to resuspend probe before use) and 5 µl Buffer B + E; dispense 24 µl reaction solution into each well. For testing large sample numbers, create a master mix of Buffers A + B, and dispense 24 µl of the A + B master mix per reaction well.

5. Sample preparation: Take 50 µl virus cell culture and place into a tube and centrifuge (benchtop microcentrifuge at $2500 \times g$ or full speed (max rpm $4000 \times g$) of Tube-Strip PicoFuge) at room temperature for 2 min.

6. Take 0.5 µl of cell culture supernatant (or 1 µl of Control RNA) for each sample and add into the appropriate reaction strip/plate well. rRT-PCR does NOT require a 95 °C heat step.

7. Place strips in real-time thermocycler and run cycles as shown below. If using a thermocycler with a rapid ramp speed, program the ramp from 44 to 60 °C for 45 s (*see* **Note 8**).

   (a) RT reaction, 42 °C, 45 min.

   (b) Inactivate RT, 95 °C, 3 min.

   (c) PCR cycles: 95 °C for 24 s, 50 °C for 30 s, then a 25 % ramp speed to 65 °C for 24 s, for 40 cycles. The end point fluorescent data is collected at the end of the anneal step.

   (d) Select the appropriate dye filter to correspond with the assay being used.

8. These Ct values are calculated automatically by the software of the real-time PCR machine, and the results are interpreted by looking for a Ct value. The Ct value cutoff is 30, with values less than 30 as positive and values more than 30 as negative (*see* **Note 9**).

3.4 Extraction of Virus RNA Using QIAamp Viral RNA Mini Column

1. Pipet 560 µl of prepared Buffer AVL containing carrier RNA into a 1.5 ml microcentrifuge tube.

2. Add 140 µl plasma, serum, urine, cell-culture supernatant, or cell-free body fluid to the Buffer AVL–carrier RNA in the microcentrifuge tube. Mix by pulse-vortexing for 15 s.

3. Incubate at room temperature (15–25 °C) for 10 min.

4. Briefly centrifuge the tube to remove drops from the inside of the lid.

5. Add 560 µl of ethanol (96–100 %) to the sample, and mix by pulse-vortexing for 15 s. After mixing, briefly centrifuge the tube to remove drops from inside the lid.

6. Carefully apply 630 µl of the solution from **step 5** to the QIAamp Mini column (in a 2 ml collection tube) without wetting the rim. Close the cap and centrifuge at 6000×*g* (or 8000 rpm) for 1 min. Place the QIAamp Mini column into a clean 2 ml collection tube and discard the tube containing the filtrate.

7. Carefully open the QIAamp Mini column, and repeat **step 6**.

8. Carefully open the QIAamp Mini column, and add 500 µl of Buffer AW1. Close the cap and centrifuge at 6000×*g* (or 8000 rpm) for 1 min. Place the QIAamp Mini column in a clean 2 ml collection tube, and discard the tube containing the filtrate.

9. Carefully open the QIAamp Mini column, and add 500 µl of Buffer AW2. Close the cap and centrifuge at full speed (20,000×*g*; or 14,000 rpm) for 3 min. Continue directly with **step 11**, or to eliminate any chance of possible Buffer AW2 carryover, perform **step 10**, and then continue with **step 11**.

10. Place the QIAamp Mini column in a clean 1.5 ml microcentrifuge tube. Discard the old collection tube containing the filtrate. Carefully open the QIAamp Mini column and add 60 µl of Buffer AVE equilibrated to room temperature. Close the cap and incubate at room temperature for 1 min. Centrifuge at 6000×*g* (or 8000 rpm) for 1 min.

3.5   VP1 Region Amplification Using RT-PCR

1. Thaw all kit reagents except enzymes, vortex and place on ice. Keep enzymes (Reverse transcriptase, RNasin, and Taq polymerase) on ice at all times. Allow RNA samples to thaw on ice and keep on ice while you are setting up the reactions.

2. Label appropriate number of 0.2 ml thin-walled, reaction tubes and place in prechilled metal cooling rack.

3. Add the following into a sterile tube. Take care to use sterile tips and avoid cross-contamination. Use new tips for each reagent and template.

Upstream primer UG1: 5′-GGG TTT GTG TCA GCC TGT AAT GA-3′.

Downstream primer UC11: 5′-AAG AGG TCT CTR TTC CAC AT-3′.

25 mM MgCl$_2$	3.0 µl
10 mM dNTPs	1.0 µl
10× PCR buffer	5.0 µl
Upstream primer (15 pmol/µl)	1.0 µl
Downstream primer (15 pmol/µl)	1.0 µl
Taq DNA polymerase (5 U/µl)	0.5 µl
RNAsin inhibitor (40 U/µl)	0.5 µl
RNA template	3.0 µl
Deionized water	33.0 µl

4. Spin the tubes briefly ($9000 \times g$ for 1 min) in a chilled microcentrifuge and immediately return the tube to the metal cooling rack.

5. Place the PCR tubes into the PCR thermal cycler and run the following PCR program:

Temperature (°C)	Time		No. of cycles
94	4 min		1
94	30 s	⎫	30
55	30 s	⎬	
72	90 s	⎭	
72	10 min		1

6. Take 5 µl of PCR product and check it on an agarose gel.

3.6 Agarose Gel Electrophoresis of PCR Products

1. Prepare 60 ml of 1.7 % agarose in 1× TAE buffer.

2. Heat in microwave for ~2 min on full power.

3. Allow to cool to about 45 °C and add 1 µl ethidium bromide (stock = 5 mg/µl) per 10 ml, giving a final concentration of 0.5 µg/ml. This can be increased to 1 µg/ml if no ethidium bromide is added to the buffer (see below).

4. Pour gel and insert comb. Allow to set on a flat surface for about 30 min.

5. Pour buffer 1× TAE buffer (containing 0.5 µg/ml ethidium bromide, i.e., 1 µl of 5 mg/ml stock to every 10 ml of buffer) into tank and remove comb from gel (*see* **Note 10**).

6. Prepare samples (1 µl 6× loading buffer + 5 µl PCR product) on Parafilm.

7. Load samples into the wells formed in the gel. It is often useful to load the molecular weight markers in both the first and last lanes.

8. Electrophorese at 100 V for 30 min.

9. View and photograph the gel on an UV-transilluminator. Use UV-safety spectacles.

3.7 Purification of PCR Products Using QIAGEN QIAquick Gel Extraction Kit

1. Excise the DNA fragment from the 1.7 % agarose gel with a clean, sharp scalpel.

2. Weigh the gel slice in a colorless tube. Add 3 volumes Buffer QG to 1 volume gel (100 mg ~ 100 μl).

3. Incubate at 50 °C for 10 min (or until the gel slice has completely dissolved), and vortex the tube every 2–3 min to help dissolve gel.

4. After the gel slice has dissolved completely, add 1 gel volume of isopropanol to the sample and mix.

5. Place a QIAquick spin column in a provided 2 ml collection tube.

6. To bind DNA, apply the sample to the QIAquick column and centrifuge for 1 min. Discard flow-through and place the QIAquick column back into the same tube. For sample volumes of >800 μl, load and spin/apply vacuum again.

7. If the DNA will subsequently be used for sequencing, add 0.5 ml Buffer QG to the QIAquick column and centrifuge for 1 min. Discard flow-through and place the QIAquick column back into the same tube.

8. To wash, add 0.75 ml Buffer PE to QIAquick column and centrifuge for 1 min. Discard flow-through and place the QIAquick column back into the same tube (*see* **Note 11**).

9. Centrifuge the QIAquick column once more in the provided 2 ml collection tube for 1 min at $17,900 \times g$ (13,000 rpm) to remove residual wash buffer.

10. Place QIAquick column into a clean 1.5 ml microcentrifuge tube.

11. To elute DNA, add 50 μl Buffer EB (10 mM Tris·Cl, pH 8.5) or water to the center of the QIAquick membrane and centrifuge the column for 1 min. For increased DNA concentration, add 30 μl Buffer EB to the center of the QIAquick membrane, let the column stand for 1 min, and then centrifuge for 1 min (*see* **Note 12**).

3.8 Sequencing Reactions

1. Take all reagents for the reaction (Ready Reaction Premix, primers, and BigDye Sequencing dilution buffer) out of the

freezer and thaw on ice or place into a PCR reaction dedicated tube rack at 4 °C until the reagents are thawed.

2. Place the reagents and PCR tubes on ice.

3. Add the following into a sterile tube. Take care to use sterile tips and avoid cross-contamination. Use new tips for each reagent and template.

BigDye Terminator v3.1	4.0 μl
5× Sequencing buffer	2.0 μl
Primer (3.2 pmol)	1.0 μl
DNA template (20–40 ng)	4.0 μl
Deionized water	9.0 μl

4. Mix the PCR reactions well by pipetting up and down. Spin the tubes briefly to remove bubbles.

5. Place the PCR tubes into the PCR thermal cycler and run the following PCR program:

Temperature (°C)	Time		No. of cycles
94	20 s		25
42	15 s		
60	4 min		
4	Forever		

3.9 Purification of Sequencing Reaction Product and Sequencing

1. Set Centri-Sep column in a 2 ml tube (*see* **Note 13**).

2. Add 900 μl of Sephadex G-50 Fine suspended in distilled water.

3. Centrifuge at $1000 \times g$, for 1 min.

4. Discard the flow-through.

5. Centrifuge at $1000 \times g$, for 2 min.

6. Transfer the column to a new 1.5 ml tube.

7. Add 25 μl of H_2O in the seq product (total 40 μl).

8. Put the diluted Seq product on the top of Sephadex G-50 Fine.

9. Centrifuge at $1000 \times g$, for 2 min.

10. Run the flow-through (about 20 μl) on a genetic sequencer.

3.10 Assay for Temperature Sensitivity (One-Step Growth Curve)

1. Titration of the virus isolates.

(a) Label seven dilution tubes (from 10^{-1} to 10^{-7}), marking each set with virus isolate number.

(b) Dispense 0.9 ml maintenance medium to each tube.

(c) Add 0.1 ml virus to the first tube (=10^{-1} dilution) using pipettor with ART tip.

(d) Take another pipette tip, mix thoroughly but gently to avoid aerosols.

(e) Transfer 0.1 ml to the second tube and discard pipette tip.

(f) Repeat dilution steps, transferring 0.1 ml into tubes 3–7.

(g) Add virus to the back titration wells of the microplate beginning at the 10^{-7} dilution in columns 1–8, rows H.

(h) Work from highest to lowest dilution, 10^{-7} to 10^{-1}.

(i) Add 100 µl of maintenance medium in columns 11–12, rows A to H, as cell control wells

(j) Trypsinize RD cells and prepare a suspension of approximately 1.5×10^5 cells per ml.

(k) Distribute 100 µl of cell suspension into each test and control well.

(l) Incubate at 36 °C using a CO_2 incubator.

(m) Continue observation and recording for development of CPE for 7 days.

(n) The virus titer is calculated by using Kärber formula [$\log CCID_{50} = L - d\,(S - 0.5)$].

(o) Perform cell counting and dilute the virus to 10 multiplicity of infection (MOI).

2. Add calculated amount of virus (MOI = 10) into the RD cells.

3. Incubate at 36 and 39 °C using two different CO_2 incubators, adsorption for 1 h.

4. Wash the cells three times using maintenance medium, and then add 1 ml maintenance medium to the tube.

5. Incubate at 36 and 39 °C using two different CO_2 incubators.

6. Harvest the virus at 0 h, 4 h, 8 h, 24 h, 48 h, and 72 h, respectively.

7. Freeze the harvested viruses at –20 °C, with repeated freeze–thaw three times.

8. Titration of viruses harvested at different temperatures and at different times as **step 1** (1.1–1.14).

9. Temperature sensitivity is expressed as logarithmic difference of the $CCID_{50}$ values at 36 and 39 °C.

3.11 Neurovirulence Testing in PVR-Tg21 Mice

1. Female and male PVR-Tg21 transgenic mice that expressed the human poliovirus receptor (CD155), aged 5–14 weeks, are bred under carefully monitored conditions.

2. $CCID_{50}$ of the virus is titrated (see Subheading 3.10).

3. Six mice (equal number of males and females) are assigned to one group and are inoculated intracerebrally with 30 μl of each virus dilution (in tenfold increments; range: 2.5–6.5 $CCID_{50}$ per mouse).

4. The mice are examined daily for 21 days after the inoculation, and the number of paralyzed or dead mice is recorded.

5. The virus titer that induced paralysis or death in 50 % of the inoculated mice (PD_{50}) is calculated by using the Kärber formula and expressed as PD_{50}/mouse.

4 Notes

1. 1+ represents up to 25 % of cells; 2+ represent 25–50 %; 3+ represents 50–75 %; 4+ represents from 75 to 100 %.

2. *Toxicity*: If cell cultures show rapid degeneration within 24 h of inoculation this may be due to nonspecific toxicity of the specimen. These tubes should be frozen at –20 °C, thawed, and 0.2 ml volumes transferred (i.e., this should be considered as a passage) into cultures of the same cell type. If toxic appearances recur, return to the original specimen extract and dilute this in PBS at 1/10 and reinoculate cultures as described above (this should be considered as inoculation of the specimen and day 0 for observation).

3. *Microbial contamination*: Contamination of the medium and cell death resulting from bacterial or fungal contamination makes detection of viral CPE uncertain or impossible. Return to the original specimen extract, *re-treat with chloroform* (*see* Chapter 6 Section 6.2 of WHO Polio Laboratory Manual) and inoculate fresh cell cultures as described above.

4. *Blind Passage*: Freeze the tubes at –20 °C, thaw, and passage 0.2 ml of culture fluid to tubes containing fresh monolayers *of the same cell type* and examine daily for at least 5 days. If cultures show no CPE by this stage, the result is regarded as negative.

5. CPE positive L20B cultures are passaged into RD cell cultures and incubated at 36 °C and observed daily, this passage is aimed at amplifying the titre of any polioviruses that may be present; and CPE positive RD cultures are passaged into L20B cells and incubated at 36 °C and observed daily, this passage is daily aimed at separating polioviruses that may be present in mixtures with other enteroviruses and amplifying the titre of any polioviruses that may be present.

6. (a) Name wells using thermocycler software for samples and controls (positive and reagent). (b) One positive control: noninfectious control RNA supplied with Polio rRT-PCR kit. (c) One reagent control: Buffer A + B with no template.

7. Do not use error correcting Taq polymerases like Pfu and Pwo, they will not work with inosine primers.

8. For the ABI 7500 machine, you can use 25 % ramp speed between the anneal and extension temperatures for all assays.

9. Samples with Ct values from 28 to 32 should be reanalyzed using extracted RNA for those samples. Samples which have a Ct value <30 but have a flat fluorescence profile, or a profile that rises just barely rise above the X axis are most likely negative, but should be repeated, using extracted RNA.

10. Take care while handling ethidium bromide; it is harmful, and gloves should be worn at all times.

11. If the DNA will be used for salt-sensitive applications (e.g., sequencing, blunt-ended ligation), let the column stand 2–5 min after addition of Buffer PE.

12. After the addition of Buffer EB to the QIAquick membrane, increasing the incubation time to up to 4 min can increase the yield of purified DNA.

13. The Centri-Sep column can be recycled after distilled water washing.

References

1. Gromeier M, Mueller S, Solecki D, Bossert B, Bernhardt G, Wimmer E (1997) Determinants of poliovirus neurovirulence. J Neurovirol 3(Suppl 1):S35–S38
2. Abe S, Ota Y, Koike S, Kurata T, Horie H, Nomura T, Hashizume S, Nomoto A (1995) Neurovirulence test for oral live poliovaccines using poliovirus-sensitive transgenic mice. Virology 206:1075–1083
3. Kew OM, Sutter RW, de Gourville EM, Dowdle WR, Pallansch MA (2005) Vaccine-derived polioviruses and the endgame strategy for global polio eradication. Annu Rev Microbiol 59:587–635
4. Kew OM, Wright PF, Agol VI, Delpeyroux F, Shimizu H, Nathanson N, Pallansch MA (2004) Circulating vaccine-derived polioviruses: current state of knowledge. Bull World Health Organ 82:16–23
5. Combelas N, Holmblat B, Joffret ML, Colbere-Garapin F, Delpeyroux F (2011) Recombination between poliovirus and coxsackie A viruses of species C: a model of viral genetic plasticity and emergence. Viruses 3:1460–1484
6. Jegouic S, Joffret ML, Blanchard C, Riquet FB, Perret C, Pelletier I, Colbere-Garapin F, Rakoto-Andrianarivelo M, Delpeyroux F (2009) Recombination between polioviruses and co-circulating Coxsackie A viruses: role in the emergence of pathogenic vaccine-derived polioviruses. PLoS Pathog 5:e1000412
7. Yang CF, Naguib T, Yang SJ, Nasr E, Jorba J, Ahmed N, Campagnoli R, van der Avoort H, Shimizu H, Yoneyama T, Miyamura T, Pallansch M, Kew O (2003) Circulation of endemic type 2 vaccine-derived poliovirus in Egypt from 1983 to 1993. J Virol 77:8366–8377
8. Kew O, Morris-Glasgow V, Landaverde M, Burns C, Shaw J, Garib Z, Andre J, Blackman E, Freeman CJ, Jorba J, Sutter R, Tambini G, Venczel L, Pedreira C, Laender F, Shimizu H, Yoneyama T, Miyamura T, van Der Avoort H, Oberste MS, Kilpatrick D, Cochi S, Pallansch M, de Quadros C (2002) Outbreak of poliomyelitis in Hispaniola associated with circulating type 1 vaccine-derived poliovirus. Science 296:356–359
9. Shimizu H, Thorley B, Paladin FJ, Brussen KA, Stambos V, Yuen L, Utama A, Tano Y, Arita M, Yoshida H, Yoneyama T, Benegas A, Roesel S, Pallansch M, Kew O, Miyamura T (2004) Circulation of type 1 vaccine-derived poliovirus in the Philippines in 2001. J Virol 78:13512–13521
10. Rakoto-Andrianarivelo M, Gumede N, Jegouic S, Balanant J, Andriamamonjy SN, Rabemanantsoa S, Birmingham M, Randriamanalina B, Nkolomoni L, Venter M, Schoub BD, Delpeyroux F, Reynes JM (2008) Reemergence of recombinant vaccine-derived poliovirus outbreak in Madagascar. J Infect Dis 197:1427–1435

11. Rousset D, Rakoto-Andrianarivelo M, Razafind-ratsimandresy R, Randriamanalina B, Guillot S, Balanant J, Mauclere P, Delpeyroux F (2003) Recombinant vaccine-derived poliovirus in Madagascar. Emerg Infect Dis 9:885–887

12. Liang X, Zhang Y, Xu W, Wen N, Zuo S, Lee LA, Yu J (2006) An outbreak of poliomyelitis caused by type 1 vaccine-derived poliovirus in China. J Infect Dis 194:545–551

13. Yan D, Li L, Zhu S, Zhang Y, An J, Wang D, Wen N, Jorba J, Liu W, Zhong G, Huang L, Kew O, Liang X, Xu W (2010) Emergence and local-ized circulation of a vaccine-derived poliovirus in an isolated mountain community in Guangxi, China. J Clin Microbiol 48:3274–3280

14. Estivariz CF, Watkins MA, Handoko D, Rusipah R, Deshpande J, Rana BJ, Irawan E, Widhiastuti D, Pallansch MA, Thapa A, Imari S (2008) A large vaccine-derived poliovirus out-break on Madura Island—Indonesia, 2005. J Infect Dis 197:347–354

15. Burns CC, Shaw J, Jorba J, Bukbuk D, Adu F, Gumede N, Pate MA, Abanida EA, Gasasira A, Iber J, Chen Q, Vincent A, Chenoweth P, Henderson E, Wannemuehler K, Naeem A, Umami RN, Nishimura Y, Shimizu H, Baba M, Adeniji A, Williams AJ, Kilpatrick DR, Oberste MS, Wassilak SG, Tomori O, Pallansch MA, Kew O (2013) Multiple independent emergences of type 2 vaccine-derived poliovi-ruses during a large outbreak in northern Nigeria. J Virol 87:4907–4922

16. Wassilak S, Pate MA, Wannemuehler K, Jenks J, Burns C, Chenoweth P, Abanida EA, Adu F, Baba M, Gasasira A, Iber J, Mkanda P, Williams AJ, Shaw J, Pallansch M, Kew O (2011) Outbreak of type 2 vaccine-derived poliovirus in Nigeria: emergence and widespread circula-tion in an underimmunized population. J Infect Dis 203:898–909

Chapter 11

Phylogenetic Analysis of Poliovirus Sequences

Jaume Jorba

Abstract

Comparative genomic sequencing is a major surveillance tool in the Polio Laboratory Network. Due to the rapid evolution of polioviruses (~1 % per year), pathways of virus transmission can be reconstructed from the pathways of genomic evolution. Here, we describe three main phylogenetic methods; estimation of genetic distances, reconstruction of a maximum-likelihood (ML) tree, and estimation of substitution rates using Bayesian Markov chain Monte Carlo (MCMC). The data set used consists of complete capsid sequences from a survey of poliovirus sequences available in GenBank.

Key words Phylogenetic analysis, Maximum-likelihood, Bayesian MCMC, Poliovirus evolution

1 Introduction

Strategically, the success of the Polio Eradication Initiative (PEI) relies on intensive surveillance of cases of Acute Flaccid Paralysis (AFP) and laboratory investigations of polio samples. Virologic surveillance depends on a well-established WHO Global Polio Laboratory Network (GPLN) where polio isolation, intratypic differentiation, and genotyping are methods shared among laboratory members. Genomic sequencing is the method with the highest resolution. Due to the rapid evolution of polioviruses (PVs) [1], phylogenetic analysis of the VP1 capsid region resolves transmission pathways [2, 3]. Phylogenetic trees are invaluable tools for monitoring the progress of immunization activities as indicated by the appearance, disappearance, or reappearance of genetic lineages. PVs generally cluster geographically on a phylogenetic tree and if sequence difference is less than 15 % they are designated as genotypes. Furthermore, groups of viruses sharing less than 5 % sequence identity are designated as clusters. Single chains of transmission are phylogenetically displayed as (usually) short branches connecting sequences; growing branches correlate with expanded transmission. In this scenario, the appearance of long, isolated branches may indicate a fragmented genetic record from orphan

Javier Martín (ed.), *Poliovirus: Methods and Protocols*, Methods in Molecular Biology, vol. 1387,
DOI 10.1007/978-1-4939-3292-4_11, © Springer Science+Business Media New York 2016

lineages (≥1.5 % sequence difference to the closest relative) indicative of possible surveillance gaps. The latest generation of rapid automated sequencers allows the routine characterization of VP1 sequences and the use of complete genome sequences for further characterization of the dynamics of PV evolution. For example, complete capsid sequences permit the use of infrequent transversion substitutions to define genetic relationships that otherwise might be obscured by saturating transition substitutions [4]. The availability of complete genomic sequences and related epidemiologic data, the development of bioinformatics and molecular evolution, and the growing computational capacity opens the door to the development of new analytical tools applied to molecular epidemiology. Here, phylodynamic methods are presented.

2   Materials

Current bioinformatics software packages are usually multiplatform; they can run on Windows, Mac, or Linux operating systems. As a recommendation, systems with at least 4 GB of RAM memory can accomplish most of the typical bioinformatics analysis described in this chapter. In order to optimize memory resources, it is best to consolidate all bioinformatics programs in a single computer and leave the rest of informatics needs (for example, word processing and email) to another computer if available. Most bioinformatics software incorporate the basic software libraries needed for running the programs. However, it is advised to follow the installation instructions provided by the software package since some may require installing or updating existing libraries (for example, Java libraries).

There are several bioinformatics packages that include phylogenetic analysis. A comprehensive list of phylogenetic resources can be found online at http://evolution.genetics.washington.edu/phylip/software.html. In this chapter the following software will be used: MEGA version 6 (http://megasoftware.net/), Geneious version 7 (http://www.geneious.com/), Seaview version 4.5 (http://doua.prabi.fr/software/seaview), BEAST version 1.8 (http://beast.bio.ed.ac.uk/), and FigTree version 1.4.2 (http://tree.bio.ed.ac.uk/software/figtree/).

2.1   General Software Packages

MEGA, Geneious, and Seaview include several approaches to phylogenetic analysis and are also useful for preparing and converting genetic data files and formats (for example, assembly and alignment). MEGA and Seaview have a long tradition of consolidating sequence editing and analysis in a single package. Geneious is a relatively new bioinformatics software package whose analytical tools extend beyond sequence and phylogenetic analysis. These programs have graphical interfaces where sequence alignments can be displayed and inspected for ambiguities and gaps. It is highly recommended to inspect the sequence alignment before proceeding with sequence analysis. All three packages infer phylogenetic trees using different methods, including Neighbor-Joining (NJ), Maximum Likelihood (ML), and Bayesian approaches.

2.2 Phylodynamics Software

BEAST (Bayesian Evolutionary Analysis Sampling Trees) is a complete software package to study viral phylodynamics [5]. For example, it has been used to infer the dates and the population dynamics of multiple emergences of circulating vaccine-derived polioviruses (cVDPV) [6]. BEAST v1.8 includes four programs: (1) BEAUTi, (2) BEAST, (3) Tree annotator, and (4) Log combiner. There are two additional programs that need to be installed separately: TRACER and FigTree. Succinctly, BEAUTi prepares the XML file that contains the sequence alignment and model specifications. BEAST takes as input the XML file and generates two outputs; a log file and a tree file. The log file can be read using TRACER while the tree file is processed using Tree annotator, which will generate a Maximum Clade Credibility (MCC) tree. The annotated MCC tree is displayed using FigTree.

3 Methods

Estimation of genetic distances is the basic building block in phylogenetic inference. Nucleotide substitutions are modeled following Markov models and estimated using substitution matrices. For in-depth study, Allman [7] and Yang [8] provide excellent background information on the statistical models used in estimation of genetic distances. Poliovirus evolution is characterized by (a) having a high substitution rate, and (b) a differential rate of accumulation of transition (substitutions between two purines A↔G or two pyrimidines T↔C) and transversion (substitutions between a pyrimidine and a purine T,C↔A,G) nucleotide changes. Genetic distances estimated using the Kimura 2-parameter model correct for both multiple hits at the same site and the differential rate of transition and transversion changes. Parameter-rich models of evolution incorporate additional substitution dynamic nuances. Fit of a particular dataset to a model of evolution can be investigated using MEGA or Modeltest [9]. In addition, substitutions are not homogenously distributed along the genome. For example, in the protein coding gene VP1, the majority of substitutions occur at the third codon position. The observed variability of substitutions across sites can be incorporated into the models of evolution by using a gamma distribution. Most phylogenetic programs incorporate this function and the user is asked to either provide the shape parameter alpha (α) or letting the program estimate it. Estimates of α for poliovirus VP1 sequence alignments gravitate around 0.3.

3.1 Genetic Distances

Dataset: FASTA alignment of WPV1 sequences available at GenBank. Accession numbers EF374000–EF374030.

1. Once the FASTA file is open in MEGA, different types of substitution models are available under the Compute Pairwise Distance menu. Three separate runs will provide estimated distances by choosing *p*-distance, Kimura 2-parameter model, and Tamura–Nei model (with gamma distributed rates among sites) respectively. Gamma distributed rates is selected by switching from Uniform rates to Gamma distributed in the Rates among Sites option.

2. Estimated distances are shown in a matrix for all pairwise comparisons and exported in tabular CSV format for use in a spreadsheet (*see* **Note 1**).

3. Comparison of genetic distances can be estimated using three different substitution models by combining each file into a single worksheet. Parameter-rich models capture increased genetic distance when sequences under comparison become more divergent. From the example dataset, estimated distances between closely related sequences (*p*-distance, 0.014 substitution per site [s/s]) have a very slight increase compared to the K2P (0.015 s/s) and TN + G (0.015 s/s) models. When moderately divergent sequences are compared (*p*-distance, e.g., 0.078 s/s), the estimated genetic distance increased by about 30 % by applying the TN + G model (~0.103 s/s).

3.2 Maximum Likelihood Tree

Dataset: FASTA alignment of WPV1 sequences available at GenBank. Accession numbers EF374000–EF374030.

1. Import the fasta alignment in Geneious v7.

2. Select Tools → Tree or click on the Tree icon.

3. The ML algorithm implemented in Geneious is PhyML [10]. It is not installed by default and should be installed as a downloadable plugin following directions in Tools → Plugins.

4. Once the PhyML plugin is installed, it will be available in the Tree menu. Click on the PhyML tab.

5. The options displayed in the PhyML menu (Fig. 1) include choice of substitution model and its parameters, branch support, and topology strategy and optimization.

   (a) In this example, the GTR model of nucleotide substitution is used (*see* **Note 2**). Next, to allow the substitution rate to vary among sites the option Number of substitution rate categories (*N*) should be >1. We set *N* to 4. The value of the Gamma distribution parameter can be fixed or estimated by the program. We set this option to Estimated. An additional option allows the proportion of invariable

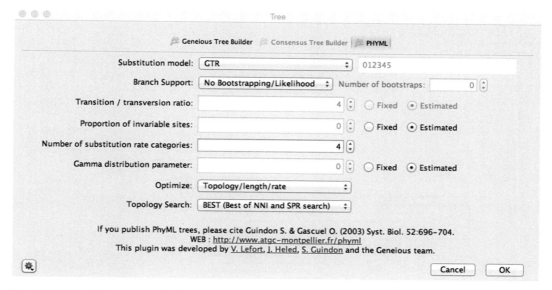

Fig. 1 PhyML window menu displayed in Geneious v7. The PhyML plugin for Geneious is available from Geneious. Model parameters and phylogenetic options for the ML run are set in this window

sites to be fixed or estimated (fixing the value to zero results in canceling this option). We set Proportion of invariable sites to Estimated.

(b) There are four options under Branch support, the most used of which is Bootstrap. This method is based on resampling with replacement from the original nucleotide sites of a sequence and inferring a new tree from each sample. Comparison of the tree from the original sequence with those arising from the resampling will indicate the level of statistical support for the branching topology. This method is computationally expensive. In this example, no bootstrap values are computed (*see* **Note 3**).

(c) Under Optimize, we set this option to Topology/length/rate.

(d) Last, the option Topology search contains three choices. We choose BEST, which combines two alternative topology strategies.

(e) After the run, PhyML will generate a tree file located in the same Geneious folder that contains the alignment under analysis. The tree file is visualized in Geneious. The tree visualization within Geneious allows further editing of the tree, including branch swapping, rooted/unrooted and circular layouts, and expansion of the tree for better visualization of trees with numerous sequences (Fig. 2). The tree can be exported in Nexus or Newick formats for further editing in other software, for example FigTree.

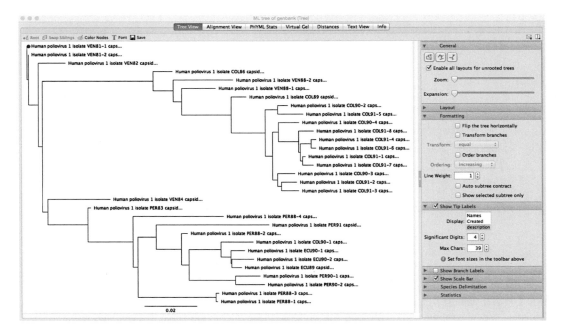

Fig. 2 Visualization within Geneious of the phylogenetic tree reconstructed by PhyML. The tree was rooted to sequence VEN-81 by selecting the sequence in the tree and pressing Root in the options shown in the *left hand corner above* the tree

3.3 Molecular Clock Analysis

This section describes the methods used to estimate substitution rates from a set of serial samples of poliovirus sequences. Rates of evolution are generally put into practice for estimating divergence times. For example, by assuming a constant rate of evolution (molecular clock) among viral lineages, it is possible to infer the dates of ancestral or source infections. Bayesian Markov chain Monte Carlo (MCMC) methods are used to infer substitution rates and are implemented in the software BEAST. In general terms, Bayesian phylogenetics are based on prior assumptions about parameters; the optimum values are obtained from a continuous distribution sample set a particular state and then proposing new states from sliding windows that will be accepted according to acceptance ratios. Development of Markov chain Monte Carlo (MCMC) algorithms provided the methods for achieving Bayesian computation.

Dataset: FASTA alignment of WPV1 sequences available at GenBank. Accession numbers EF374000–EF374030.

1. The first step is incorporating the temporal data associated with each sequence in the sequence name (tip dates). Source of temporal data includes specimen date or onset date. BEAST can recognize several date formats. In this example, a decimal format is used. For example, when the date is 1 July 1981, we

Fig. 3 BEAUTi window menu showing the options for incorporating numerical fields (tip dates) from the sequence name in the analysis

search the day of the year for 1 July (in this example, the 182nd day) and then divide it by 365. The corresponding decimal format is 1981.49. This process can be easily obtained using Excel.

2. Once all sequence names are labeled incorporating temporal data, the corresponding alignment (in Nexus format) is imported into BEAUTi.

3. BEAUTi prepares an XML file for BEAST. Once the file is imported in BEAUTi, choose the Tip tab and check Use tip dates option. Pressing Guess dates will open a new menu for recognizing the numerical field for each sequence name (Fig. 3) (*see* **Note 4**). If successful, the Date column will be populated with the corresponding numerical values.

4. The next tab to be modified is the Site tab. Under Substitution model, choose GTR and under Site heterogeneity, choose Gamma + Invariant sites.

5. In the clocks tab, select Strict clock and Estimate. Alternatively, the rate can be fixed to a value consistent with the units of time.

6. For the purpose of this example, the tab Trees can be left at default values (*see* **Note 5**).

7. The Priors tab shows every parameter of the model selected and its corresponding prior distribution. The priors that appear in red should be set. In this example, clock.rate. A prior selection window appears after clicking on the parameter. Select uniform distribution (*see* **Note 6**).

8. The Operators tab is set to Auto Optimize.

9. The MCMC tab tunes the MCMC chain. Length of chain depends on the size of the data set and the models chosen. For initial tests, choose a value of 1,000,000. The Log parameters option specifies how often a sample is recorded. It is recommended a final sampling of no more than 10,000 samples. The value in this field can be calculated as Length of chain/10,000. The option Echo state is related to the amount of information displayed on the screen and it is recommended to follow the value calculated from the Log parameters option. The remaining options can be set as default or adjusted per user requirement.

10. Click on Generate BEAST file. If any parameter has improper priors, it will be shown in the window before saving the XML file.

11. Run BEAST. The XML file generated in BEAUTi can be chosen from the dialog box. Click on Run. Progress of the run will be displayed on the screen. Once finished, two files will be generated; log (file with extension .log) and tree files (file with .trees extension).

12. The log file is analyzed using the program Tracer.

    (a) Import the log file into Tracer.

    (b) The mean substitution rate estimated from BEAST is displayed under clock.rate. Confidence intervals around the mean are displayed in the Estimates tab as 95 % HPD (highest posterior density) intervals (Fig. 4).

    (c) Assessment of sample autocorrelation is checked in the ESS (Effective Sample Size) column. Low ESS values are displayed in red or yellow and generally are indicative of short runs and the chain length needs to be adjusted accordingly.

13. Processing of the tree file is performed in TreeAnnotator. The tree file contains all sampled trees (in this example, 10,000 trees). TreeAnnotator summarizes in a single tree the information from all sampled trees.

    (a) Burnin (as the number of trees) is the number of excluded trees from the summary. In general is set to 10 % of the total number of sampled trees (in this case, 1000).

    (b) Nodes are annotated according to the Posterior probability limit set in TreeAnnotator. If set to zero, all nodes will incorporate a summary of the annotations stored during the BEAST run.

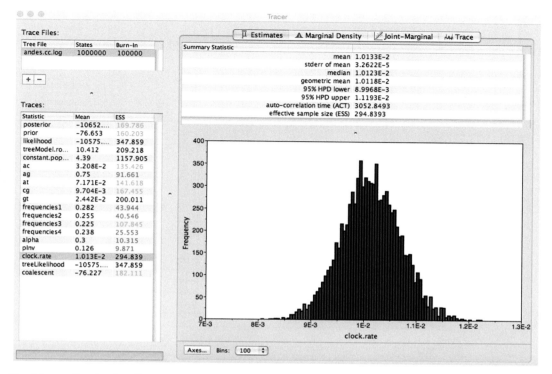

Fig. 4 Inspecting the log file in TRACER. The *left hand side* column displays the traces of each parameter including the mean and ESS values. On the *right*, statistics about parameters (including 95 % HPD) are shown in the *upper* window and a graphical visualization is displayed in the *lower* window. Each sampled value during the MCMC run can be graphically inspected in the tab labeled Trace

(c) Select Maximum clade credibility (MCC) tree in the Target tree type option for obtaining a tree that has the highest product of the posterior probability of all of its nodes. Node heights can be summarized as Mean or Median values.

(d) Choose the tree file generated by BEAST and give a file name for the resulting tree.

(e) The MCC tree generated by TreeAnnotator can be visualized in FigTree.

14. In order to generate a time scale in FigTree, open the Time Scale tab and in the Offset by option enter the date of the most recent sample (as displayed in the sequence name) and in the Scale factor option change the default to −1 (negative one). In the Scale axis tab, check the option Reverse Axis. Check the Node Labels tab and select Node Ages from the Display option. All nodes will display estimates of the divergence dates in time scale values (Fig. 5). 95 % HPD values can be summarized in bars across nodes by checking Node Bars and selecting the parameter of interest for display.

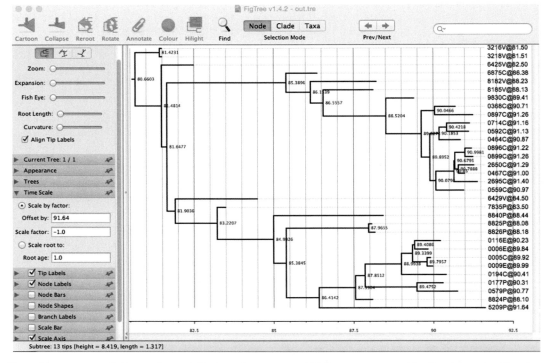

Fig. 5 The program FigTree edits and displays the annotated MCC tree generated by TreeAnnotator. The time scale on the tree is displayed by setting the Time scale and Scale axis options as shown in the figure

4   Notes

1. Pairwise distances displayed in MEGA can be exported in Excel and CSV formats. In addition, the order of sequences can be changed before exporting the matrix by holding the sequence name and moving it up or down the column.

2. In order to fix the Transition/Transversion ratio to a specific value (for example, fixed to 10), choose TN93, HKY85, or K80 models of evolution.

3. The number of bootstraps is dependent on the size of the data set. It is recommended choosing a minimum of 100 replicates when selecting this option.

4. It is recommended to include a tag (prefix) before the date for easier detection of the numerical field in the sequence name. For example, VEN81-1@1981.49 When a date is unknown, for example the sequence from Sabin vaccine, BEAST can estimate the date by choosing Tip date sampling (sampling with individual priors) in the Tips menu. The sequence is first chosen in the Taxon sets tab and it becomes available in the Apply to taxon set option.

5. There are several models for investigating population size changes under coalescent models, including coalescent exponential growth and Bayesian skyline. There are numerous sources of information dealing with specific coalescent models, including the BEAST-users mailing list (https://groups.google.com/forum/#!forum/beast-users).

6. For polioviruses, a reasonable initial value for clock.rate is 0.011 and upper and lower values in the range of 0.1 and 0.001 respectively. Alternatively, the Gamma distribution can be set as prior distribution.

References

1. Jorba J, Campagnoli R, De L, Kew O (2008) Calibration of multiple poliovirus molecular clocks covering an extended evolutionary range. J Virol 82:4429–4440

2. Kew OM, Mulders MN, Lipskaya GY, da Silva EE, Pallansch MA (1995) Molecular epidemiology of polioviruses. Semin Virol 6:401–414

3. Kew OM, Pallansch MA (2002) The mechanism of polio eradication. In: Semler BL, Wimmer E (eds) Molecular biology of picornaviruses. ASM Press, Washington, DC, pp 481–491

4. Al-Hello H, Jorba J, Blomqvist S, Raud R, Kew O, Roivainen M (2013) Highly divergent type 2 and 3 vaccine-derived polioviruses isolated from sewage in Tallinn, Estonia. J Virol 87:13076–13080

5. Volz EM, Koelle K, Bedford T (2013) Viral phylodynamics. PLoS Comput Biol 9:e1002947

6. Burns CC, Shaw J, Jorba J, Bukbuk D, Adu F, Gumede N, Pate MA, Abanida EA, Gasasira A, Iber J, Chen Q, Vincent A, Chenoweth P, Henderson E, Wannemuehler K, Naeem A, Umami RN, Nishimura Y, Shimizu H, Baba M, Adeniji A, Williams AJ, Kilpatrick DR, Oberste MS, Wassilak SG, Tomori O, Pallansch MA, Kew O (2013) Multiple independent emergences of type 2 vaccine-derived polioviruses during a large outbreak in northern Nigeria. J Virol 87:4907–4922

7. Allman ES, Rhodes JA (2004) Mathematical models in biology: an introduction. Cambridge University Press, New York

8. Yang Z (2006) Computational molecular evolution. Oxford University Press, Oxford

9. Posada D, Crandall KA (1998) MODELTEST: testing the model of DNA substitution. Bioinformatics 14:817–818

10. Guindon S, Dufayard JF, Lefort V, Anisimova M, Hordijk W, Gascuel O (2010) New algorithms and methods to estimate maximum-likelihood phylogenies: assessing the performance of PhyML 3.0. Syst Biol 59:307–321

Chapter 12

Generation of Infectious Poliovirus with Altered Genetic Information from Cloned cDNA

Erika Bujaki

Abstract

The effect of specific genetic alterations on virus biology and phenotype can be studied by a great number of available assays. The following method describes the basic protocol to generate infectious poliovirus with altered genetic information from cloned cDNA in cultured cells.

The example explained here involves generation of a recombinant poliovirus genome by simply replacing a portion of the 5′ noncoding region with a synthetic gene by restriction cloning. The vector containing the full length poliovirus genome and the insert DNA with the known mutation(s) are cleaved for directional cloning, then ligated and transformed into competent bacteria. The recombinant plasmid DNA is then propagated in bacteria and transcribed to RNA in vitro before RNA transfection of cultured cells is performed. Finally, viral particles are recovered from the cell culture.

Key words Recombinant poliovirus, cDNA cloning, Restriction cloning, Bacterial transformation, In vitro transcription, RNA transfection, Virus recovery

1 Introduction

Poliovirus has a positive sense single stranded RNA genome which functions directly as mRNA once it has entered a host cell. The viral genome is translated to produce the viral proteins including the structural (capsid) proteins and the non-structural proteins which are required for replication. The genome then serves as a template for the synthesis of complementary negative strand used in replication of further positive strands to produce infectious virions by packing them with capsid proteins. For this reason the naked RNA of poliovirus is infectious on delivery into the host cell. The in vitro manipulation of cDNA clones to introduce defined changes to the viral genomes and their recovery as infectious virus proved to be an essential research tool to study the roles of specific sequences since it was first carried out with poliovirus in 1981 [1].

Javier Martín (ed.), *Poliovirus: Methods and Protocols*, Methods in Molecular Biology, vol. 1387,
DOI 10.1007/978-1-4939-3292-4_12, © Springer Science+Business Media New York 2016

The protocol explained here involves the generation of recombinant poliovirus cDNA by simply replacing a portion of the poliovirus genome with a synthetic gene containing the altered nucleotide sequences by directional cloning. Alternatively, the mutations can be introduced individually by two-step overlap PCRs, using mutagenesis primers, before restriction cloning [2]. Restriction cloning of synthetic sequences offers a great flexibility in the length of the altered sequence and also in the number of changes introduced simultaneously within that.

The example shown in this protocol has the original Sabin 2 poliovirus domain V sequence in the 5′ noncoding region replaced with S15domV, the domain V region of S15/2, a genetically engineered Sabin 2 strain containing stabilized attenuating mutations [3]. S15domV, a 482 bp long synthetic sequence was ordered from a commercial company and was received in the form of purified plasmid DNA. The insert DNA was designed to fit within two defined unique restriction enzyme recognition sites present in Sabin 2 so as the new fragment could be easily cloned into the place of the excised original sequence.

The full length Sabin 2 poliovirus cDNA is cloned into a pBR322 plasmid vector with a T7 promoter and a ribozyme sequence upstream of the viral genome. The use of the phage T7 RNA polymerase transcription system permits in vitro transcription of viral RNAs just as it would occur in cells naturally infected with virus. A T7 promoter in the construct also allows transcription of the recombinant cDNA by the use of T7 DNA dependent RNA polymerase [4]. A hammerhead ribozyme sequence between the promoter and the viral genome ensures an accurate 5′ end for efficient replication [5]. An essential 3′ poly(A) tail of a minimum of eight As is required downstream of the viral genome so it gets recognized by the host cell as mRNA. An appropriate antibacterial resistance gene in the plasmid vector is used for selection of transfected bacteria.

2   Materials

2.1   Equipment

1. Vortex.
2. Microcentrifuge.
3. Water bath.
4. Gel electrophoresis apparatus consisting of power pack, tank, and casting mold with combs.
5. UV transilluminator with digital camera.
6. Tissue culture microscope, inverted phase-contrast.
7. T25 cm² and T75 cm² tissue culture flasks.
8. CO_2 monitored Incubator, at 37 °C and 5 % CO_2.
9. Dry bath, heating block taking 1.5 ml tubes.

10. 1.5 ml microcentrifuge tubes.

11. Pipettes and compatible disposable filter-tip pipette tips.

12. 90 mm diameter petri dishes.

13. Bacterial spreader.

14. Microwave oven.

15. 15, 250 ml centrifuge tubes.

16. Centrifuge with variable inserts to take 15–200 ml centrifuge tubes.

17. Thermocycler.

18. MSC class 2.

19. 7 ml glass bijoux bottles.

20. Fluorometer or spectrophotometer for nucleic acid quantification.

21. –20 and –80 °C freezers.

22. 4 °C fridge.

23. 1 l measuring cylinder.

24. Scalpel or razor blade.

25. Shaker-incubator with fittings to take 2, 15 ml tubes and 500 ml glass bottles.

26. 250 ml glass bottle with screw cap.

27. Parafilm or 96-well plate with conical bottom for mixing samples with loading dye before electrophoresis.

2.2 Reagents

2.2.1 Cleavage of Vector and Insert DNA with Restriction Endonucleases for Directional Cloning

1. Insert DNA: 1 µg of S15domV plasmid DNA (482 bp in this case, provided in a 2.5 kb pEX-A vector) (*see* **Note 1**).

2. Vector DNA: 1 µg of pT7RbzS2 plasmid DNA (11.1 kb vector).

3. Restriction enzymes MluI and SacI, stored at –20 °C.

4. Endonuclease buffer, 10× provided with the enzymes, stored at –20 °C.

5. Bovine serum albumin (BSA), 10× diluted from the 10 mg/ml, 100× stock provided with the enzymes, stored at –20 °C (*see* **Note 2**).

6. Nuclease-free molecular biology grade water.

2.2.2 Separation and Analysis of DNA Digests by Gel Electrophoresis

1. Insert and vector DNA digests.

2. TAE working solution, diluted from TAE 50× stock prepared with 242 g Tris base, 57.1 ml glacial acetic acid, and 100 ml 0.5 M EDTA topped up to 1 l with water.

3. Agarose, electrophoresis grade.

4. Ethidium bromide, molecular biology grade, 10 mg/ml solution.

5. DNA loading dye, 6×.

6. Molecular weight DNA marker: 100 bp and 1 kb ladders.

2.2.3   Purification of DNA Fragments

(a) With commercially available spin-column gel extraction kit.

    1. Double digests of insert and vector DNA.

(b) With specialist electrophoresis system.

    1. Double digests of insert and vector DNA.

    2. SYBR Green.

    3. DNA marker.

    4. Nuclease-free molecular biology grade water.

2.2.4   Ligation of Insert and Vector DNA

1. Purified DNA fragments.

2. T4 DNA ligase.

3. T4 DNA ligase buffer 10×.

4. Nuclease-free molecular biology grade water.

2.3   Propagation of Recombinant DNA

2.3.1   Transformation of Competent Bacterial Cells with Recombinant DNA

1. 5 μl ligated plasmid DNA.

2. 1 shot/50 μl DH5α High Efficiency Competent *E. coli* cell suspension, kept frozen.

3. 950 μl SOC, Super Optimal Broth (SOB) with Catabolite repression medium, usually provided with competent cells.

4. Luria agar.

5. Tetracycline, stock made to 12.5 mg/ml in ethanol and kept in −20 °C freezer, used at 10 μg/ml in liquid culture or 12.5 μg/ml in agar plates (or other appropriate antibiotic in suitable concentration depending on the vector used).

2.3.2   Screening of Colonies, Plasmid DNA Purification from Overnight Cultures

1. Agar plate with bacterial colonies.

2. Luria broth.

3. Tetracycline.

4. Commercial spin-column kit for plasmid DNA purification (Miniprep kit).

5. Commercial Maxiprep kit.

6. Commercial dsDNA quantification kit using fluorometer (*see* **Note 3**).

2.4   In Vitro Transcription of Viral RNA

2.4.1   Linearization of cDNA Plasmid

1. 2 μg cDNA plasmid.

2. 1 μl endonuclease enzyme Hind III at least 5 units.

3. 2 μl BSA 10× working stock.

4. 2 μl Restriction buffer.

5. Nuclease-free molecular biology grade water.

Also, reagents for electrophoresis as in Subheading 2.1, **item 2**.

2.4.2 Transcription
of Viral RNA

1. 10 µl (1 µg) of linearized plasmid DNA.

2. 10 µl 5× buffer.

3. 2 µl DTT (1 M).

4. 2 µl rNTP (10 mM).

5. 1. 25µl RNAse inhibitor (50 U).

6. 1.5 µl T7 RNA polymerase (30 U).

7. 23.25 µl DEPC-treated nuclease-free molecular biology grade water.

Also, reagents for electrophoresis as in Subheading 2.2.2.

2.5 RNA Transfection
and Virus Recovery

2.5.1 RNA Transfection

1. 2×25 cm^2 tissue culture flasks of confluent Hep2C cell cultures (one per sample plus one for transfection control).

2. PBS-A solution (PBS solution without Ca^{2+} and Mg^{2+}).

3. ~50 µl of RNA from in vitro transcription.

4. 1 ml HBSS (10×) made up from 50 g HEPES, 80 g NaCl, 3.7 g KCl, and 1.25 g Na$_2$HPO$_4$·2H$_2$O per liter in water, pH set to 7.05.

5. 1 ml DEAE (10×) dextran of stock made up from 5 g DEAE-dextran per liter of water.

6. 100 µl D-glucose (10 % D-glucose in water).

7. 8 ml water.

8. Maintenance media made up of EMEM cell culture media supplemented with 1 % serum, 1 % Penicillin-Streptomycin.

2.5.2 Virus Recovery

1. Tissue culture flasks containing the virus.

---

3 Methods

3.1 Construction
of Recombinant DNA
Molecule

The first step in constructing a recombinant DNA molecule is producing insert and vector DNA with cohesive ends by restriction endonuclease digest. The digests are then analyzed by gel-electrophoresis and the desired fragments can be subsequently purified. Most commercially available spin-column kits work by solubilizing the agarose then binding the nucleic acid to silica beads in the presence of a chaotropic salt. An alternative to spin-column purification is the use of a specialist bufferless electrophoresis gel apparatus which allows collection of the DNA fragment after separation directly from the gel. The two DNA molecules are joined by T4 ligase enzyme using the cohesive or sticky ends produced by restriction digest.

3.1.1 Cleavage of Vector and Insert DNA with Restriction Endonucleases

1. Defrost Endonuclease buffer and BSA 10× and vortex them briefly then quick-spin along with enzymes to collect droplets from the lid. Place enzymes on ice.

2. Prepare double digest of insert by combining 1 μg of insert DNA, 2 μl endonuclease buffer, 1 μl BSA 10× solution, and 1 μl of each enzyme in a microcentrifuge tube with water to 20 μl final volume (*see* **Notes 4–7**).

3. Set up double digest of vector similarly as in **step 2** using 1 μg of vector DNA (*see* **Note 1**).

4. Incubate 1–1.5 h at 37 °C in a water bath.

3.1.2 Separation and Analysis of Digested DNA by Gel Electrophoresis

1. To prepare 1 % (w/v) agarose gel for two mini gels add 1 g of agarose to 100 ml of 1× TAE buffer in a 250 ml glass bottle with a screw cap and heat with a loose cap in a microwave oven until the agarose is fully dissolved (*see* **Note 8**). Allow to cool to about 50 °C then add 8 μl of ethidium bromide and mix. Pour the still molten agar into the gel casting tray with the combs in place. Check for air bubbles around the teeth of the comb and remove any with a fine tip pipette tip.

2. Let the gel set at room temperature for about 20 min. When the gel has solidified, immerse it in the electrophoresis tank with 1× TAE buffer and remove the comb. Unused gels can be stored in the fridge for a few days wrapped in cling film to avoid drying out.

3. Load 1 μl of the insert and vector digests mixed with 6× Gel loading dye mixed together in individual droplets on a piece of Parafilm or in a 96-well plate with conical bottom wells, mix a few microliters of water to the sample for loading if necessary to bulk up volume. Include one well of molecular weight marker.

4. Electrophorese at 130 V for 30 min or until the loading dye reaches near the bottom of the gel. The ethidium bromide allows the gel to be examined under UV illumination at any time during the electrophoresis.

5. Examine and photograph under UV light.

3.1.3 Purification of DNA Fragments

(a) with commercial spin-column gel purification kit.

   1. Cut the bands of interest out of the gel with a clean scalpel and transfer gel slices into separate 1.5 ml tubes.

   2. Perform gel clean up by following the protocol in the kit and elute DNA in 30–50 μl elution buffer.

(b) with specialist electrophoresis system.

   1. Prepare, run and collect digest samples according to instructions of the manufacturer (*see* **Note 9**).

3.1.4 Ligation of Insert and Vector DNA

1. Check approximate concentration of aliquots of purified digested vector and insert with spectrophotometer or by agarose gel electrophoresis.

2. Add 50–100 ng digested vector and threefold molar excess of digested insert in a 0.5 ml tube. Make up volume to 8 μl with water if necessary.

3. Mix with pipetting up and down and incubate content at 45 °C for 5–10 min.

4. Add 1 μl 10× T4 ligation buffer and 1 μl T4 DNA ligase enzyme and mix gently.

5. Incubate mixture overnight at room temperature.

3.2 Propagation of Recombinant DNA

To obtain sufficient amounts of the recombinant DNA molecule competent bacteria are transformed with the plasmid DNA. The simpler method of chemical transformation is perfectly adequate to be used here, but if the equipment is available, electroporation can be performed as a quick and highly efficient method. Transformed bacteria are selected by their resistance to appropriate antibiotics. Single colonies, originating from single transformed cells can then be used to inoculate bacterial cultures, again containing antibiotic. A few milliliters of bacterial culture usually yield enough plasmid DNA for initial manipulations and screening. Bacterial cultures can be upscaled to produce sufficient quantities to be used for transcription and subsequent transfection of mammalian cell cultures.

3.2.1 Transformation of Competent Bacterial Cells with Recombinant DNA

1. Thaw 1 tube/50 μl of competent *E. coli* suspension on ice for 10 min.

2. Add 5 μl ligation reaction and carefully tap the tube with a finger to mix content gently. Store remaining ligation reaction at –20 °C.

3. Incubate cells on ice for 30 min.

4. Heat shock bacteria in 42 °C water bath for 20 s (*see* **Note 10**).

5. Add 900 μl room temperature SOC to the cells and incubate at 37 °C with shaking at 250 rpm for 2 h.

6. Prepare Luria agar plates by melting Agar in the microwave and letting it to cool to ~50 °C in a water bath. Transfer agar to MSC and add Tetracycline stock to 1 in 1000 dilution. Swirl to mix then pour about 20 ml per petri dish.

7. Let agar plates to set and agar surface to dry with open lids for about 20 min in the MSC.

8. Warm two plates per transformation in a 37 °C incubator.

9. Vortex bacterial culture briefly then plate 20 and 100 μl onto plates by pipetting 20 or 100 μl culture onto the plate and spreading it over the whole surface with a spreader. Continue spreading until the liquid is absorbed by the agar (*see* **Note 11**).

10. Incubate agar plates overnight, upside down in 37 °C incubator for colonies to appear.

1. Pick 3–6 individual colonies with a sterile 200 µl pipette tip and use it to inoculate 3 ml Luria Broth supplemented with antibiotic in a 15 ml centrifuge tube.

2. Incubate at 37 °C overnight with shaking at 250 rpm.

3. To purify plasmid DNA mini-prep follow the kit manufacturer's instructions. Usually up to 20 µg plasmid DNA can be purified on one mini-prep spin column.

4. Quantitate plasmid DNA and confirm presence of insert by digesting them with the appropriate restriction enzymes or by DNA sequencing.

5. Store plasmid DNA at 4 °C for short term, at –20 °C for storage over a year.

6. When more plasmid DNA is needed, use an aliquot of the mini-prep bacterial culture to inoculate a maxi-prep plasmid culture of Luria broth supplemented with antibiotics. 150 ml culture of low copy number plasmids and 250 ml of plasmids with high copy number is usually used in a 500 ml glass bottle. Follow the plasmid DNA maxi-prep kit manufacturers' instructions for purification of up to a few hundred micrograms plasmid DNA.

3.3 In Vitro Transcription of viral RNA

To obtain infectious poliovirus RNA the plasmid DNA purified from the bacterial culture needs to be linearized by a restriction enzyme whose unique restriction site is downstream of, and as close as possible to the 3′ end of the viral genome so the sequence starting from the T7 promoter to the poly-A tail is accessible for the T7 RNA polymerase to produce RNA. Linearization prevents the production of concatemers that would arise from the copying of a circular template.

3.3.1 Linearization of cDNA Plasmid

1. Set up single digest of 2 µg purified plasmid DNA with 1 µl HindIII, 2 µl endonuclease buffer with water in a 20 µl reaction.

2. Incubate in 37 °C water bath for an hour and then check for completion a 0.5 µl aliquot diluted with water and mixed with loading dye on an agarose gel. Incubate digest further or add enzyme if necessary to complete digest.

3.3.2 Transcription of Viral RNA

1. Defrost reagents for transcription. Keep enzymes and rNTP on ice until use.

2. Prepare 50 µl reaction for transfection of one T25 flask of cells. Combine 10 µl 5× Transcription buffer, 2 µl 0.1 M DTT, 2 µl 10 mM rNTP and 1 µg (10 µl) of linearized plasmid DNA template with 23.25 µl DEPC treated water.

3. Add 1.25 µ RNAse inhibitor (50 U), 1.5 µl of T7 RNA polymerase (30 U) and tap the tube to mix the solution gently.

4. Incubate reaction for 1 h in a 37 °C water bath and run an aliquot of 1 μl on agarose gel to check for presence of RNA (*see* **Note 12**).

3.4 RNA Transfection and Virus Recovery

RNA transfection of cultured mammalian cells is performed mediated by DEAE-Dextran [4]. The cationic polymer binds the negatively charged RNA and the complex is then taken up by the cells via endocytosis. The virus can then be recovered from the cell culture.

3.4.1 RNA Transfection

1. Seed 3×10^6 cells into T25 cm² tissue culture flasks the day before and grow overnight to reach confluency. Have one flask for transfection control, where the following steps are carried out without addition of RNA.

2. Prepare transfection media by mixing 1 ml HBSS (10×), 1 ml DEAE-dextran (10×), 0.1 ml D-glucose with 8 ml water. Keep on ice until use.

3. Mix the freshly transcribed RNA (~50 μl) with 200 μl of transfection media by vortexing briefly and incubate on ice for 20–30 min.

4. Wash cell sheet 2× with PBS-A, and once with serum-free media.

5. Remove final wash and add 250 μl of RNA-Dextran complex. Swirl around the flask gently to cover the whole cell sheet. Add 250 μl of transfer media without RNA to control flask.

6. Incubate for 2–30 min at RT and add 5 ml culture media with 1 % serum and transfer flasks to CO_2 incubator at 35 °C to incubate until full CPE is detected or virus recovery will be attempted.

3.4.2 Virus Recovery

1. Transfer tissue culture flask to −20 or −70 °C freezer and freeze completely. When virus is recovered before full CPE was reached, freeze-thaw cells three time to release virus.

2. Thaw cells completely at RT and transfer everything to a suitable centrifuge tube.

3. Centrifuge for 5 min at $2500 \times g$ to pellet cell debris and decant supernatant to a fresh container.

4 Notes

1. In case more insert DNA is needed than is available it can be retransformed by following the present method from Subheading 3.2 before preparing the recombinant viral cDNA.

2. Some endonucleases are sensitive to low protein concentration so require the addition of BSA to avoid denaturation. These

enzymes are usually supplied with BSA already in the endonuclease buffer or provided separately so it can be diluted and stored as 10× working stock to be added when setting up the reaction.

3. There are several easy to use spectrophotometric or fluorometric systems available for routine quantification of nucleic acids, most of which is accurate enough to use in this protocol.

4. Restriction enzymes are usually provided in 50 % glycerol so make sure that the total volume of enzyme does not exceed 10 % of the final reaction volume so enzyme activity is not affected by the high glycerol content.

5. One unit of restriction enzyme should completely digest 1 μg of substrate in 1 h in 50 μl reaction, but it is common practice to use a few times more to overcome DNA prep variations.

6. When doing double digest adjust number of enzyme units added or incubation time according to activity rating in the buffer used.

7. When there is no buffer in which the two enzymes exhibit >50 % activity then set up reaction with the enzyme that has the recommended buffer with the lower salt concentration first. Incubate to completion by checking an aliquot if all the DNA got linearized, then perform a second digest sequentially after adjusting the buffer conditions. Otherwise purify DNA after first digest on a spin-column then set up reaction with the second enzyme.

8. A linear DNA fragment of a given size migrates at different rates through gels with different agarose concentrations: 0.5 % agarose gel is efficient for separation of fragments of 0.5–20 kb and 2 % gel for 0.1–2 kb. 1 % agarose gel suites most applications for fragments within the range of 0.4–10 kb.

9. Some gel-electrophoresis systems require dilution of double digest reactions to lower salt content prior to loading. Enhanced visibility of DNA bands can be achieved by incubation of samples with SYBR Green prior to loading.

10. For best efficiency of bacterial transformation, follow the instructions provided with the competent cells as incubation times and temperatures might differ slightly from those recommended here.

11. It is advised to leave the transformed bacterial culture at RT until confirmation of growth on the agar plates so repeated plating of bacteria can be done if necessary. To counteract the low transformation efficiency increase the number of bacteria to be plated by centrifuging the cells to concentrate, removing media then resuspending in 100 μl SOC to be plated on new plates.

12. Transcribed RNA ran in 1 % agarose TAE gel will degrade and produce a smeary band and might also show some remaining plasmid DNA if linearization was incomplete. Transfection of poor quality RNA may still produce virus and transcription reactions can be left to proceed for additional time if necessary, but the 1.5 h total incubation normally results in appropriate amount of RNA to transfect cells.

References

1. Racaniello VR, Baltimore D (1981) Cloned poliovirus complementary DNA is infectious in mammalian cells. Science 214:916–919
2. Burrill CP et al (2013) Poliovirus: generation and characterisation of mutants. Curr Protoc Microbiol. Chapter 15:Unit 15H.2
3. Macadam AJ et al (2006) Rational design of genetically stable, live-attenuated poliovirus vaccines of all three serotypes: relevance to poliomyelitis eradication. J Virol 80: 8653–8663
4. Van der Werf S et al (1986) Synthesis of infectious poliovirus RNA by purified T7 polymerase. Proc Natl Acad Sci U S A 83:2330–2334
5. Herold J, Andino R (2000) Poliovirus requires a precise 5′ end for efficient positive-strand RNA synthesis. J Virol 74:6394–6400

Chapter 13

A Rapid Method for Engineering Recombinant Polioviruses or Other Enteroviruses

Maël Bessaud, Isabelle Pelletier, Bruno Blondel, and Francis Delpeyroux

Abstract

The cloning of large enterovirus RNA sequences is labor-intensive because of the frequent instability in bacteria of plasmidic vectors containing the corresponding cDNAs. In order to circumvent this issue we have developed a PCR-based method that allows the generation of highly modified or chimeric full-length enterovirus genomes. This method relies on fusion PCR which enables the concatenation of several overlapping cDNA amplicons produced separately. A T7 promoter sequence added upstream the fusion PCR products allows its transcription into infectious genomic RNAs directly in transfected cells constitutively expressing the phage T7 RNA polymerase. This method permits the rapid recovery of modified viruses that can be subsequently amplified on adequate cell-lines.

Key words Enterovirus, Poliovirus, Genetic engineering, Recombinant virus

1 Introduction

Studying unnatural viruses arisen from modified genomes is a method commonly used to address numerous questions regarding virus characteristics, such as replication cycle, tropism, virulence, pathogenic power, or interactions between viral proteins and host factors. Modifications of viral sequences range from a unique nucleotide (nt) change made by site-directed mutagenesis to large-scale nt substitutions. Additionally, nt deletions can be operated within genomic sequences to decipher the importance and the role of a given nt sequence or of the protein encoded by this sequence. Nt stretches can also be inserted in genomics sequences, particularly to introduce tags into viral proteins. These tags are short amino-acidic patterns that are used subsequently for purification or detection of the tagged proteins [1].

Unnatural viruses are generally obtained by modifying infectious cDNA clones, i.e., reverse-transcribed RNA genomic sequences that have been cloned into bacterial plasmids. The modifications can be done by using site-directed procedures and/or by digestion and

Javier Martín (ed.), *Poliovirus: Methods and Protocols*, Methods in Molecular Biology, vol. 1387,
DOI 10.1007/978-1-4939-3292-4_13, © Springer Science+Business Media New York 2016

ligation methods. These procedures have the advantage of giving rise to new plasmids that contain and preserve the infectious cDNA of the modified viral genome, which can be the basis for subsequent new modifications. Nonetheless, the cloning of large virus sequences is often labor-intensive and time-consuming because of plasmid instability in bacteria. This pitfall constitutes an obstacle for the production of large panels of unnatural viruses.

In order to circumvent this problem, we have developed a PCR-based method that allows the generation of modified full-length enterovirus genomes [2]. This method can be used to construct chimeric genomes from different parental viruses (Fig. 1) and to produce genomes with nt insertions or highly modified sequences (Fig. 2). This method relies on fusion PCR, which

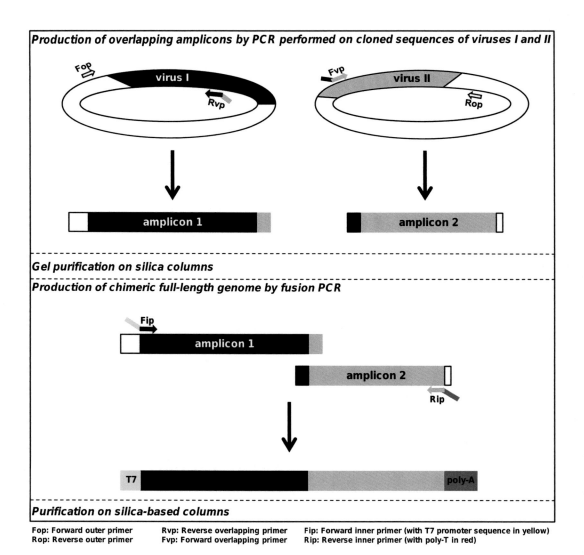

Fig. 1 Overview of the method used to engineer a 2-component chimeric genome

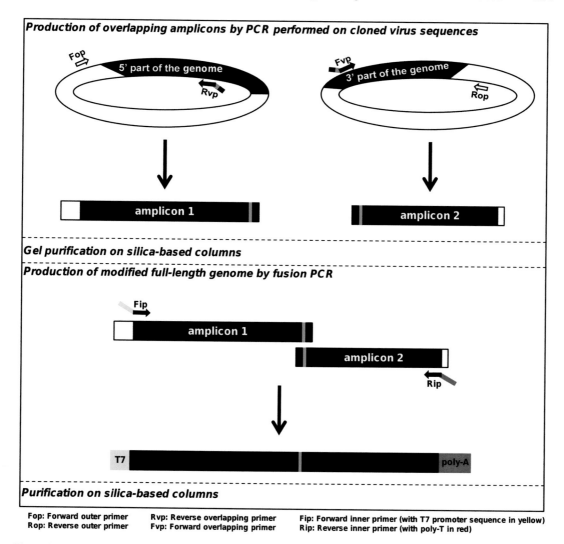

Fig. 2 Overview of the method used to insert a nucleotide stretch in a poliovirus genome

enables the concatenation of several DNA amplicons produced separately [3]. Two adjacent amplicons overlap each other through non-priming 5'-flags of the PCR primers. A T7 promoter sequence added upstream the genome allows the transcription of the fusion PCR product into RNA molecules that act as genuine virus genomic RNA to initiate the viral cycle in transfected cells. In order to avoid the use of in vitro transcription kits, the fusion PCR product is used to transfect cells that constitutively express the T7 RNA polymerase. Enterovirus genome is particularly adapted for this method as it is relatively short (<7500 nt) and does not require 5'-cap to be translated.

Our method consists of five main steps: (1) Design of primers that allow the production of amplicons with overlapping extremities. This step is crucial as the primer sequences determine the location of the breakpoints or insertions. Besides, primers targeting the 5′ and 3′ extremities of the unnatural genome must contain the T7 promoter sequence and a poly-T, respectively. (2) Generation of overlapping amplicons by PCR and purification on silica-based columns. (3) Production of a full-length DNA genome by fusing all the overlapping amplicons by a single PCR and then purified on silica-based columns. (4) Transfection of HEK-293T-T7 cells with the full-length DNA genome to recover the virus (1 week or less post-transfection). (5) Amplification of the virus on HEp-2c cells or other permissive cells according to the produced enteroviruses. The steps (2)–(4) can be performed in 2 working days only and several unnatural viruses can be produced in parallel.

The protocol described below presents the procedures to obtain a chimeric full-length genome with a unique breakpoint or to obtain a full-length genome with a unique nt insertion. These procedures can be adapted and/or mixed to obtain complex chimeric genomes, i.e., genomes with multiples breakpoints, multiple parental strains and multiple insertions. By using this method, we produced tens of modified poliovirus genomes that associate genomic sequences of polioviruses and genomic sequences of other enteroviruses species C. Some genomes contain five breakpoints and were produced by fusing 6 DNA amplicons in a single fusion PCR round [2]. This protocol was also used to create polioviruses whose 3A protein was tagged with the FLAG-tag sequence [4].

2 Materials

2.1 Cell Lines

1. Human HEp-2c cells provided by the Dutch National Institute of Public Health and the Environment (RIVM, Bilthoven, The Netherlands).

2. Human embryonic kidney HEK-293T-T7 cells (provided by Pierre Charneau, Institut Pasteur, Paris, France) are modified cells derivative from HEK-293T cells that constitutively express the phage T7 RNA polymerase (*see* **Note 1**).

2.2 Cell Media

1. Medium A: Dulbecco Modified Eagle Medium (DMEM) supplemented with 10 % fetal calf serum and 2 mM L-glutamine.

2. Medium B: DMEM supplemented with 2 % fetal calf serum and 2 mM L-glutamine.

3. Medium C: DMEM supplemented with 5 % newborn calf serum and 2 mM L-glutamine.

2.3 Plasmids Containing Viral Sequences of Interest

Plasmids with viral sequences that encompass the regions of interest obtained by cloning amplicons produced by RT-PCR with any TA or blunt-end cloning kits and purified following any miniprep procedures (*see* **Notes 2** and **3**). Alternately, such plasmids can be ordered to specialized companies that ensure chemical synthesis and cloning of the required viral sequences.

All plasmids should be sequenced to ensure that they do not contain mutations or deletions.

2.4 Enzymes and DNA Purification Kits

PCR polymerase: Phire Hot Start II DNA polymerase (Thermo Scientific, F-122S or L) (*see* **Note 4**).

DNA purifications are performed on silica-based Wizard SV Gel columns. Purified DNA is eluted in 50 μl of DNase- and RNase-free water.

2.5 Transfection Reagent

Purified T7 promoter tagged-DNA genomes are transfected in permissive cells with a DNA transfection reagent (FuGENE 6 Transfection Reagent from Roche Diagnostics or jetPRIME® from Polyplus Transfection), following the manufacturer recommendations (*see* **Note 5**).

3 Methods

3.1 Outer and Inner Primer Design

1. Forward outer primer (Fop)/Reverse outer primer (Rop) must be designed as 18 to 24-mer targeting plasmid sequences located ~50 nt upstream/downstream the cloned viral sequences (*see* **Note 3**).

2. Forward inner primer (Fip) contains the T7 RNA polymerase promoter sequence followed by the 24 first nucleotides of virus I genome (GGGTAATACGACTCACTATAGGGxxxxxxxxxxxxxx xxxxxxxxxx).

3. Reverse inner primer (Rip) contains 24T followed by 24 nt corresponding to the complementary sequence of the very last nucleotides of virus II genome preceding poly(A) (TTTT TTTTTTTTTTTTTTTTTTTTxxxxxxxxxxxxxxxxxxxxxxxx).

3.2 Design of Overlapping Primer Used to Construct Chimeric Genomes (See Note 6)

1. Forward overlapping primer (Fvp) contains the 50 nt-long sequence of virus I located upstream the breakpoint followed by the 24 nt-long sequence of virus II located downstream the breakpoint (Fig. 3a).

2. Reverse overlapping primer (Rvp) contains the 50 nt-long complementary sequence of virus II located downstream the breakpoint followed by the 24 nt-long complementary sequence of virus I located upstream the breakpoint (Fig. 3a).

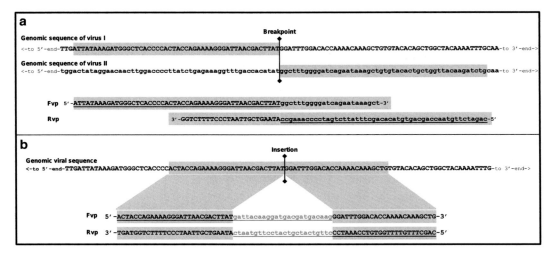

Fig. 3 Design of the overlapping primers to create a chimeric breakpoint (**a**) and to insert a short nucleotide stretch (**b**). The non-priming region of each primer is *underlined*. The inserted nucleotide stretch is indicated in *green*. The Rvp sequences are indicated in the 3′ → 5′ direction

3.3 Design of Overlapping Primer Used to Insert Short Nucleotidic Stretches

1. Forward overlapping primer (Fvp) consists of the inserted sequence flanked by the ~30 nt-long viral sequences located upstream and downstream the insertion point (Fig. 3b).

2. Reverse overlapping primer (Rvp) consists of the sequence complementary to the Forward overlapping primer (Fig. 3b).

3.4 Generation of Purified Overlapping Amplicons

1. The overlapping amplicons are generated by PCR by using the Phire Hot Start II DNA polymerase. PCR is carried out in a final volume of 50 μl that includes 50 ng of plasmid containing viral sequence of interest, 10 μl of 5× Phire Reaction Buffer, 200 μM of each high quality dNTP, 2 nM of each primer (Fop/Rvp or Fvp/Rop) (*see* Subheading 3.2) and 1 μl of Phire Hot Start II DNA polymerase. The thermocycler profile is 30 s at 98 °C followed by 30 cycles including 5 s at 98 °C, 1 min at 60 °C, and elongation at 72 °C (elongation time was adjusted depending on the fragment sizes).

2. Perform an agarose gel electrophoresis of the PCR products (*see* **Note 7**).

3. Excise the bands corresponding to the expected size for each amplicons.

4. Purify each amplicon on silica-based column (*see* **Note 8**).

5. Quantify spectrophotometrically each purified amplicon.

6. Store the purified amplicons at −20 °C until use.

3.5 Fusion PCR

1. Equimolar amounts of the purified overlapping fragments are mixed and joined by fusion PCR performed in a final volume

of 50 μl that includes 10 μl of 5× Phire Reaction Buffer, 200 μM of each high quality dNTP, and 1 μl of Phire Hot Start II DNA polymerase without any primers; the thermocycler profile is 30 s at 98 °C followed by ten cycles including 5 s at 98 °C, 1 min at 56 °C, and elongation at 72 °C (elongation time was adjusted depending on the fragment sizes).

2. After these ten cycles, 2 nM Fip and Rip primers (*see* Subheading 3.1) are directly added in the mixture and PCR resumed with twenty cycles of 5 s at 98 °C, 5 s at 65 °C, and 3 min at 72 °C (*see* **Note 9** for an alternative method).

3. Check the fusion PCR efficiency by loading 5 μl of each PCR product in a 0.8 % agarose gel and perform electrophoresis (*see* **Note 7**).

4. Purify the fusion PCR products on silica-based columns (*see* **Notes 8** and **10**).

3.6 Cell Propagation

1. HEK-293T-T7 and HEp-2c cells are cultivated in T flasks in medium A and C (*see* Subheading 2.2), respectively. Cell passaging can be done by using a dilution factor ranging from 1/20 to 1/10.

2. As HEK-293T-T7 cells are poorly adherent, neither trypsinization nor cell scratching is required: these cells can be detached by gently pipetting fresh propagation medium onto the monolayer. HEp-2c monolayers must be detached by trypsinization.

3. For both cell lines, flasks must be incubated at 37 °C with 5 % CO_2.

3.7 Transfection of HEK-293T-T7 Cells to Recover Viruses

As transfection is expected to give rise to infectious virions, the following steps must be conducted in biosafety laboratories (*see* **Note 11**).

1. Dilute 10 μl of purified fusion PCR into 50 μl jetPRIME® buffer. Mix by vortexing.

2. Add 1 μl jetPRIME®, vortex for 10 s, spin down briefly.

3. Incubate for 10 min at RT.

4. Pour the transfection mixture directly into empty well. One well must receive transfection mixture without purified fusion PCR to serve as control.

5. Add 1 ml of medium B (*see* Subheading 2.2) containing 2×10^5 cells per ml.

6. Incubate the plates in an incubator at 37 °C under 5 % CO_2 atmosphere.

7. Each day, observe each well under a microscope to check cytopathic effects.

8. When cytopathic effects are observe, or 5 days post-transfection if no cytopathic effects have occurred, collect the supernatant in a sterile tube by pipetting up and down to detach the monolayer.

9. Centrifuge the tubes at $1000 \times g$ for 5 min.

10. Collect the supernatants in new sterile tubes and discard the cell pellets.

11. Use 500 µl of clarified supernatant to infect HEp-2c cells (*see* Subheading 3.8) and store the remaining supernatant at −80 °C.

3.8 Virus Amplification in HEp-2c Cells

1. The day before, seed HEp-2c cells (or other permissive cells according to the produced enteroviruses) in T25 flasks. Prepare one flask per transfected well plus an additional flask to serve as non-infected control. The cells must be subconfluent on the day of infection.

2. Remove the T25 medium.

3. Pour 500 µl of the clarified transfection supernatant in each flask. Add 500 µl of medium B in the control flask.

4. Incubate 30–60 min at 37 °C under 5 % CO_2 atmosphere.

5. Add 6 ml of medium B (*see* Subheading 2.2) in each flask and place in an incubator at 37 °C under 5 % CO_2 atmosphere.

6. Each day, observe each well under a microscope to check cytopathic effects.

7. When the monolayer has been destroyed, collect the supernatant in a sterile tube and clarify by centrifugation at $1000 \times g$ for 5 min. Aliquot the supernatant and store at −80 °C until use.

3.9 What If No Cytopathic Effects Have Been Observed on HEp-2c?

The modifications made in viral genomes can have deleterious effects, so leading to non-functional genomes or to viruses that grow very slowly. To determine whether the absence of cytopathic effects observed for a given genome must be attributed to a complete lack of functionality of this genome in the one hand, or to a decrease in the replication speed of the virus derived of this genome on the other hand, an additional passage on HEp-2c must be conducted, following the procedure described below.

1. Collect the clarified infection supernatant of each well where no cythopathic effects have been observed 7 days post-transfection in a sterile tube.

2. Clarify by centrifugation at $1000 \times g$ for 5 min.

3. Prepare a new sub-confluent T25 HEp-2c flask per supernatant to be tested and an additional one as control.

4. Aspirate the supernatant and add 500 µl of the clarified infection supernatant.

5. Incubate 30–60 min at 37 °C under 5 % CO_2 atmosphere.

6. Aspirate the inoculums and rinse twice with 3 ml PBS (*see* **Note 12**).

7. Add 6 ml of medium D (*see* Subheading 2.2) in each flask and place in an incubator at 37 °C under 5 % CO_2 atmosphere.

8. Each day, observe each well under microscope to check cytopathic effects.

9. If cytopathic effects are observed, collect the supernatant in a sterile tube and clarify by centrifugation at $1000 \times g$ for 5 min. Aliquot the supernatant and store at −80 °C until use.

10. If no cytopathic effects are observed 7 days after the second round of infection, perform three freeze–thaw cycles by placing the flasks into a freezer until icing and into an incubator until complete defrosting, alternately.

11. Collect the supernatant in sterile tubes and centrifuge at $1000 \times g$ for 5 min.

12. After RNA extraction by using any dedicated procedures (*see* **Note 13**), presence or absence of viral RNA in passage two cell cultures can be tested by using any molecular methods (*see* **Note 14**).

4 Notes

1. We developed our protocol by using HEK-293T-T7 cells, which are not commercially available. Any cell lines that express the T7 RNA polymerase, could probably replace these cells, provided that these cells are permissive to the replication of the enteroviruses produced by PCR fusion.

   Alternately, the T7 promoter can be replaced by another promoter (such as the CMV promoter for instance) to allow the expression of the genome in any mammalian cell lines permissive to polioviruses in which this promoter is active. As the CMV promoter length is >500 nucleotides, it cannot be inserted in the Forward inner primer but can be included as an overlapping amplicons during the fusion PCR step.

   HEK-293T-T7 cells are permissive to poliovirus infection but cannot be infected by some non-polio enteroviruses (for instance coxsackieviruses A11, A13 and A17). However, these cells allowed us to recover recombinant non-polio enteroviruses species C by achieving the generation of RNA from the transfected fusion PCR products and a single round of production of infectious particles. The viruses were subsequently amplified by infecting HEp-2c cells, which are permissive to enterovirus C infection. In this case, no CPE was observed after tranfection of HEK-293T-T7 (*see* 3.7.8).

2. In order to avoid the undesirable production of parental viruses, the plasmids used as matrix must not be able to promote viral expression in HEK-293T-T7 cells. Therefore, we used plasmids in which only virus partial genomic sequences have been cloned [2, 4]. Plasmid containing infectious viral cDNA (complete viral genomic sequence) cannot be used since it can lead to infectious viruses following transfection in permissive cell lines, even without any promoter sequences. In all cases, the ability of the used plasmids to promote viral expression can be tested easily by transfection in HEK-293T-T7 cells, by following the procedure exposed in Subheading 3.8. If required, after producing the overlapping amplicons, plasmids can be eliminated specifically by using DpnI, a restriction enzyme that targets methylated DNA [5].

3. In our protocol, the amplicons are generated by PCR by using cloned viral sequences as matrix. In pilot experiments, we successfully produced full-length genomes by using amplicons generated by RT-PCR directly from viral RNA extracted from cell culture supernatant [2]. This procedure avoids the cloning steps and gives rise to amplicons that probably reflect better the genetic diversity of the viral quasi-species. However, RT-PCR performed to amplify long fragments (>4500 nt) has frequently a low yield. Producing overlapping amplicons directly by RT-PCR does not require neither Forward outer primer nor Reverse outer primer: the RT-PCR step is performed with the Forward inner and Reverse inner primers.

4. Fusion PCR (described in Subheading 3.5) requires a polymerase with a high processivity. In pilot experiments performed to test different polymerases, the Phire Hot Start II DNA polymerase gave the best results. For optimal results always use 200 µM of each high quality dNTP (Thermo Scientific, R0191). The PCR performed to generate the overlapping amplicons (described in Subheading 3.4) can be done by using any polymerase, especially high-fidelity ones.

5. Classical transfection on cells monolayer or the reverse transfection protocols can be used. In reverse transfection protocol, nucleic acid is added to plate wells, followed by transfection reagent. Cells in suspension are added subsequently, after the formation of the complexes in the wells. In pilot experiments, Lipofectamine 2000 (from Life Technologies) was found toxic for cells. Its use thus requires to remove the medium containing the lipofectamine/DNA complexes after a few hours of incubation. As HEK-293T-T7 are poorly adherent, this step often leads to a disruption of cell monolayers. By contrast, no toxicity was observed by using FuGENE 6 or jetPRIME.

6. In order to avoid frame-shifting due to truncated or incomplete primers, the overlapping primers Fvp and Rvp can be ordered as PAGE-purified primers.

7. Agarose gel electrophoresis of long DNA fragments takes a long time. Check that the temperature of the electrophoresis buffer does not increase too much, in order to avoid DNA denaturation that would compromise BET-staining of the amplicon.

8. Recovery of amplicons >4500 nt in length is generally low after purification on silica-based columns. We tested two different kits from two providers: the Qiaquick Purification Kit (Qiagen) and the Wizard SV Gel and PCR Clean-Up System (Promega). In our hands, the second one gave better results for long amplicons. In order to increase the recovery of long amplicons, similar PCR can be performed in parallel and pooled onto the same column.

9. All the genomes were obtained successfully by a single round of fusion PCR regardless of the number of overlapping fragments. However, alternatively, PCR fusion resulting from a mixture of overlapping amplicons with highly heterogeneous sizes can be optimized: pour 2 μl of the solution obtained at step 1 (after the ten cycles without primers used for PCR fusion) into a new tube and add 2 nM of Fop and Rop primers in a final volume of 50 μl that included 10 μl of 5× Phire Reaction Buffer, 200 μM of each high quality dNTP, and 1 μl of Phire Hot Start II DNA polymerase. Then perform PCR with 20 cycles of 5 s at 98 °C, 5 s at 65 °C, and 3 min at 72 °C.

10. Fusion PCR product must not be purified after gel excision. In pilot experiments, we found that cell transfection with fusion PCR products that have been gel-excised before purification did not allow the recovery of viruses.

11. Wild polioviruses must be handled in biosafety level 2 laboratories. Unnatural polioviruses arisen from modified genomes must be handled in convenient biosafety level 2 or 3 laboratories, according to local regulation for genetically modified microorganisms.

12. Rinsing with PBS is crucial to remove DNA from transfection that could remain in the supernatant, so as to avoid false-positive detection of viral RNA during the following steps.

13. Any methods suitable for viral RNA extraction can be used.

14. Many real-time or end-point molecular methods have been developed to detect specifically enterovirus genomes. We use a real-time RT-PCR that targets conserved sequences within 5′-UTR [6].

Acknowledgments

This work was supported by the Transverse Research Program PTR-276, the *Agence Nationale pour la Recherche* (ANR 09 MIEN 019), the *Fondation pour la Recherche Médicale* (FRM DMI20091117313).

References

1. Terpe K (2003) Overview of tag protein fusions: from molecular and biochemical fundamentals to commercial systems. Appl Microbiol Biotechnol 60:523–533
2. Bessaud M, Delpeyroux F (2012) Development of a simple and rapid protocol for the production of customized intertypic recombinant polioviruses. J Virol Methods 186:104–108
3. Horton RM, Hunt HD, Ho SN, Pullen JK, Pease LR (1989) Engineering hybrid genes without the use of restriction enzymes: gene splicing by overlap extension. Gene 77:61–68
4. Teoule F, Brisac C, Pelletier I, Vidalain PO, Jegouic S, Mirabelli C, Bessaud M, Combelas N, Autret A, Tangy F, Delpeyroux F, Blondel B (2013) The golgi protein ACBD3, an interactor for poliovirus protein 3A, modulates poliovirus replication. J Virol 87:11031–11046
5. Nelson M, McClelland M (1989) Effect of site-specific methylation on DNA modification methyltransferases and restriction endonucleases. Nucleic Acids Res 17(Suppl):r389–r415
6. Monpoeho S, Dehee A, Mignotte B, Schwartzbrod L, Marechal V, Nicolas JC, Billaudel S, Ferre V (2000) Quantification of enterovirus RNA in sludge samples using single tube real-time RT-PCR. Biotechniques 29:88–93

Chapter 14

Methods to Monitor Molecular Consistency of Oral Polio Vaccine

Konstantin M. Chumakov

Abstract

Replication of viruses leads to emergence of mutations and their content in viral populations can increase by selection depending on growth conditions. Some of these mutations have deleterious effect on vaccine safety, such as neurovirulent reversions in the 5′-UTR of attenuated Sabin strains of poliovirus. Their content in vaccine batches must be tightly controlled during vaccine manufacture to ensure safety of the product. This chapter describes a quantitative molecular procedure called mutant analysis by PCR and restriction enzyme cleavage (MAPREC) that is used to monitor content of neurovirulent revertants in Oral Polio Vaccine (OPV). The method can be used for quantitative analysis of any other mutation in a viral population.

Key words Reversion to virulence, Quality control of Oral Polio Vaccine, Quantitative analysis of mutations, PCR, Deep-sequencing

1 Introduction

Low fidelity of replicases of RNA viruses result in emergence of all possible mutations in viral populations during their growth in vitro and in vivo. Lethal mutations and mutations lowering fitness of the virus are eliminated by selection, while the content of those mutations that provide virus an advantage under particular growth conditions increases. Some mutations can be enriched even if they do not increase virus fitness. This can happen if a neutral or near-neutral mutation occurs in the same RNA molecule with an unrelated fitness-increasing mutation that drives the selection. Therefore any change in the mutational composition is undesirable because it could lead to unpredictable phenotypic changes. Consequently, to ensure consistency of manufacture of viral vaccines it is important to monitor profiles of mutations. Mutations that are known to lead to increased virulence must be closely monitored and not allowed to exceed a certain limit that was found to be safe in prior clinical tests. Vaccine batches containing excessive amount of mutations

Javier Martín (ed.), *Poliovirus: Methods and Protocols*, Methods in Molecular Biology, vol. 1387, DOI 10.1007/978-1-4939-3292-4_14, © Springer Science+Business Media New York 2016

shall not be used for vaccination. For example, mutations in Sabin strains of poliovirus that have been shown to affect attenuation as measured by monkey neurovirulence test (MNVT) [1] are $472_{U \to C}$ in type 3 [2, 3], $481_{A \to G}$ in type 2 [4], and $480_{G \to A}$ or $525_{U \to C}$ in type 1 [5]. All these mutations were found in OPV bulks and if their contents exceeded a certain level, the bulks were found to fail MNVT [6, 7]. Therefore these mutations are monitored by MAPREC to ensure that neurovirulence of newly manufactured bulks was within acceptable limits.

All other mutations for which no biological function is known or no connection to safety of vaccine was demonstrated could be used to monitor consistency to reveal potential breaches in manufacturing process. Unexpected increase of the content of such mutations does not necessarily mean that the vaccine batch cannot be used for immunization. However, it must trigger investigation and troubleshooting of the manufacturing process [6].

Monitoring content of mutations may be very helpful during vaccine development. First, it can reveal genetic instabilities that may or may not be linked to adverse properties of the product. If mutations are found to be undesirable, their content must be controlled. In contrast, if a selectable mutation does not adversely affect vaccine properties, it could be incorporated into vaccine the strain to increase viral fitness and yield, and to prevent further genetic changes by means of passive selection (see above). Identification of such selectable fitness-increasing mutations can be done by serial passaging under conditions of vaccine manufacture followed by quantitative analysis of mutations.

There is a number of methods that could be used for the discovery and quantification of mutants in viral populations [8–12]. As a rule of thumb, methods that allow screening of the entire genome tend to be less sensitive, while sensitive methods focus only at a small part of the genome or one mutation only. To date, MAPREC is possibly the most sensitive method for quantification of mutants but it can detect only one mutation at a time. It is not suitable for discovery of unknown mutations, but can be successfully used for accurate quantification of very small quantities of mutants in viral populations. The method had been elaborated in great detail, and its utility for vaccine quality control evaluated in the International Collaborative Study organized by the World Health Organization (WHO). It has been recommended by the WHO Expert Committee on Biological Standardization as an in vitro method of choice for lot release of OPV of all three serotypes [1]. The main focus of this chapter will be on describing procedure for conducting MAPREC tests.

Other procedures that could be used for monitoring quantities of mutants in batches of viral vaccines include quantitative real-time PCR. The protocols for these methods will not be described here because they were not fully validated yet. Recently a new approach for monitoring genetic consistency of viral vaccines was proposed. It is based on deep sequencing of cDNA prepared from vaccine batches and allows each nucleotide in viral genome to be read hundreds of thousands times. The method combines screening of the entire genome in search for new mutations with their accurate quantification. Deep sequencing platforms continue to evolve rapidly, and no established validated protocol for using deep sequencing exists so far. Therefore in this chapter we only provide a detailed technical protocol for MAPREC test.

1.1 MAPREC Principle and Design

MAPREC principle is similar to restriction fragment length polymorphism (RFLP) analysis, except for it is quantitative and employs PCR primers modified to create recognition site(s) for restriction endonucleases. In case of type 2 and type 3 polioviruses only one mutation is analyzed (481_G and 472_C, respectively), and only one primer is modified to create a restriction site affected by the mutation of interest. In case of type 1 poliovirus two mutations (480_A and 525_C) located across from each other in a hairpin structure of viral RNA are tested. To enable this, both primers are modified, forward primer to create a site affected by mutation at nucleotide 480, and reverse primer modified to test for 525_C. Schemes of MAPREC procedures for all three serotypes are shown on Figs. 1, 2, and 3.

Since double-stranded DNA is repeatedly denatured—reannealed during PCR procedure, complementary strands with different mutations may reassort to form heteroduplexes that may or may not be susceptible to restriction enzyme digestion. To avoid strands reassortment, PCR amplification is conducted with significant excess of one primer so that predominantly single-stranded DNA is synthesized. Next, one-step second-strand DNA synthesis is performed using radiolabeled primer so that the resulting double-stranded DNA is composed of completely complementary stands one of which is end-labeled with radioisotope.

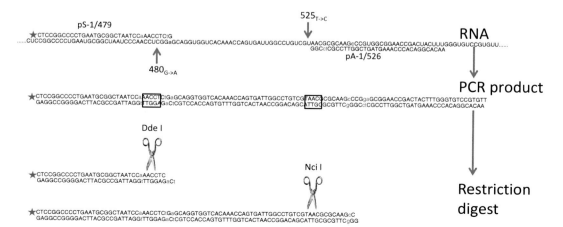

Fig. 1 Scheme of MAPREC assay for 480-A and 525-C in batches of type 1 OPV

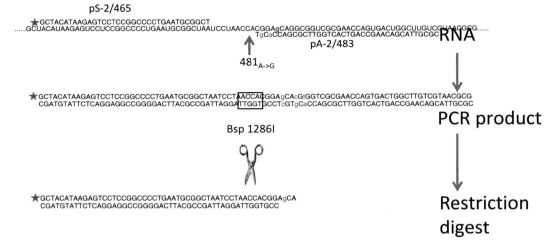

Fig. 2 Scheme of MAPREC assay for 481-G in batches of type 2 OPV

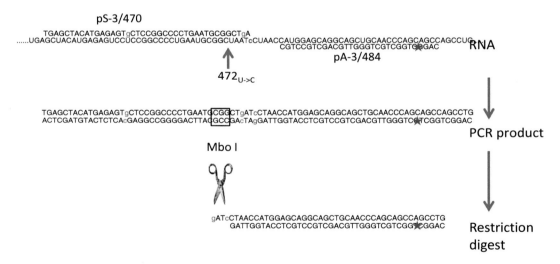

Fig. 3 Scheme of MAPREC assay for 472-C in batches of type 3 OPV

MAPREC test includes several control and reference samples that are used for test validation and evaluation of its results. The controls include a blank sample (with no viral cDNA added) to test for potential contamination, and 100 % mutant control (synthetic DNA with 100 % 472_C) to test for completeness of restriction enzyme digestion. To calibrate mutants content a synthetic DNA sample with mutants level equal to the target content for vaccine batches is included. The content of mutants in this International Standard (IS) sample (that may be tested in duplicate to increase accuracy) is used as a denominator for calculating ratio that is used for evaluation and interpretation of MAPREC results. If the ratio is greater than 1.0 by a statistically significant margin (MAPREC procedure is performed at least five times to enable statistical evaluation of the data), the sample contains more excessive amount of mutants. The test also includes two viral controls—one "passed" viral control and one "failed" viral control that represent virus stocks with mutant contents that are below and above the acceptable level. The samples serve for validation of test performance, and are expected to pass and fail, respectively. Alternatively, the test is considered to be invalid because it did not correctly identify these samples.

2 Materials

In addition to commercially available reagents, MAPREC procedure uses custom-synthesized PCR primers as well as reference materials and standards available from WHO reference laboratory at the National Institute for Biological Standardization and Control (NIBSC) in Potters Bar, UK.

Standard MAPREC procedure uses primers labeled with radioactive isotope ^{32}P (see **Note 1**). Details of radioactive imaging of polyacrylamide gels and determining radioactivity in DNA bands depend on the equipment that is available. The simplest solution is to use X-ray film for imaging and location of radioactive bands in the gel followed by their excision and determining radioactivity using a liquid scintillation counter. There are a number of flat-bed radioactive imagers (e.g., phosphorimagers) that allow accurate quantification of radioactivity in two-dimensional media such as slab polyacrylamide gels used in MAPREC. They can be used if their dynamic range covers the level of radioactivity present in MAPREC gels.

1. General reagents and disposables used for MAPREC include the following materials or their equivalent from other vendors: Eagle MEM (Life Technologies), Sephadex spin-columns (Millipore PD-10), SDS (Sigma), buffer-saturated phenol (Life Technologies), isopropanol (Sigma), ultrapure DEPC-treated water (Life Technologies), reverse transcriptase kit (SuperScript III, Life Technologies), 10 mM dNTP solutions (Life Technologies), T4 polynucleotide kinase (Life Technologies), Restriction endonucleases Dde I, Nci I, Bsp1286I, Mbo I (Life Technologies), acrylamide–bis-acrylamide solution (Protogel, National Diagnostics), 0.5× TBE (Life Technologies), Hybond N+ membrane (GE Healthcare Life Sciences).

2. RNA from viral sample can be isolated using phenol saturated with pH 7.5 tris–HCl buffer with 1 % SDS or special kits for RNA isolation (e.g., Qiagen RNeasy kit).

3. Mo-MuLV Reverse transcriptase, 200 U/μl (e.g., Life Technologies SuperScript III or equivalent). Random hexa-

nucleotide primer is used in cDNA reactions. A mixture of four dNTPs for cDNA reaction must contain dATP, dCTP, dGTP, and dTTP at 10 mM concentration.

4. T4 polynucleotide kinase is used to label 5′-end of one primer. The concentration is 10 U/μl. Gamma-[^{32}P]-adenosine-triphosphate (γATP, 6,000 Ci/mmol, 10 mCi/ml) serves as the donor of radioactive phosphate group added to the 5′-end of the primer (see Note 2).

5. MAPREC primers are designed to amplify a short region of poliovirus genome with the nucleotide of interest located near the 3′-end of one primer. The recognition site of restriction endonuclease used to discriminate between vaccine and mutant sequences is composed from the 3′-end of primer and newly synthesized DNA chain, and the primer part may contain a nucleotide substitution depending on the substrate specificity of the restriction enzyme. Nucleotide sequences of MAPREC primers used to analyze mutations in all three serotypes of vaccine poliovirus are listed in Table 1.

Nucleotides shown in lowercase letters were modified compared to the viral sequence to create or destroy recognition sites for restriction endonucleases. Primers marked with an asterisk are those used for radiolabeling. Primers must be purified by gel electrophoresis, HPLC, or another procedure that ensures that they contain only oligonucleotides of intended size and sequence. Sense primer solutions are used at 30 μg/ml, antisense at 3 μg/ml to ensure that predominantly single-strand cDNA is synthesized during RT reaction.

Table 1
PCR primers used for MAPREC test

Serotype	Polarity	Name	Start	Nucleotide sequence	End
1	Sense	pS-1/479*	445	CTCCGG CCCCTGAATG CGGCTAATCC aAACCTCtG	479
1	Antisense	pA-1/526	560	AACACGGACA CCCAAAGTAG TCGGTTCCGC tcCGG	526
2	Sense	PS-2/465*	431	GCTACATAAG AGTCCTCCGG CCCCTGAATG CGGCT	465
2	Antisense	PA-2/483	520	CGCGTTACGA CAAGCCAGTC ACTGGTTCGC GACCaCgT	483
3	Sense	pS-3/470	431	TGAGCTACAT GAGAGTgCTC CGGCCCCTGA ATGCGGCTgA	470
3	Antisense	pA-3/484*	513	CAG GCTGGCTGCT GGGTTGCAGC TGCCTGC	484

Table 2
Restriction endonucleases used in MAPREC test

Serotype	Nucleotide	Enzyme
Type 1	480-A (revertant)	Dde 1
	480-G (vaccine)	BstX 1
	525-C (revertant)	Nci 1
	525-U (vaccine)	Mro 1
Type 2	481-G (revertant)	Bsp 1286I
	480-A (vaccine)	Afl 3
Type 3	472-C (revertant)	Mbo 1
	472-U (vaccine)	Hinf 1

6. Taq polymerase is used for PCR and for second-strand synthesis using radiolabeled primer. Perkin Elmer AmpliTaq® Kit or similar enzyme can be used according to the manufacturer's protocol and using reagents from the kit.

7. Restriction endonucleases used to detect mutant and vaccine nucleotides in three serotypes of Sabin viruses are listed in Table 2 (see **Note 3**). Buffers and other reagents used in restriction enzymes reactions must be a part of respective reagent kits, and all conditions should be according to the manufacturer's protocol.

8. Acrylamide–bis-acrylamide mixture (20:1) is used. The gels are prepared in tris–borate–EDTA buffer and polymerized using TEMED and ammonium persulfate. 30×40 cm glass plates are treated with silicone solution (see **Note 4**), and used with 0.2 mm teflon spacers and combs to create 4–5 mm wells separated by ~2 mm walls. A mixture of tracer dyes bromophenol blue and xylene cyanol in glycerol is added to each sample prior to loading on the gel.

9. Pattern of radioactive DNA band in the gel can be revealed by exposure on X-ray film (Kodak) or phosphorimager (Storm 860, Molecular Dynamics).

10. Several types of references and control samples are used for MAPREC. Two controls represent synthetic DNA of the sequence similar to the PCR amplicon. The first serves as a 100 % revertant control to check for completeness of restriction endonuclease digestion. The second DNA (called International Standard) contains about 1 % of mutants (see **Note 5**), and serves for ratiometric calibration of results providing the denominator for calculating ratios. In addition to DNA controls, two viral references are used. The contents of

Table 3
Reference materials used in MAPREC test

	Type 1	Type 2	Type 3
Passed virus reference	00/416	97/756	96/572
Failed virus reference	00/422	98/596	96/578
100 % revertant control	00/410	98/524	94/790
International standard (IS)	00/418	97/758	95/542

mutants in the virus references is above and below the threshold of mutants in acceptable vaccines (*see* **Note 6**). They serve for determining validity of the test: if the failed virus reference does not fail or the passed virus reference does not pass, the test is considered invalid because of the failure to correctly identify these samples. Table 3 lists the control and reference samples used for all three serotypes of OPV. They are available on request from WHO/NIBSC.

3   Methods

3.1   RNA Extraction

1. Take 0.45 ml virus aliquot and, if necessary, clarify by low-speed centrifugation. If the virus medium contains alcohol-insoluble salts (e.g., phosphates), remove them prior to extraction by dialysis against Eagle MEM or passing though Sephadex spin-columns (*see* **Note 7**).

2. Add 50 μl of 10 % SDS and 500 μl of buffer-saturated phenol and vortex for 30–60 s.

3. Centrifuge for 5 min at $15,000 \times g$ at room temperature, collect the upper phase and extract it one more time with phenol.

4. Collect the upper phase after the second extraction and add 2 volumes of –20 °C isopropanol, and keep at –20 °C overnight. The alcohol suspension of RNA can be stored at –20 °C for at least 1 year.

3.2   Reverse Transcription

1. Take 0.3 ml of RNA suspension in isopropanol from Subheading 3.1 and centrifuge for 15 min at $15,000 \times g$ at +4 °C (*see* **Note 8**).

2. Carefully remove the supernatant and wash the RNA pellet with 0.3 ml 70 % ethanol to remove traces of phenol and centrifuge again.

3. Carefully remove the supernatant and dry the pellet in vacuum for 5–10 min.

4. Prepare the reaction mixture containing the following components (per one reaction): 12.8 μg of deionized DEPC-treated water, 4 μl of 5× buffer supplied in the reverse transcriptase kit, 0.2 μl of 0.1 M DTT, 0.25 μl of each dNTP (10 mM), 1 μl of 50 ng/μl random dN_6, and 1 μl of reverse transcriptase (200 U/μl).

5. Reconstitute the RNA pellet in the cDNA reaction mixture and incubate for 60 min at 37 °C.

6. Incubate for 5 min at 94 °C to inactivate residual RTase (*see* **Note 9**). cDNA prepared in this way can either be used directly, or stored at –20 °C for at least 1 year.

3.3 PCR Amplification

1. Besides the test sample, each MAPREC test includes the following controls and references:

   • Blank with no cDNA (water or mock-cDNA made without RNA).

   • 100 % revertant control.

   • Two replicates of International Standard.

   • Passed viral reference.

   • Failed viral reference.

   The references listed in Table 3 are available on request from WHO/NIBSC.

2. Prepare the PCR mixture containing the following components (per one reaction): 26.3 μl of deionized DEPC-treated water, 4.5 μl of 10× buffer supplied with Taq polymerase kit, 1 μl of each dNTP (10 μM), 5 μl of sense and 5 μl of antisense primer, and 0.2 μl of AmpliTaq® DNA polymerase. The names and concentrations of primers for all three types of poliovirus are given in Table 4.

3. Put 45 μl of this mixture into each PCR tube, add 5 μl of cDNA, and overlay with mineral oil if required by the thermocycler manual (*see* **Note 10**).

4. Incubate in the thermocycler for 40 cycles, each consisting of 30 s at 94 °C, 15 s at 55 °C, and 3 min at 65 °C, followed by

Table 4
Concentrations of primers used for asymmetric PCR reaction

Type	Sense primer	Antisense primer
1	pS-1/445 (3 μg/ml)	pA-1/526 (30 μg/ml)
2	pS-2/431 (3 μg/ml)	pA-2/483 (30 μg/ml)
3	pS-3/470 (30 μg/ml)	pA-3/484 (3 μg/ml)

indefinite incubation at 4 °C. Other conditions may be used if shown to produce comparable result. For instance, two-step PCR amplification consisting of melting for 15 s at 94 °C and extension for 1 min at 65 °C were successfully used on Perkin/Elmer thermocycler 9600 (*see* **Note 11**).

3.4 Radiolabeling of Oligonucleotide Primer

1. Prepare the reaction mixture by adding 2.67 μl of 10× polynucleotide kinase buffer (a part of T4-PNK kit), 2.67 μl of 1 U/μl T4 polynucleotide kinase, and 17.33 μl of a primer at 3 μg/ml.

2. Incubate for 30 min at 37 °C.

3. The labeled primer can be used immediately, or stored frozen for few days.

3.5 Second-strand DNA Synthesis

1. Prepare dNTP mixture by mixing 30.8 μl of deionized DEPC-treated water, 6.5 μl of 10× polymerase buffer (from Taq polymerase kit), and 1.5 μl of each dNTP (10 mM) to give the total volume of 50 μl.

2. Add 26.7 μl of labeled primer from Subheading 3.4 and 2.67 μl of AmpliTaq® DNA polymerase (5 U/μl).

3. Take 10 μl from each PCR reaction (*see* Subheading 3.3) and add 2 μl of the radiolabeled primer from the previous step.

4. Incubate for 5 min at 72 °C.

3.6 Restriction Endonuclease Digestion

1. Take two 5 μl aliquots from each reaction (see previous step).

2. Add 1 μl of restriction endonuclease to one sample in each pair, and leave the second one as a control. The following enzymes are used: Dde I for quantification of 480-A and Nci I of 525-C in type 1 poliovirus, Bsp1286I for 481-G in type 2, and Mbo I for 472-C in type 3 (*see* **Note 12**).

3. Incubate both aliquots for 60 min at 37 °C.

3.7 Polyacrylamide Gel Electrophoresis

1. Prepare 33 cm×42 cm×0.2 mm 7.5 % polyacrylamide gel (20:1 acrylamide–bis-acrylamide) containing 1× TBE buffer.

2. Install the gel into electrophoresis apparatus with 0.5× TBE electrode buffer.

3. Add 1 μl of tracer dye solution in 50 % glycerol to each aliquot from Subheading 3.6 and load into the wells.

4. Perform electrophoresis for about 60 min at 60 W (ca. 1500 V) until xylene cyanol tracer migrated about 18 cm.

5. Remove top glass plate, place DE81 DEAE-paper or Hybond N+membrane over the gel, and take the gel with paper from the bottom glass plate.

6. Cover the gel with plastic film (e.g., Saran wrap) and dry the gel in vacuum at 80 °C.

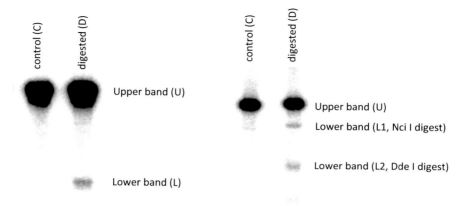

Fig. 4 Examples of electrophoretic separartion of restriction digests of PCR products in MAPREC assay for type 2 and type 3 OPV (left panel) and type 1 OPV (right panel)

7. Expose the gel to X-ray film or use other radio-imaging equipment.

8. If using autoradiography, cut out pieces of polyacrylamide gel containing DNA bands and determine radioactivity using liquid scintillation counter. If using phosphorimager or other direct imaging equipment use it to determine radioactivity in the DNA bands.

3.8   Data Analysis

Control (C) and enzyme-digested (D) portions of each sample are electrophoresed in adjacent wells of polyacrylamide gel. Figure 4a, b show DNA bands representing PCR-amplified DNA and its fragment(s) produced by restriction endonuclease digestion for polioviruses of type 2 and 3 (Fig. 4a) and type 1 (Fig. 4b).

1. To calculate percentage of 481-G in type 2 and 472-C in type 3 samples use the following formulas:

$$F_D = L_D / (L_D + U_D)$$ Radioactivity fraction in digested band

$$F_C = L_C / (L_C + U_C)$$ Similar fraction in control lane (background)

$$\text{Percent of mutants} = (F_D \quad F_C) \times 100$$

2. To calculate percentage of 480-A and 525-C mutants in type 1 samples use similar formulas using radioactivity in L1 and L2 bands:

$$F1_D = L1_D / (L1_D + L2_D + U_D)$$

$$F2_D = L2_D / (L1_D + L2_D + U_D)$$

$$F1_C = L1_C / (L1_C + L2_C + U_C)$$

$$F2_C = L2_C / (L1_C + L2_C + U_C)$$

$$\text{Percent of 525 C mutants} = \left(Fl_{\text{D}} \quad Fl_{\text{C}}\right) \times 100$$

$$\text{Percent of 480 A mutants} = \left(F2_{\text{D}} \quad F2_{\text{C}}\right) \times 100$$

3. Calculate ratios between the percentage of mutants in each sample (both test articles and references) and the average between two replicates of International Standard.

3.9 Test Validity Criteria

1. The cDNA and PCR controls should be negative (no full-length PCR product seen in the gel).

2. Apparent mutant content determined in the 100 % control should be consistent with the past performance in this laboratory (*see* **Note 13**).

3. Standard deviation of MAPREC data should not be greater than 0.3, the level found to be feasible to achieve in most laboratories.

4. Values obtained for International Standard must be consistent with past performance (*see* **Note 14**).

5. The values obtained for the duplicate IS samples (A and B) should not differ significantly from each other as determined by a paired t test of % mutation values (*see* **Note 15**).

6. The Failed viral reference must fail the test, and the Passed viral reference must pass the test (*see* **Note 16**).

3.10 Interpretation of Results

Pass/fail decisions are made based on ratiometric analysis of MAPREC data. For serotypes 2 and 3 481_{G} and 472_{C} mutations are considered, while for serotype 1 analysis is performed based on the combination of 480_{A} and 525_{C} mutations. Apparent content of mutants in test samples is divided by the content of mutants determined in International Standard (average of duplicate determinations). The fiducial upper limit (FUL) on the ratio between mutant content in IS replicates A and B is calculated as described in **Note 15**. If the ratio between mutant content in a test sample and the average mutant content in IS replicates A and B exceeds the FUL, the sample fails the test, otherwise it passes MAPREC test.

4   Notes

1. Standard MAPREC protocol uses primers radioactively labeled with ^{32}P. Alternative methods of labeling such as the use of infrared fluorescent dyes were also shown to be acceptable [13].

2. In nonradioactive MAPREC protocol infrared fluorescent dyes are covalently linked to the 5′-end of one primer during or after its synthesis.

3. Other isoschizomers can be substituted for these enzymes after appropriate validation.

4. One glass plate is covered with repel-silane while another with bind-silane.

5. The content of mutants in the International Standard was adjusted to represent the level that is considered the maximum acceptable for batches of OPV. For type 3 OPV it was determined from analysis of batches of vaccine that passed and failed monkey neurovirulence test (MNVT). For other two types it was determined by analysis of multiple batches of vaccine that passed MNVT and experimental samples with increased content of mutants that were also tested in MNVT.

6. The contents of mutants in the Passed and Failed virus references were adjusted so that they pass and fail, respectively, when tested together with International Standard used as the denominator to calculate ratio (*see* **Note 5**).

7. When using other procedures for RNA isolation (e.g., Qiagen RNeasy kit) follow the manufacturer's protocol.

8. Some protocols used with RNA isolation kits do not include alcohol precipitation step and result in RNA solution in deionized water. In this case skip **steps 1–3** and use the RNA solution to prepare reverse transcriptase reaction mixture.

9. Residual reverse transcriptase can inhibit PCR and stimulate primer-dimer formation [14].

10. Hot-start PCR performed by separating polymerase from other reagents by wax prior to the beginning of polymerase reaction was successfully used to minimize primer-dimer formation.

11. The quantity and quality of cDNA may affect MAPREC results. The amount of cDNA in each reaction must be at least at a 100- to 1000-fold excess compared to the minimum required for amplification. It is highly advisable to perform preliminary amplification at varying doses of cDNA (titration) to select the optimal concentration.

12. Other isoschizomers can be used if demonstrated to produce similar results.

13. Restriction enzymes cannot digest 100 % of PCR-amplified DNA and the residual undigested DNA band normally contains about 2–6 % of radioactive material (94–98 % digestion efficiency). It depends on the type and the source of restriction enzyme and its target value must be determined based on historical performance of this enzyme under optimal conditions.

14. Data on determinations of the IS DNA sample should be accumulated and a mean (μ) and standard deviation (S.D.) calculated.

95 % Fiducial lower and upper limits (FLL and FUL) on the values obtained for the IS can be calculated as:

$$FLL = \mu - 1.96 \times S.D.$$

$$FUL = \mu + 1.96 \times S.D.$$

which represents roughly two "sigmas" below or above the mean value. If a more than one determination for the IS falls beyond this range, the test is considered invalid and must be repeated.

15. The 99 % fiducial upper limit (FUL) on the ratio variability is calculated as:

$$FUL = S.D. \times SQUARE\ ROOT\left(X^2 P, N - 1 \div (N - 1) \right)$$

where $X^2 P$, $N-1$ is a value taken from a statistical table for a number of tests N and probability P (0.01 for 99 % confidence).

16. Fiducial upper limit on the ratio of two replicates of IS (A and B) is calculated as described in **Note 15**. If the average ratio between the mutant content in a sample and the average content determined for IS replicated A and B exceeds the FUL, the sample is considered failed the MAPREC test.

References

1. WHO (2012) Recommendations to assure the quality, safety and efficacy of live attenuated poliomyelitis vaccine (oral). WHO technical report series, vol 904. World Health Organization, Geneva, Switzerland. pp 1–87

2. Evans DM et al (1985) Increased neurovirulence associated with a single nucleotide change in a noncoding region of the Sabin type 3 poliovaccine genome. Nature 314(6011):548–550

3. Cann AJ et al (1984) Reversion to neurovirulence of the live-attenuated Sabin type 3 oral poliovirus vaccine. Nucleic Acids Res 12(20):7787–7792

4. Macadam AJ et al (1991) The 5′ noncoding region of the type 2 poliovirus vaccine strain contains determinants of attenuation and temperature sensitivity. Virology 181(2):451–458

5. Christodoulou C et al (1990) Mapping of mutations associated with neurovirulence in monkeys infected with Sabin 1 poliovirus revertants selected at high temperature. J Virol 64(10):4922–4929

6. Chumakov K et al (1992) RNA sequence variability in attenuated poliovirus. In: Brown F, Chanock R, Ginsberg H, Lerner R (eds) Vaccines 92. Cold Spring Harbor Laboratory Press, Cold Spring Harbor, NY, pp 331–335

7. Chumakov K et al (1993) Assessment of the viral RNA sequence heterogeneity for control of OPV neurovirulence. Dev Biol Standard 78:79–89, discussion 88–79

8. Punia P, Cane P, Teo CG, Saunders N (2004) Quantitation of hepatitis B lamivudine resistant mutants by real-time amplification refractory mutation system PCR. J Hepatol 40(6):986–992

9. Troy SB et al (2011) Use of a novel real-time PCR assay to detect oral polio vaccine shedding and reversion in stool and sewage samples after a Mexican national immunization day. J Clin Microbiol 49(5):1777–1783

10. Laassri M, Bidzhieva B, Speicher J, Pletnev AG, Chumakov K (2011) Microarray hybridization for assessment of the genetic stability of chimeric West Nile/dengue 4 virus. J Med Virol 83(5):910–920

11. Amexis G et al (2001) Quantitative mutant analysis of viral quasispecies by chip-based matrix-assisted laser desorption/ionization time-of-flight mass spectrometry. Proc Natl Acad Sci U S A 98(21):12097–12102

12. Proudnikov D et al (2000) Analysis of mutations in oral poliovirus vaccine by hybridiza-

tion with generic oligonucleotide microchips. Biologicals 28(2):57–66

13. Bidzhieva B, Laassri M, Chumakov K (2011) MAPREC assay for quantitation of mutants in a recombinant flavivirus vaccine strain using

near-infrared fluorescent dyes. J Virol Methods 175(1):14–19

14. Chumakov KM (1995) Reverse transcriptase can inhibit PCR and stimulate primer-dimer formation. PCR Methods Appl 4:62–64

Chapter 15

Methods for the Quality Control of Inactivated Poliovirus Vaccines

Thomas Wilton

Abstract

Inactivated poliovirus vaccine (IPV) plays an instrumental role in the Global Poliovirus Eradication Initiative (GPEI). The quality of IPV is controlled by assessment of the potency of vaccine batches. The potency of IPV can be assessed by both in vivo and in vitro methods. In vitro potency assessment is based upon the assessment of the quantity of the D-Antigen (D-Ag) units in an IPV. The D-Ag unit is used as a measure of potency as it is largely expressed on native infectious virions and is the protective immunogen. The most commonly used in vitro test is the indirect ELISA which is used to ensure consistency throughout production.

A range of in vivo assays have been developed in monkeys, chicks, guinea pigs, mice, and rats to assess the potency of IPV. All are based on assessment of the neutralizing antibody titer within the sera of the respective animal model. The rat potency test has become the favored in vivo potency test as it shows minimal variation between laboratories and the antibody patterns of rats and humans are similar. With the development of transgenic mice expressing the human poliovirus receptor, immunization-challenge tests have been developed to assess the potency of IPVs. This chapter describes in detail the methodology of these three laboratory tests to assess the quality of IPVs.

Key words ELISA, Rat potency, Neutralization titer, Back titration, Transgenic mice, Human poliovirus receptor, Immunization-challenge, Biosensor, Surface plasmon resonance

1 Introduction

In the late nineteenth century improvements in hygiene standards in some countries delayed transmission of poliovirus in new born infants until maternal antibodies had waned. This delay in transmission consequently changed the pattern of poliomyelitis in those countries from endemic to epidemic. The occurrence of epidemics involving large number of children, many of whom became paralyzed, prompted a research based response focussed on developing effective vaccines against poliovirus. It was demonstrated that poliovirus could be propagated in human cells in culture [1], as well as a range of cells from tissues of nonhuman primates. The monkey kidney became the source of tissue for vaccine

Javier Martín (ed.), *Poliovirus: Methods and Protocols*, Methods in Molecular Biology, vol. 1387,
DOI 10.1007/978-1-4939-3292-4_15, © Springer Science+Business Media New York 2016

production, allowing the development of two vaccines; an inject-
able (killed) inactivated poliovirus vaccine (IPV) by Salk et al. [2],
and a live attenuated oral poliovirus vaccine (OPV) by Sabin [3].
This chapter focuses on IPV and the methods used to assess the
quality of the vaccine.

IPV is a mixture of three PV strains, one from each of the three
serotype, made by harvesting cell culture supernatants, purifying
and inactivating them using formaldehyde. In the early 1950s, following several technical developments, Salk and colleagues were
able to grow large quantities of the three PV serotypes and analyze
the kinetics of inactivation with formaldehyde (HCHO) [4]. It was
concluded that if filtration was used to remove aggregates of PV,
the virus could be inactivated at a constant rate, allowing complete
elimination of infectivity provided that sufficient time was allowed
for complete inactivation. Several trivalent vaccine pools were prepared and found to be safe and immunogenic in both monkeys and
humans [5], based on these laboratory findings large-scale field
trials in humans were conducted by Thomas Francis and his associ-
ates in 1954. The vaccine was found to be safe and effective at
60–70 % for serotype 1 and 90 % for serotype 2 and 3. Potency,
based on the antibody response in children, correlated with effectiveness of the vaccine [6] and IPV was licensed shortly after.

Following several technical advances in the late 1960s and
1970s, enhanced-potency IPV (eIPV) was developed. This second
generation IPV differed from the first generation IPV as the virus
harvest was concentrated and purified before inactivation by ultrafiltration and column chromatography. In addition, to improve
virus yield, they increased the density of cells by growing them on
microbeads in large fermenters. In 1967 van Wezel developed a
microcarrier system which could be applied to 100 l fermenters [7],
either secondary or tertiary subcultures of kidneys from pathogen-
free monkeys or human diploid cell strains or the Vero African
green monkey kidney cell line were used as the cell substrate. The
use of well characterized cells ensures that IPV is free of extraneous
contaminating agents. This trivalent eIPV acts as the conventional
IPV (cIPV) and is administered either alone or in combination
with other vaccines, including diphtheria, tetanus and acellular
pertussis, hepatitis B and/or Hemophilus influenzae b. These IPV
preparations are highly immunogenic in infants following three
doses at 2, 3, and 4 months of age [5].

The wild-type PV strains used by Salk et al. for the first generation IPV are still used by most modern manufacturers. These
include Mahoney (serotype 1; Brunenders is used in Sweden and
Denmark); MEF-1 (serotype 2); and Saukett (serotype 3).
Although HCHO inactivation has been noted to modify the antigenic site 1 of PV serotypes 2 and 3 [8], immunization with IPV
can induce high titers of neutralizing antibodies protective against
all PV strains.

One of the key means for the quality control of IPV is the assessment of the immunogenicity or potency of the vaccine. When the IPV was developed by Salk in the 1950s, the potency of IPVs was not assessed; instead each dose of IPV was designed to be the equivalent of a specific volume of harvest fluid from PV-infected primary monkey kidney tissue cells. Inactivated poliovirus vaccines assessed by this method showed variable immunogenicity in humans. In response to the Cutter incident in which vaccine recipients were paralyzed by the use of incompletely inactivated IPV, filtration steps were introduced to remove aggregates [9]. While these steps led to increased safety, the antigenicity of the IPVs fell as a result and this, in turn, lowered the immunogenicity of the vaccine. Consequently, potency assays for the final products were required.

Both in vitro and in vivo potency assays have been developed to assess the immunogenicity of IPV. Previously the in vivo assays were used as the official batch release tests carried out on the final IPV product, while the in vitro assays were principally used for in-process monitoring. This situation has since changed and due to the requirements of the European convention on the protection of vertebrate animals used for experimental and other scientific purposes, it is possible to waive the in vivo assay and assess the potency solely by in vitro assays, should certain conditions be met [10]. In vitro assays measure the potency of IPV preparations by D-Ag units. In the 1960s the D-Ag and C-Ag units were established following characterization of purified virus preparations by sucrose gradient centrifugation where two bands were identified [11]. The D-Ag unit is used as a measure of potency as it is largely expressed on native infectious virions and is the protective immunogen. The most commonly used in vitro test is the indirect ELISA. This assay was first developed in 1980 and replaced the gel diffusion method as it was found to be more sensitive and less time consuming [12]. It is used to assess the D-Ag content of IPVs and ensure consistency throughout production. This assay is described in detail below.

A range of in vivo assays have been developed in monkeys, chicks, guinea pigs, mice and rats. All are based on assessment of the neutralizing antibody titer within the sera of the respective animal model. Initially the monkey potency test was the standard in vivo assay to determine the potency of IPV preparations [13]. In this test at least ten simians are inoculated, using a three dose schedule comparable to that used for human recipients. The geometric mean titer of neutralizing antibodies for each of the three serotypes is compared with a reference trivalent antiserum. An IPV is deemed acceptable if the mean titer is above that of the reference antiserum.

While the monkey potency test can distinguish qualitatively between good and poor IPVs, it is not sufficient for obtaining quantitative data about vaccine potency, as it uses a reference serum rather than a reference vaccine. In addition, this test requires many expensive primates. These concerns led to a search for another animal model to assess the potency of IPVs. Animals which are nonsusceptible to PV are known to be capable of forming specific neutralizing antibodies following parenteral administration of live or inactivated PV. Consequently, a range of nonsusceptible animal models have been developed, two of which are the guinea pig/ chick models [14, 15]. These two models are based on a potency tests in which the extinction end-point of the median effective dose (ED50) is calculated after the animals have been inoculated with a single dose of the vaccine. Both the guinea pig and chick potency test have been adopted as a single test within the European Pharmacopoeia. In this test, three or more dilutions of an IPV are used to immunize either guinea pigs or 3-week-old chicks (ten animals/dilution). After 5–6 days, animals are bled and sera are diluted to one in four. Poliovirus (100 TCID50) is mixed with the diluted serum and incubated at 37 °C for 4.5–6 h and then 5 ± 3 °C for 12–18 h. Mixtures are added to cell cultures for up to 7 days, to detect unneutralized PV. For each group of animals, the number of sera which have neutralizing antibodies is noted and the dilution of the IPV which gives an antibody response in 50 % of the animals (the end-point of the ED50) is calculated. An IPV is acceptable if a dilution of 1 in 100 or more produces an antibody response for each of the three serotypes of PV in 50 % of the animals.

Although an accepted technique of the European Pharmacopoeia, the guinea pig/chick potency test is limited in that the calculated single dose ED50 can only provide a qualitative measure of the immunogenicity of an IPV. This test lacks the sensitivity and reproducibility to distinguish between IPVs which differ in antigenic content as much as fourfold to eightfold. In addition, this test is a poor predictor of human immune response. In particular, this test can only measure IgM instead of IgG which is typically measured following vaccination in humans. In response to these limitations the Rijks Instituut voor de Volksgezondheid developed an alternative potency test in which the titer of the antibody response was measured rather than the extinction titer (end-point). The antibody response of guinea pigs and various strains of rats and mice were assessed to identify the animal which would produce a good dose-related titer response after a single injection, preferably in the IgG class. Rats were found to give the highest titers, a good dose-related titer response in the IgG class and were consistent across different strains. Consequently, a rat potency test was developed to assess the immunogenicity of a range of IPVs in relation to a reference IPV of proven efficacy in humans. Comparable distribution of antibody titers for reference and sample IPVs was found, allowing accurate assessment of potency. Antibody patterns of rats and humans were similar, indicating that the rat maybe a suitable model to assess the immunogenicity of IPV in humans [16]. An international collaborative study compared the use of the chick/guinea pig and the rat potency tests to determine the immunogenicity of six trivalent IPVs. Wide variation was found between laboratories using the chick/guinea pig potency test. The rat potency test was found to be far less variable between laboratories [17]. The results of this study led to the validation of the rat potency test by manufacturers and national control laboratories as an alternative means of determining the immunogenicity of IPVs. The rat potency test is now included within the European Pharmacopoeia [18]. Currently the European Pharmacopoeia guidelines require that the potency of an IPV is assessed either in vivo by the chick/guinea pig or rat tests or by an in vitro method (following a waiving of in vivo tests). Due to its benefits, the rat potency test is considered the in vivo method of choice. The rat potency test is described in detail below.

Recently an immunization-challenge test using transgenic (Tg) mice to further evaluate the immunogenicity and protective properties of IPVs has been developed independently by the US Food and Drug Administration (FDA) and NIBSC [19, 20]. The identification and isolation of genes which encode human and primate PVRs has allowed Tg mice lines which express the human PVR (TgPVR mice) to be established. Transgenic mice expressing the human PVR can be infected with PV serotypes by various routes and develop clinical signs of paralysis and morphological lesions in the CNS. As the TgPVR mice can clinically manifest paralysis, it is possible to determine the 50 % end-points for paralysis (PD_{50}) or lethality (LD_{50}), if a suitable dose range of a PV is used.

Previous research by the FDA and the NIBSC has identified suitable immunization-challenge regimes with TgPVR mice for assessing the immunogenicity of IPV preparations. A single or booster vaccination (1 week apart) followed by a challenge was sufficient to model protection. In Tg21PVR mice both protection and the level of neutralizing antibody elicited by vaccination with IPV have been found to be dose-dependent. A good correlation has been found between the neutralizing antibody titers in blood samples from Tg21PVR mice and the immune protection conferred. Immunization-challenge regimes with TgPVR mice allow a more direct analysis of the protection conferred by IPV preparations than other potency test which assess the immunogenicity on the basis of seroconversion. The immunization-challenge test with TgPVR mice assesses the protection conferred by other aspects of the immune response (e.g., cellular immunity) in addition to the neutralizing antibody response. The immunization-challenge assay with TgPVR mice is described in detail below.

---

2   Materials

2.1   ELISA Potency Test

2.1.1   Equipment

- +4 °C refrigerator.
- –20 °C freezer.
- –70 °C freezer.
- Incubator set at 37 °C.
- 96-well microtiter plates for ELISA.
- 2 ml microtubes.
- Pipettors in calibration.
- Multichannel pipette and suitable tips.
- Sterile 1, 5, 10, and 25 ml serological pipettes.
- Multipipette and tips.
- Multidrop Combi Reagent Dispenser.
- Spectrophotometer for 96-well plates with 492 nm (nominal) filter.
- Balance.
- Plastic box.
- Vortex mixer.

2.1.2   Reagents

- Test vaccines—stored at either 2–8 °C (refrigerator) or –20 °C or –70 °C (Freezer).
- Dulbecco's 6 salt PBS.
- Dried milk powder.
- Tween 20.
- Carbonate coating buffer (storage at 2–8 °C)—6.36 g sodium carbonate and 11.72 g sodium hydrogen carbonate made up to 4 l with deionised H_2O.
- Wash Buffer—Dulbecco's 6 Salt PBS containing 2.0 % dried milk and 0.5 % Tween 20—prepare on day of assay and discard any unused buffer after use.
- Assay diluent—Dulbecco's 6 Salt PBS containing 2.0 % dried milk powder—prepare on day of assay and discard any unused buffer after use.
- Serotype specific capture antisera (*see* **Note 1**).
- Monoclonal antibodies (MAbs) (*see* **Note 1**).
- Reference vaccine (with assigned potency in D-Ag units). Any reference vaccine used in the study should have been calibrated using an International Standard for IPV. The current WHO International Standard (12/104) has an assigned potency of 277, 65 and 248 D-Ag units for types 1, 2 and 3 respectively (stored at –70 °C). The current European Pharmacopoeia biological reference preparation (Ph Eur BRP Batch 2) has an

assigned potency of 320-67-282 D-Ag units for types 1, 2 and 3 respectively.

- Monitor vaccine samples.

A previously validated monitor IPV sample must be included in all batch release assays.

- Peroxidase conjugated anti-mouse IgG whole molecule (Sigma-Aldrich, store at -20 °C) diluted in assay diluent to a working dilution of 1:400.

- Substrate buffer—Mix 12.15 ml 0.1 M citric acid, 12.85 ml 0.2 M Na_2HPO_4, and 25 ml distilled H_2O. Prepare immediately before use.

- Substrate reagents:

0.1 M citric acid—19.2 g made up to 1 l with distilled H_2O in a measuring cylinder or
0.1 M citric acid·H_2O—21.0 g made up to 1 l with distilled H_2O in a measuring cylinder (Storage—room temperature).
0.2 M Na_2HPO_4—28.4 g made up to 1 l with distilled H_2O in a measuring cylinder or
0.2 M Na_2HPO_4·$12H_2O$—71.63 g made up to 1 l with distilled H_2O in a measuring cylinder (Storage—room temperature).

- OPD substrate—Prepare in a 50 ml centrifuge tube: 1×30 mg o-phenylenediamine dihydrochloride substrate tablet (Sigma-Aldrich) + 50 ml substrate buffer + 50 µl hydrogen peroxide (30 %, Sigma-Aldrich). Use within 1 h of addition of tablet to buffer and add H_2O_2 immediately before use (Store in dark).

- Stop solution—1 M H_2SO_4 (in the proportion of 1 ml concentrated acid in 17 ml H_2O) (Storage—room temperature).

2.2 Rat Potency Test

2.2.1 Equipment

- Class II Microbiological safety cabinet.
- Refrigerated centrifuge.
- Sterile universal bottles.
- +4 °C refrigerator (+2–8 °C).
- −20 °C freezer.
- −80 °C freezer.
- Multipipette and sterile tips.
- Half test tubes in racks.
- Sterile 1, 5, 10 and 25 ml pipettes.
- Microtiter plates (flat bottomed sterile tissue culture grade).
- Pressure sensitive film (PSF).
- Sterile blotting paper.
- Multidrop Combi Reagent Dispenser.
- Inverted microscope.

- Incubator at $+35+1$ °C.
- Gilson and electronic pipettes in calibration + sterile tips.
- Multichannel pipettes in calibration + sterile tips.
- Sterile Multidrop dispensing cassettes.

2.2.2 Reagents

- Assay diluent for inoculum to be immunized to rats (100 ml volume):

Eagles Minimum Essential Medium (MEM) 10× concentration	10 ml
NaHCO$_3$ (7.5 %) (Sigma-Aldrich)	2.9 ml
7 % Bovine albumin in MEM	2 ml
Sterile distilled water	Up to 100 ml

Assay diluent is made up before use using Class II Microbiological safety cabinet and used immediately without storage.

- Assay diluent for neutralization assay (500 ml volume):

MEM ×1 concentration	Up to 500 ml
Or MEM 10 concentration	50 ml
NaHCO$_3$ (7.5 %) (Only required with ×10 MEM)	8.85 ml
Penicillin + Streptomycin (Sigma-Aldrich)	5 ml
(With 10,000 units' penicillin and 10 mg streptomycin/ml in 0.9 % NaCl)	
Amphotericin B solution (250 µg/ml) (Sigma-Aldrich)	5 ml
Fetal calf serum	20 ml
l-Glutamine (200 mM) (Sigma-Aldrich)	5 ml
Sterile water (only required with ×10 MEM)	Make up to 500 ml

- Assay diluent is made up before use and used immediately without storage.
- 0.25 % trypsin—EDTA solution (Sigma-Aldrich).
- HEp2-C cells. Use HEp2-C 75 cm^3 flask stock cultures up to 7 days from the last subculture. (No maximum passage level has been established for this assay method however cells are not used for more than 3 months before returning to original stocks.)
- Plastic stain bath containing naphthalene black—6 % acetic acid (100 % glacial), 1.3 % sodium acetate, 1 % naphthalene black.

- Reference vaccine.

  A previously validated vaccine reference has to be used in the assay. Ideally, this reference should be traceable to clinical studies showing suitable immunogenicity in humans. An established International Standard for IPV should be used to validate the assay.

- Monovalent reference polioviruses for each serotype.

- In-house reference positive and negative sera of known neutralization titer.

2.3 Immunization-challenge assay in TgPVR mice.

2.3.1 Equipment

- Class II microbiological safety cabinet.
- +4 °C refrigerator (+2-8 °C).
- –20 °C freezer.
- –80 °C freezer.
- 2 ml microtubes.
- Gilson and electronic pipettes in calibration.
- Sterile 1, 5, 10, and 25 ml serological pipettes.
- Vortex.
- All equipment needed to house, handle, and inoculate TgPVR mice is not listed here.

2.3.2 Reagents

- Transgenic mice expressing the human poliovirus receptor so they are susceptible for poliovirus infection and paralysis.

- Eagles Minimum Essential Medium (MEM) 1× concentration is used as assay diluent to prepare solutions containing vaccines and virus challenges Assay diluent is handled in a sterile manner in a Class II microbiological safety cabinet and used immediately without storage.

- Reference vaccine.

  A previously validated vaccine reference has to be used in the assay. Ideally, this reference should be traceable to clinical studies showing suitable immunogenicity in humans. An established International Standard for IPV should be used to validate the assay.

3 Methods

3.1 ELISA Potency Test

3.1.1 Preparation of Assay

1. In a class II safety cabinet, dilute serotype specific capture antibody in carbonate coating buffer as required (*see* **Note 1**). Add 50 μl to each well of a 96-well ELISA plate. One plate per poliovirus serotype.

2. Incubate overnight at 2–8 °C in a box with a humidified atmosphere.

3. Wash each ELISA plate 4× with wash buffer. Leave plate containing last wash at room temperature for at least 30 min.

4. Prepare independent series of 4, twofold dilutions of reference and test vaccines in assay diluent.

5. Start at 1 in 20 for the IRR and BRP. For non-desorbed single human dose vaccines (40:8:32, respective virus types) use a starting dilution of 1 in 3. Predilute vaccines of different formulations to approximately this concentration in assay diluent.

6. Add 50 μl of vaccine or reference dilution to a minimum of 2 wells per dilution for each independent series of dilutions. Add diluent only to row 1 and 12 to serve as a blank.

7. Seal plates and incubate for at least 2 h at 35–38 °C in a box with a humidified atmosphere.

8. Wash 3× with wash buffer.

9. Add 50 μl of monoclonal antibody (*see* **Note 1**) diluted in assay diluent to each well, including blanks.

10. Seal plates and incubate for at least 1 h but not more than 75 min at 35–38 °C in a box with a humidified atmosphere.

11. Wash 3× with wash buffer.

12. Add 50 μl peroxidase conjugated anti-mouse diluted in assay diluent to each well.

13. Seal plates and incubate for at least 1 h but not more than 75 min at 35–38 °C in a box with a humidified atmosphere.

14. Wash 3× with PBS.

15. Add 50 μl OPD substrate to each well. Leave at room temperature in the dark.

16. Stop reaction after 30 min by addition of 50 μl of 1 M H_2SO_4 per well.

17. Read optical density at 492 nm as soon as possible.

3.1.2 Analysis of Results

1. Parallel line analysis is performed on the assay data by using a statistical program such as CombiStats. This program gives the potency of the test vaccine relative to the concurrently tested reference. The dilutions and the corresponding OD values are used to calculate the D-Ag content of each vaccine relative to the reference (*see* **Note 1**).

2. For an assay to be valid the dose response curves should be linear and parallel. A minimum of three dilutions on the linear section of the dose response curve should be used in the analysis. Tests for linearity and parallelism are performed by the computer program.

3. The levels of significance of deviation from linearity and parallelism for the reference, monitor and all individual vaccine samples should be 1 % or below.

4. The D-Ag content of the monitor sample should be within specifications assigned for that sample.

3.1.3 Interpretation of Results

Vaccines tested in assays that meet all the assay validity criteria should be assessed with regard to their D-Ag content/dose. Vaccines must have D-Ag within the given release specification for a vaccine to be recommended for release. Recent research has been directed to assessing alternative methods to the current ELISA protocol for establishing the D-Ag content of IPV (*see* **Note 2**).

3.2 Rat Potency Test

This potency test involves inoculating trivalent test vaccine/s plus a reference into rats (*see* **Note 3**). The rats are bled between 20 and 22 days and the sera is then tested for neutralizing antibody to all three polio virus types using a fixed virus varying serum, 50 % end point technique in a microtiter system. A vaccine is satisfactory if the potency is not significantly less than that of the reference preparation. This testing is undertaken in line with European Pharmacopoeia (Ph Eur) potency test for IPV.

3.2.1 Inoculation of Rats

1. Prepare assay diluent for inoculation of rats using a class II cabinet.

2. Remove the vaccines from +4 °C and the appropriate reference preparation from either −80 °C or +4 °C.

3. Assess the appearance of the vaccine (described in the product licence). Assess the sample against a light background with light coming from the side and also against a matt black background. This is to ensure it meets manufacturer's specifications and does not contain any extraneous particles. If the vaccine fails to meet the criteria then the test is stopped.

4. Label sterile universal bottles with the sample name, test number, dilution, and test date. Transfer of sample from the original container to the first container of the dilution series is witnessed and countersigned by a second person.

5. Prepare dilutions of test vaccine and reference as detailed in Table 1.

6. Inoculate each group of rats with 2×0.25 ml of diluted material into each hind leg.

7. Following 20–22 days post-inoculation anesthetize rats and bleed out by cardiac puncture.

8. Allow blood samples to clot at +4 °C for 1–2 days.

9. Centrifuge all blood samples at $800–1000 \times g$ in a refrigerated centrifuge for 5 min to aid separation of the blood clot and sera.

Table 1
Preparation of dilutions of test vaccine and pre-diluted reference

	Dilution	Volume diluent	Vol. sample (ml)	No. of rats
Vaccine + reference	Undiluted	NA	12	10
	1/2	6 ml	6	10
	1/4	6 ml	6	10
	1/16	6 ml	2	10
Ref EP BRP or IS 91/574	[a](See below)	12 ml	3	10
	1/2	6 ml	6	10
	1/4	6 ml	6	10
	1/16	6 ml	2	10

[a]The first dilution made is in fact a 1:5 for these references. This dilution is required to bring the references to single human dose as they are 10× more concentration than the test vaccine. This dilution will be treated as undiluted as the reference now equals a single human dose as per the vaccine section

10. Transfer sera in a sterile manner to tubes labeled with the test number, rat cage number, a unique letter and the date.

11. Store the sera at –20 or –80 °C until required for testing.

3.2.2 Neutralizing Antibody Assay

Preparation of Assay

1. Obtain a culture of Hep-2C cells. Check the culture microscopically for condition of monolayer and absence of overt bacterial contamination. Use only cultures which are confluent and in good condition.

2. Remove rat sera from –20/–80 °C storage and thaw overnight at +4 °C.

3. Thaw positive/negative reference sera from –20 °C overnight at +4 °C.

4. Prepare sufficient assay diluent for the test in the class II cabinet.

5. Prepare a worksheet in which code number assigned to each serum are recorded. Note the volumes used to make up the challenge virus on the worksheet.

6. Label all tubes and plates with the serum or sample number or the uniquely assigned code letter or number from the worksheet.

Dilution of Challenge Viruses to 100 Tissue Culture Infective Dose 50 % End point (TCID$_{50}$)/0.05 ml

1. Remove relevant challenge virus strains from –80 °C storage and thaw on ice.

2. Dilute the challenge virus strains to 100 Tissue Culture Infective Dose 50 % end point (TCID$_{50}$)/0.05 ml. This dilution is based on previously established titer of the challenge virus strains. The dilution to a challenge dose of 100 TCID$_{50}$/0.05 ml is done in a series of steps.

3. Adjust the volume of the last dilution step so that it is sufficient for the relevant number of plates.

4. Incubate 1.0 ml of each virus challenge with 1.0 ml of diluent in half test tubes in the same test conditions as the test plates to perform the back-titration of the virus challenge.

Titration of Sera

1. Prepare sufficient plates for the number of sera to be tested. Each serum will be added to three plates so that a different plate is used for each virus type.

2. A different starting dilution might be required for each serotype based on preliminary tests, label all plates with the serum number or unique code, the virus type and the assay date.

3. Use the plate in the 12 across 8 down orientation allowing six sera to be titrated per plate using this layout.

4. Three further plates will be needed for the in-house rat positive sera and one plate (or 6 wells) is needed for the in-house rat negative sera.

5. Using Titertek Multidrop dispensing machine dispense 0.05 ml of diluent into each well of a microtiter plate. Keep each plates covered as much as possible during manipulations using sterile blotting paper.

6. Add 0.05 ml of the first serum sample to columns 1 and 2 in Row A of each of three plates. Because an equal volume of medium has been added to each well.

7. Add 0.05 ml of the second serum to columns 3 and 4 in Row A of each of three plates and repeat this process for other serum samples until all samples are added.

8. Add 0.05 ml of the in-house rat negative sera to 2 wells per virus type. This can either be on a separate plate or at the end of the test sera.

9. Make a 1:1 dilution of the in-house rat positive sera by adding 1 ml of assay media to the 1 ml aliquot (so starting dilution on plate of 1/4). Then add 0.05 ml of the in-house rat positive sera to wells A-H in row 1 (this plate is in the 8–12 orientation). One plate will be used per type (8 wells per virus type).

10. Make twofold serial dilutions of all samples using a multichannel pipette with disposable tips. Do this by aspirating manually or electronically to mix and withdraw 0.05 ml volumes from each well in Row A and transfer these to Row B. Continue through to Row H and discard the last 0.05 ml.

Addition of Virus Challenge and Neutralization Step

1. Add 0.05 ml of each virus challenge to the appropriate plates using a suitable piece of equipment that is calibrated to dispense 50 µl.

2. Seal all plates with pressure sensitive film (PSF).

3. Incubate the assay, including the 1 + 1 virus challenge, overnight at +4 °C then for 3 h at 35 °C.

Back Titration of Virus Challenge

1. After incubation, dispense 4.5 ml of diluent into 3 half test tubes for each virus type. Back titrate each 1 + 1 virus challenge sample a further 3× tenfold steps.

2. Transfer and mix 0.5 ml of 1 + 1 virus challenge into 10^{-1} tube. Use a fresh sterile 1 ml pipette or tip each time. Continue serial dilution for 10^{-2} and 10^{-3}.

3. Inoculate 0.1 ml of each titration into eight wells of a microtiter plate beginning at most dilute and ending with the most concentrated dilution.

4. Repeat for each virus challenge.

Addition of HEp2-C Cells

1. Prepare sufficient HEp-2C cell suspension for the number of plates in the assay, (10 ml of cell suspension per plate plus about 10 % extra) in class II safety cabinet as follows:

   (a) Decant medium from 75 cm² culture flasks.

   (b) Wash cell monolayer with PBS.

   (c) Add 5 ml of 0.25 % trypsin-EDTA solution.

   (d) Spread this over the cell sheet and decant.

   (e) Place culture at 35 °C for 5 min.

   (f) When cells begin to detach from the flask suspend in approx. 10 ml of diluent, aspirating well.

   (g) Make up to 100 ml using assay diluent. The cell suspension is now ready for use (approx cell count $1–2 \times 10^5$/ml).

2. Remove plates from the 35 °C incubator and remove PSF. Keep each plate covered with sterile blotting paper.

3. Using Titertek Multidrop dispensing machine add 0.1 ml of cell suspension into each well in all plates. Seal the plates thoroughly with PSF. Incubate the plates at 35 °C for 5–7 days. Ensure at least two rows or columns of cell controls are included in each test.

4. After incubation check plates visually for condition of cells and for absence of bacterial contamination. If contamination is suspected plates are read microscopically.

5. Uncontaminated plates are then fixed and stained as below and results are recorded. Contaminated plates are read microscopically by eye.

Fixing and Staining

1. Remove pressure sensitive film from plates.

2. Discard medium from wells into 5 % Microsol.

3. Immerse plates in stain bath containing naphthalene black stain. Leave for at least 1 h or overnight.

4. Rinse in tap water, drain and dry in incubator.

5. Only those wells with a confluent monolayer of cells will be stained blue. These are scored negative. Those wells where polio cytopathic effect (CPE) has completely destroyed the monolayer will remain colorless. These are scored positive.

Assessment of the Assay

1. The test must be assessed for tissue condition: the cell controls must be in such condition that clear distinction can be made between the presence and absence of polio CPE. Negative wells on the test plates may also be used as guide to tissue condition.

2. Serum titrations with significant incidental losses of more than three wells must be repeated.

3. The titer of the challenge virus must be in the range 1.0–3.0 log10 TCID50.

Analysis of Results

1. The serum antibody titer is calculated as the highest dilution which protects 50 % of the cultures against 100 $TCID_{50}$ of challenge virus.

2. The raw data from the rat assays are converted to animals responding or not responding. The number of animals positive for neutralizing antibody out of the number inoculated is scored for both test vaccine and reference at each dilution for each of the polio virus types. The cut off values for assessing whether a serum is positive should be established before to define the dilution with best dynamic range. These are usually different for each of the three poliovirus serotypes. The geometric mean neutralization titers generated by each of the vaccines against different poliovirus challenge strains can also be determined and compared.

3. The number of animals positive for neutralizing antibody out of the number inoculated is scored for both test vaccine and reference at each dilution for each of the polio virus types. These results are used for determining if the response to test vaccine/s is significantly different to that of the reference vaccine.

4. The dose response for each preparation should be linear, and the dose-responses of the reference and test preparations should be parallel. Tests for linearity and parallelism are performed by computer analysis.

5. A potency value with confidence intervals is obtained for each test vaccine relative to the reference vaccine.

6. The "Upper limit" and "Lower limit" ($P = 0.950$) for vaccine potency must be within 25–400 % of the value.

7. The 50 % effective dose (ED50) is calculated and must lie between the largest and the smallest doses for both the reference and the test preparation for the assay to be valid.

Interpretation of Results

A vaccine is considered suitable for release if the potency of each of the polio types is not significantly less than that of the reference preparation. The release criteria is that the upper fiducial limit (UFL) of the relative potency for the vaccine must be greater or equal to 1.0 (>1.0). This specification is noted in the Ph Eur monograph for IPV.

3.3 Immunization-Challenge Assay in TgPVR Mice

3.3.1 Inoculation of TgPVR Mice

1. Prepare assay diluent for inoculation of TgPVR mice using a class II cabinet.

2. Remove the vaccines from either $-80\ ^{\circ}C$ or $+4\ ^{\circ}C$.

3. Assess the appearance of the vaccine (described in the product licence). Assess the sample against a light background with light coming from the side and also against a matt black background. This is to ensure it meets manufacturer's specifications and does not contain any extraneous particles. If the vaccine fails to meet the criteria then the test is stopped.

4. Label sterile bijou with the sample name, test number, and test date.

5. Prepare dilutions of test vaccine/s and reference to the required D-Ag dosage concentrations.

6. Inoculate groups of eight TgPVR mice of equivalent age and gender (6–8 weeks) by the intra-peritoneal route (0. 2 ml per mouse) with each test/reference vaccine or diluent medium.

7. Following 14 days, booster the mice with the same dose by the same route.

8. After a further 21 days tail bleed all mice.

9. Allow blood samples to clot at $+4\ ^{\circ}C$ for 1–2 days.

10. Centrifuge all blood samples at $800-1000 \times g$ in a refrigerated centrifuge for 5 min to aid separation of the blood clot and sera.

11. Transfer sera in a sterile manner to tubes labeled with the test number, TgPVR mice cage number, a unique letter and the date.

12. Store the sera at -20 or $-80\ ^{\circ}C$ until required for testing. The sera can be tested for neutralization titers against different poliovirus strains following a neutralization assay similar to the one used for rat sera described above.

13. After obtaining tail bleeds prepare the virus inoculum containing the selected challenge PV strain at a dosage $10-50 \times PD_{50}/0.05$ ml to inoculate each TgPVR mouse.

14. Inoculate the mice by the intramuscular route with a paralyzing dose of the selected challenge PV strain.

15. Monitor the mice for signs of paralysis for 14 days. Clinically score daily— a minimum of partial paralysis in either hind limb must be apparent before mice are culled.

3.3.2 Analysis of Results

1. The number of paralyzed mice is recorded for each group of mice.

2. For the assay to be valid, a significant number of animals in the control group (mock-immunized) should be paralyzed.

3. Survival curves showing the proportion of surviving animals during the 14-day observation period can be obtained using specialized software.

4. Statistical tests such as the Log-rank (Mantel–Cox) Test or the Gehan–Breslow–Wilcoxon Test can be used to determine if there are significant differences between survival curves of animal groups immunized with different vaccines and/or challenged with different poliovirus strains.

3.3.3 Interpretation of Results

This assay allows comparison between vaccines for protection against challenge with paralytic poliovirus. It will also be possible to assess whether different vaccines protect differently against various poliovirus strains such as vaccine, wild or vaccine-derived isolates. These results should be compared to those measuring the neutralization antibody titers against different poliovirus strains sera from mice prior to the virus challenge.

4 Notes

1. A variety of combinations of capture-detection antibodies are used in different laboratories. These include both monoclonal and polyclonal antibodies obtained in different animals and used for either step and in all possible combinations. The use of such different platforms in different laboratories highlights the need for International Standards to calibrate laboratory references used in these assays. It is critical to calibrate any antibody reagents to specifically detect D-Ag potency.

2. One such method is the biosensor-based analytical system which is based on Surface Plasmon Resonance (SPR) technology. Surface Plasmon Resonance technology is an optical method which measures the refractive index of very thin layers of material adsorbed on a metal (such as gold). A detailed explanation of SPR technology can be found elsewhere [21, 22]. The SPR technology has been developed by various research institutions to study molecular interactions, including

GE Healthcare which has developed the Biacore biosensor. Preliminary work by Kersten et al. [23] using the Biacore biosensor indicates that the biosensor approach could be used to determine the D-Ag content of IPV. More recent research [24] has developed a biosensor-based protocol which established comparable D-Ag/ml estimates to those obtained by the current ELISA method, indicating that it could offer an alternative means to assess the potency of IPVs. Variation within assays and laboratories could be lessened by the automated nature of biosensors. The variability in polyclonal and monoclonal antibodies could be eliminated with the use of a biosensor system, as this would allow the rapid characterization (within days) of antibodies, which would enable laboratories to rapidly screen and select the optimal antibodies.

3. A good titer response has been found in both outbred (such as Wistar) and inbred strains (including Sprague Dawley, Lewis, and others). There is little difference in immune response between the rat strains [16].

References

1. Enders JF, Weller TH, Robbins FC (1949) Cultivation of the Lansing strain of poliomyelitis virus in cultures of various human embryonic tissues. Science 109(2822):85–87

2. Salk JE, Krech U, Youngner JS, Bennett BL, Lewis LJ, Bazeley PL (1954) Formaldehyde treatment and safety testing of experimental poliomyelitis vaccines. Am J Public Health Nations Health 44(5):563–570

3. Sabin AB (1985) Oral poliovirus vaccine: history of its development and use and current challenge to eliminate poliomyelitis from the world. J Infect Dis 151(3):420–436

4. Furesz J (2006) Developments in the production and quality control of poliovirus vaccines—historical perspectives. Biologicals 34(2):87–90

5. Robbins FC (2004) The history of polio vaccine development. In: Plotkin SA, Orenstein WA (eds) Vaccines. Saunders, Philadelphia, pp 17–30

6. Plotkin SA, Vidor E (2008) Poliovirus vaccine—inactivated. In: Plotkin SA, Orenstein WA, Offit PA (eds) Vaccine. Saunders Elsevier, Philadelphia, pp 605–629

7. van Wezel AL (1967) Growth of cell-strains and primary cells on micro-carriers in homogeneous culture. Nature 216(5110):64–65

8. Ferguson M, Wood DJ, Minor PD (1993) Antigenic structure of poliovirus in inactivated vaccines. J Gen Virol 74(Pt 4):685–690

9. Nathanson N, Langmuir AD (1963) The Cutter incident: poliomyelitis following formaldehyde-inactivated poliovirus vaccination in the United States during the spring of 1955. II. Relationship of poliomyelitis to Cutter vaccine. Am J Epidemiol 78:29.6–60.6

10. European Pharmacopoeia, 2.7.20 (2011) In vivo assay of poliomyelitis vaccine (inactivated). In: European Pharmacopoeia 7.0. pp 226–227

11. Beale AJ, Mason PJ (1962) The measurement of the D-antigen in poliovirus preparations. J Hyg (Lond) 60:113–121

12. Souvras M, Montagnon BJ, Fanget B, Van Wezel AL, Hazendonk AG (1980) Direct enzyme-linked immunosorbent assay (ELISA) for quantification of poliomyelitis virus D-antigen. Dev Biol Stand 46:197–202

13. United States Department of Health Education and Welfare (1968) Biological Products Public Health Service, in 42, Public Health Service (ed). Division of Biological Standards National Institutes of Health, Maryland. pp 32–44

14. Gard S, Wesslen T, Fagraeus A, Svedmyr A, Olin G (1956) The use of guinea pigs in tests for immunogenic capacity of poliomyelitis virus preparations. Arch Gesamte Virusforsch 6(5):401–411

15. Timm EA, Rope EZ, Mc LI Jr (1958) Chick potency tests of poliomyelitis vaccine: basic studies on response. J Immunol 80(5):407–414

16. van Steenis G, van Wezel AL, Sekhuis VM (1981) Potency testing of killed polio vaccine in rats. Dev Biol Stand 47:119–128

17. Wood DJ, Heath AB (1995) A WHO collaborative study of immunogenicity assays of inactivated poliovirus vaccines. Biologicals 23(4):301–311

18. European Pharmacopoeia (2014) In vivo assay of poliomyelitis vaccine (inactivated): test in rats. In: European Pharmacopoeia 8.0. Strasbourg, France

19. Martin J, Crossland G, Wood DJ, Minor PD (2003) Characterization of formaldehyde-inactivated poliovirus preparations made from live-attenuated strains. J Gen Virol 84 (Pt 7):1781–1788

20. Taffs RE, Chernokhvostova YV, Dragunsky EM, Nomura T, Hioki K, Beuvery EC, Fitzgerald EA, Levenbook IS, Asher DM (1997) Inactivated poliovirus vaccine protects transgenic poliovirus receptor mice against type 3 poliovirus challenge. J Infect Dis 175(2):441–444

21. Jason-Moller L, Murphy M, Bruno J (2006) Overview of Biacore systems and their applications. Curr Protoc Protein Sci. Chapter 19:Unit 19 13

22. Pattnaik P (2005) Surface plasmon resonance: applications in understanding receptor-ligand interaction. Appl Biochem Biotechnol 126(2): 79–92

23. Westdijk J, Brugmans D, Martin J, Van't Oever A, Bakker WA, Levels L, Kersten G (2011) Characterization and standardization of Sabin based inactivated polio vaccine: proposal for a new antigen unit for inactivated polio vaccines. Vaccine 29(18):3390–3397

24. Kersten G, Hazendonk T, Beuvery C (1999) Antigenic and immunogenic properties of inactivated polio vaccine made from Sabin strains. Vaccine 17(15–16):2059–2066

Chapter 16

Measuring Poliovirus Antigenicity by Surface Plasmon Resonance. Application for Potency Indicating Assays

Janny Westdijk, Larissa van der Maas, Rimko ten Have, and Gideon Kersten

Abstract

The D-antigen ELISA is the commonly accepted test for release of inactivated poliovirus containing vaccines. However, this test has a few drawbacks regarding the many variations in the method to quantify the D-unit. The result may depend on method and reagents used which makes standardization of inactivated polio vaccines, based on D-units, to a real challenge. This chapter describes a surface plasmon resonance based method to quantify D-units. The advantage of the calibrated D-antigen assay is the decrease in test variations because no labels, [no incubation times] and no washing steps are necessary. For standardization of both IPV and Sabin IPV, the calibration free concentration analysis could be an improvement as compared to ELISA or other SPR methods because this method combines quantity (particle concentration) and quality (antigenicity) in one assay.

Key words D-antigen ELISA, Surface plasmon resonance, Calibrated D-ag assay, Calibration free concentration assay, Polio vaccine, Sabin, IPV

1   Introduction

For release and stability studies the active component in polio vaccines is measured by a sandwich ELISA in which D-ag specific antibodies are used to determine the D-ag concentration [1]. Also the rat potency test [2, 3] is used. Since 2005 the European Pharmacopoeia allows waiving of the rat potency and to rely exclusively on the ELISA as a potency indicating test for release of inactivated poliovirus containing vaccines [4].

However, the ELISA has a few drawbacks. The most important is that there is no common inter-laboratory method to quantify the D-antigen unit [5]. The methods differ in the type of (catching and detecting) antibodies. Regarding the antibodies, which could be polyclonal or monoclonal, infinite combinations of detecting and catching antibodies are possible. If the detecting antibody is labeled, even more alternatives are possible. The method can also

Javier Martín (ed.), *Poliovirus: Methods and Protocols*, Methods in Molecular Biology, vol. 1387,
DOI 10.1007/978-1-4939-3292-4_16, © Springer Science+Business Media New York 2016

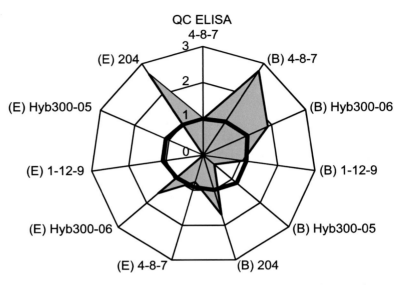

Fig. 1 Antigenic fingerprint of Sabin and wt IPV Type 3. The *right half* of the radarplot represents the ratio D-antigen concentration obtained by Biacore (marked (B)) to the D-antigen concentration obtained by QC-ELISA with mab 4-8-7. At the *left half* of the radarplot D-antigen ratio's by ELISA are shown (marked (E)). Reproduced from Vaccine, 2011 [6] with permission from Elsevier

differ in buffers used, incubation times, the sort of conjugation, blocking steps, and washing steps. For instance, differences in incubation time and temperature (2 h at 37 °C versus 2 h at 37 °C and overnight at 4 °C) resulted in a twofold difference in the apparent D antigen concentration [6]. Furthermore, many different monoclonal antibodies (mabs) are available for each serotype. These mabs could also be the reason for a variation in D-antigen concentration. With regard to inactivated poliomyelitis vaccine (IPV) this variation is not that pronounced; however, the D-antigen concentration of Sabin IPV (sIPV) depends strongly on both method and mabs, when using anti-Salk mabs (Fig. 1) Because of the many variations in the method to quantify the D-antigen unit, which may result in considerable inter-laboratory variability [7], standardization of IPV remains under discussion. If, in the foreseeable future, Salk IPV will be replaced by sIPV, which could be a serious option [8], standardization of sIPV will even be more of a challenge, considering the dependence of the D-unit on method and mabs.

Surface plasmon resonance (SPR) technology does not solve the sIPV dependency on mabs which is also illustrated in Fig. 1. SPR calibrated concentration assays use the same format as ELISA's for the quantification of antigens: first an anti-mouse antibody is immobilized to the surface of a chip, then a complex is formed by subsequent injections of an antigen specific mouse mab and a captured antigen. However, the advantages of SPR-measurements are that the measurements are label-free where the ELISA requires a conjugated antibody. The ELISA requires also several washing steps, Biacore detects the vaccine directly. Furthermore, the detection is in real-time: the Biacore monitors continuously each binding step in the assay, allowing visual monitoring each step, where the result of the ELISA is only seen after the final step. Last but not least, the hands-on time of an ELISA is quite longer than that of a Biacore assay. Besides, the inter-laboratory variability might decrease substantially because of the lacking of incubation steps, intermediate washing steps and different sets of polyclonal/monoclonal antibodies. Finally, results from Biacore correlate well with D-units from ELISA (Fig. 2). Also the lower limits of quantification are more or less the same: 3–0.4–1.1 DU/ml (type 1, 2, and 3, respectively) for Biacore and 1.5–0.4–0.7 DU/ml for ELISA.

Biacore measurements rely on SPR. The underlying complex theory of SPR is accurately described in the Biacore concentration analysis handbook [9]. It suffices to know that SPR-based instruments use an optical method to measure the refractive index near (within ~300 nm) a sensor surface. In the Biacore this surface forms one side of a small flow cell, through which a solution (the running buffer) passes under continuous flow. In order to detect an interaction one molecule (the ligand) is immobilized onto the sensor surface. Its binding partner (the analyte) is injected in buffer through the flow cell, also under continuous

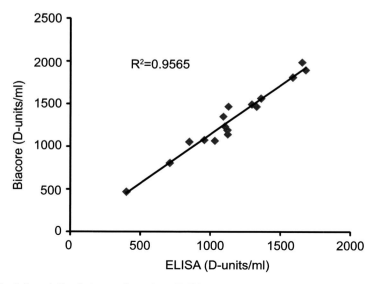

Fig. 2 Correlation between D-antigen ELISA and the calibrated Biacore D-antigen assay of 16 monovalent bulk vaccines of either type 1, 2, or 3

flow. As the analyte binds to the ligand, the accumulation of protein on the surface results in an increase in the refractive index. This change in refractive index is measured in real time, and the result plotted as response or resonance units (RUs) versus time (a so-called sensorgram) [10].

For the calibrated D-antigen assay it comes down to immobilization of anti-mouse IgG Fc-specific antibodies on the dextran layer of a sensorchip by primary amine coupling. Next, a detecting antibody is injected p, followed by IPV. The sensor chip is regenerated with 10 mM glycine-HCl, pH 1.5, preparing the surface for the next run, varying the sample or reference dilution. Assay data are analyzed by a four-parameter logistic curve fitting using the Biacore evaluation software. Antigenicity is calculated relative to the international IPV reference. This SPR method is robust and considerably more accurate as compared to ELISA. CV values below 5 % are easily achieved by this classical SPR method.

A few years ago, Biacore introduced a second approach, the calibration-free concentration analysis (CFCA). CFCA allows the determination of the absolute antigen concentration without the use of a reference. The analysis relies on measurement of the binding rate during sample injection under partially or complete mass transport limited conditions. Mass transport is a diffusion phenomenon that describes the movement of molecules from a higher concentration (e.g., in the bulk fluid in the center of a capillary) to a lower concentration (e.g., in the stagnant layer of fluid near the surface of a capillary). Mass limitations arise when either analyte binds to the surface faster than it diffuses from the solution during injection, or analyte does not diffuse fast enough from the surface during dissociation, leading to rebinding. To calculate the analyte concentration in the bulk solution a relation between initial binding rate and analyte concentration in the bulk is used. On a sensor surface with a high immobilization level the initial binding rate (slope) can be described as a function of the molecular weight of the analyte, the mass transport coefficient and the concentration of the analyte. The mass transport coefficient is calculated using flow cell dimensions, flow rate, and diffusion coefficient. The sample is run over the surface at two flow rates (5 and 100 µl/min), the initial binding rate (dR/dt) is measured and the software resolves the concentrations based on the aforementioned parameters. As long as the affinity of antigen and antibody is high enough ($k_a < 5 \times 10^4/\text{M}^{-1}\,\text{s}^{-1}$ and/or $K_D > 10^{-6}$ M), the calculated concentration is independent of the antibody used. The dynamic range of the method is approximately 0.05–5 µg/ml [9].

Indeed, results of CFCA with sIPV show that the active concentration measurement of type 1, 2 and 3 IPV (wt IPV and sIPV) is independent of the mab (see Fig. 3). The active particle concentration of polio vaccines also correlate with the virus concentrations of the vaccines calculated from the absorbance at 260 nm [6].

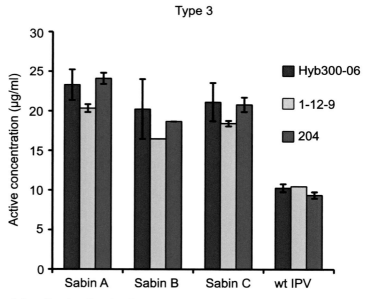

Fig. 3 Results of calibration free concentration analysis of Sabin/wt IPV type 3, resp. 15/50 DU/ml; D-units based on QC-ELISA. Error bars represent SD at $n = 3$; where error bars are missing, the concentration is an average of a duplicate measurement. Reproduced from Vaccine, 2011 [6] with permission from Elsevier

Because the D-antigen unit is not well defined, IPV quantification by using protein or virus concentrations is attractive from a standardization point of view. However, use of a protein assay is only valid if almost 100 % of the protein is poliovirus and has the native conformation, i.e., D-antigenicity. Measurement of the active particle concentration by CFCA is an attractive alternative as it combines quantity and quality in one assay. CFCA requires no calibration curve and the measurement is highly specific. Differences between different mabs do not exist; all high affinity mabs give the same results. Implementation of this assay as an in vitro measurement of IPV potency would require additional validation and an international harmonization study, but it may be worth doing considering the increasing role of IPV in the world.

2 Materials[1]

1. CM3-chip (GE Healthcare) (*see* **Note 1**). Storage 4–8 °C.

2. Amine coupling kit (GE Healthcare): N-hydroxysuccinimide (NHS), 115 mg; 1-ethyl-3-(3-dimethylaminopropyl) carbodiimide hydrochloride (EDC), 750 mg; 1 M ethanolamine

---

[1] Identification of particular products is provided as a guide to aid in the selection of equivalent, suitable products.

hydrochloride-NaOH pH 8.5. Dissolve the EDC and NHS by adding 10.0 ml of filtered, deionized water (Milli-Q) to each vial. Cap vials tightly and agitate until the solids are completely dissolved. Dispense the EDC and NHS solutions separately in 100 μl aliquots for storage at −18 °C or below. For the aliquots use the plastic vials Ø 0.7 mm and cap the vials with the type 3 rubber caps. Use aliquots within 2 months (*see* **Note 2**).

3. Acetate 5.0 (GE Healthcare): 10 mM sodium acetate, pH 5.0. Storage 4–8 °C.

4. HBS-P buffer 10x (GE Healthcare): 0.1 M HEPES; 1.5 M NaCl; 0.5 % v/v surfactant P20 will yield pH 7.4 when diluted: supplement 900 ml of Milli-Q with 100 ml 10× HBS-P buffer 10×. Store at room temperature. This buffer will be used as running buffer and dilution buffer. Filter the diluted buffer through a 0.2 μm filter before running a method.

5. Glycine 1.5 (GE Healthcare): 10 mM glycine-HCl pH 1.5. Storage 4–8 °C.

6. Goat anti-mouse (GaM) IgG F_c fragment specific (Thermo Scientific): antibody concentration 2.3 mg/ml in 0.01 M sodium phosphate and 0.25 M NaCl, pH 7.6. Store at 4 °C. Expiration date 1 year from date of product receipt. Prepare a ten times dilution (200 μg/ml GaM IgG Fc) in 100 μl acetate 5.0 buffer in a Ø 7 mm plastic vial (*see* also Subheading 3.1 **step 11**).

7. Reference preparation poliomyelitis vaccine inactivated trivalent (Pu91-01, RIVM): 430-95-285 DU/ml for type 1, type 2, and type 3, respectively. Storage ≤−70 °C. Calibration curves of trivalent Pu91-01 are twofold dilution series of 8 dilutions, starting with 1:1 in HBS-P buffer (*see* also Table 1). Prepare fresh before use.

8. IPV serotype 1, 2, and 3 specific mouse monoclonal antibodies (mabs) Hyb295-17 (IgG2b/k); Hyb 294-06 (site 1 specific, IgG2a/k); and Hyb300-06 (site 1 specific, IgG2a/k), respectively (BioPorto Diagnostics A/S): each mab 1 mg/ml ± 15 % in 0.01 M Phosphate Buffered Saline (PBS), pH 7.0; 0.5 M NaCl and 15 mM Sodium Azide. Storage 4–8 °C (*see* **Note 3**). For the calibrated D-antigen assay, dilute each mab to 1 μg/ml in HBS-P buffer. For the calibration-free assay the mabs are diluted to 4 μg/ml in HBS-P. Prepare fresh before use.

9. Inactivated polio vaccine samples for testing (*see* **Note 4**). The vaccines must be stored at 4–8 °C. The polio vaccines are stable for 24 months at 4–8 °C. Prepare appropriate dilutions (within the range of the calibration curves) fresh before use.

10. Microplates 96-well (GE Healthcare): polystyrene microplates.

Table 1
Typical D-antigen concentrations used for calibration curves of type 1, 2, and 3 of the international reference Pu91-01[a]

Standard nr.	D-antigen concentration (DU/ml)		
	Type 1	Type 2	Type 3
1	1.7	0.4	1.1
2	3.3	0.7	2.2
3	6.7	1.5	4.5
4	13.4	3.0	8.9
5	26.9	5.9	17.8
6	53.8	11.9	35.6
7	107.5	23.8	71.3
8	215.0	47.5	142.5

[a]Twofold dilution series of Pu91-01 with starting dilution 1:1 in HBS-P buffer

11. Microplate foils (96-well) (GE Healthcare): self-adhesive, transparent plastic foils.

12. Glass Vials, Ø 16 mm (GE Healthcare): 4.0 ml borosilicate screw top glass vials

13. Rubber caps, type 2 and 3 (GE Healthcare): penetrable cap made of Kraton G. Ventilated.

14. Plastic vials, Ø 7 mm (GE Healthcare): 0.8 ml rounded polypropylene microvials (*see* **Note 5**).

15. Plastic vials, Ø 11 mm (GE Healthcare): 1.5 ml polypropylene vials with wide opening that allows a pipette to reach the bottom.

16. SFCA Serum Filter Unit (Thermo Scientific): 500 ml bottle, 0.2 μm pore size

3   Methods

Although the calibrated D-antigen assay can be performed on all available Biacore instruments, whereas the CFCA method runs at the X100 and T200 Biacore, the procedures below are based on working on a T200-instrument (Biacore control software version 2.0) with T200 evaluation software (version 1.0).

3.1   Immobilization

Before performing the calibrated D-antigen assay or the calibration free assay, the sensor chip must be prepared enabling to capture the anti-polio mouse monoclonal antibodies. For this purpose

Goat-anti mouse IgG Fc specific antibodies are covalently bound to a CM3 sensorchip. In case of performing a calibration-free concentration assay, a second flow cell must be prepared with a blank immobilization.

1. Open the Biacore T200 control software (*see* **Note 6**).

2. Click on **insert sensor chip** icon. A window appears: **this will eject the sensor chip**.

3. Click on **Eject Chip** (this will take a minute).

4. A dialog box **Insert Chip** appears. Click on where appropriate, either **new chip** or **reuse**. In case of a new chip, select the chip type **CM3**, enter a chip id and chip lot number. In case of a reused chip, select the chip id from the pull-down menu. Close the sensor chip port cover and click **Dock Chip** to dock the chip in the instrument.

5. Click **File**, then **Open/new wizard template**, and then click **Immobilization** from the **Surface preparation** directory.

6. Click on **New** or **Open** an existing immobilization file.

7. The dialog box **Immoblization setup** appears (Fig. 4). Select **CM3** as **Chip type**. Select flow cell **1, 2, 3, or 4** (calibrated D-antigen assay)or in case of CFCA flow cells **1** or **3 blank** and **2** or **4 anti-mouse IgG Fc**. Select as **Method: Amine (CFCA)** or **Amine + Regeneration 3×** (calibrated D-antigen assay) (*see* **Note 7**). Enter the **ligand** Goat anti Mouse IgG Fc (for **CFCA** in either flow cell 2 or 4). Tick off **Specify contact time and flow rate. Contact time** 420 (s), **flow rate** 10 (µl/min). Click in case a CFCA is performed, in either flow cell 1 or 3 **blank immobilization**. Press **Next**.

8. The window **System preparations** appears. Tick off **Prime before run**. Enter **analysis temperature**, 25 °C, **compartment temperature** 25 °C. Press **Next**.

9. The window **Rack positions** appears. This dialog box shows where samples and reagents are placed in **reagent** rack 2 or **sample and reagent** rack 1 (*see* **Note 8**). Positions are color-coded according to sample and reagent categories. Positions are described by tool tips (hold the cursor over the position for a couple of seconds to display the tool tip). Empty positions show the position capacity and dead volume. Used positions show in addition the content name and the volume that will be used.

10. If the rack is already placed in the compartment, click in the dialog box **Eject rack**. The rack tray is ejected.

11. Place all necessary samples in the required volumes and concentrations (*see* **Note 9**) in the appropriate positions

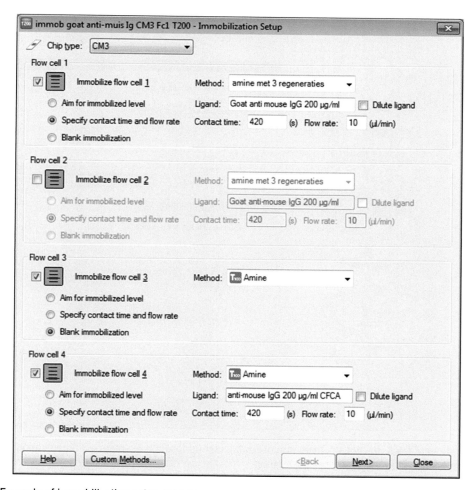

Fig. 4 Example of immobilization setup screen

according to the dialog box: frozen aliquots of EDC and NHS, 130 μl Ethanolamine, Goat-anti mouse IgG (*see* also heading 2 **item 6**) in acetate pH 5 buffer, and glycine-HCl (Fig. 5).

12. Click again **Eject rack**. The rack compartment opens. Place the rack into the compartment and click **OK**. Click **Next**.

13. A window **Prepare Run Protocol** appears. Perform the instructions in the dialog box (*see* **Note 10**). Click on **Save as** and enter a name for the immobilization method. Click on **Run** and enter a file name for the immobilization results.

14. After the immobilization run is finished (Fig. 6), a window **Immobilization Results** appears. This dialog box summarizes the results of the immobilization (*see* **Note 11**).

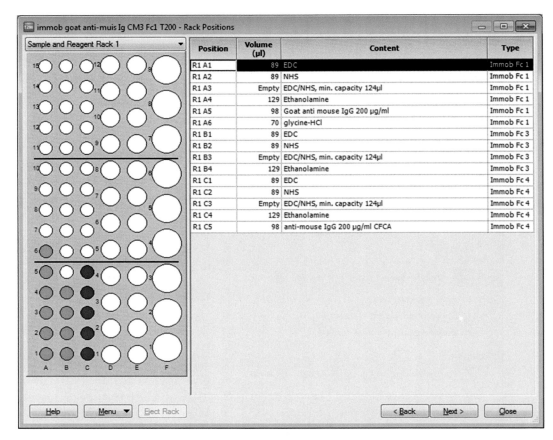

Fig. 5 Example of Rack positions screen. Immobilization set up for Row A: amine + 3 regenerations method; row B: blank immobilization; row C: amine method

3.2 Calibrated D-Antigen Assay for One Serotype and One mab

If samples consist of just one serotype, e.g., monovalent bulk, the concentration assay can be performed following the T200 concentration wizard. If samples are trivalent or it is feasible (considering the number of samples, *see* **Note 12**) to measure three serotypes in a row, it is better to build up a method (*see* Subheading 3.3).

1. Open the Biacore T200 control software.

2. Click **File**, then **Open/new wizard template**, and then click **Concentration Analysis** from the **Assay** directory.

3. Click on **New** or **Open** an existing concentration analysis file.

4. The window **Injection Sequence** appears. Choose the flow path (refers to the flow cell that has been immobilized) for the experiment. Choose the chip type (CM3) for the experiment. Select the sequence of injections for the assay: **Ligand capture** (refers to the use of anti-polio serotype specific mab), **Sample** (refers to the IPV sample injection), **Regeneration** (refers to the injection of regeneration solution to remove bound

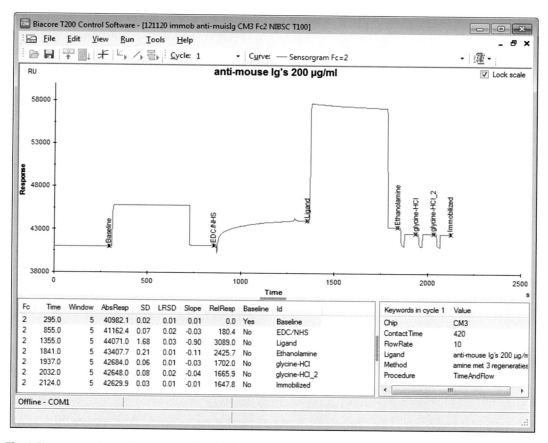

Fig. 6 Sensorgram immobilization Goat anti-Mouse IgG Fc by amine coupling with three regenerations

analyte) (the complex of mab and vaccine) from the surface. Click 1 injection for the regeneration. Click **Next**.

5. The window **Injection Sequence** appears. Click **Run startup cycles** to include dummy analysis cycles at the start of the run to make sure the system is stable. Enter HBS-P for the solution. Enter 5 for the number of cycles. Click **Next**.

6. The window **Injection Parameters** appears. Enter successively: **ligand name** the name of the serotype specific mab, **Contact time** 120 (s), **flow rate** 10 (μl/min), **Stabilization period** 0 (s), **Sample Contact time** 120 (s), **flow rate** 10 (μl/min), **Regeneration solution** Glycine-HCl 1.5, **Contact time** 30 (s), **flow rate** 30 (μl/min), **Stabilization period** 0 (s). Click **Next**.

7. The window **Calibration Parameters** appears. Enter the name of the International Reference used for the calibration curve. The option **Run First** is always checked, and indicates that a calibration curve will always be run before the first sample or control sample cycle. Optional: Check **Repeat calibration every … sample cycles** and enter an interval to repeat the calibration curve at the specified interval throughout the assay.

Check **Run last** to include an additional calibration cycle at the end of the assay. This will ensure that calibration trends can be used correctly in the Evaluation Software to compensate for drift in calibration responses during the assay. Enter the D-antigen concentration for each point in the calibration curve (*see* Table 1 for D-antigen concentrations of International Reference Pu91-01). You can change the concentration unit for the analyte samples in the table header. Click **Next**.

8. The window **Control Samples** appears. Skip the **Control Samples** window and add the control vaccine samples to the sample table (*see* **Note 13**). Click **Next**.

9. The **Sample Table** appears together with a **Sample and Position Import table**. Enter cancel. Enter the sample names and an optional dilution factor (*see* **Note 14**). The dilution factor is used to calculate the concentration in the original sample (e.g., if you enter a dilution factor 5, a measured concentration of 3 μM in a sample will be reported as a calculated concentration of 15 μM). If you leave the dilution factor blank, a value of 1 will be used. Each row in the table represents one cycle. If you want to run replicate samples, enter the same sample name on different rows. Click **Next**.

10. The window **System preparations** appears. Tick off **Prime before run**. Enter **analysis temperature**, 25 °C, **compartment temperature** 25 °C. Press **Next**.

11. The window **Rack positions** appears. This dialog box shows where samples and reagents are placed in the microplate and rack (*see* **Note 12**). You can change sample and reagent positions manually: Click on the sample or reagent in the sample plate and rack illustration and drag it to a new (empty) position. You cannot drag to a position that does not have sufficient capacity for the required volume of sample or reagent. Positions can also be reorganized using **Automatic positioning** under the **Menu** button (*see* **Note 15**).

12. Prepare samples, mabs, calibration curves, and controls (*see* Subheading 2.1, **items 7–9**) in the required volumes and concentrations (*see* **Note 16**) and place them in the appropriate positions according to the dialog box. If necessary, eject first the required rack from the T200.

13. Click again **Eject rack**. The rack compartment opens. Place the rack into the compartment and click **OK**. Click **Next**.

14. A window **Prepare Run Protocol** appears. Perform the instructions in the dialog box (*see* **Note 10**). Click on **Save as** and enter a name for the concentration method. Click on **Run** and enter a file name for the result file.

15. The sensorgram window appears. This window displays the sensorgram in real time for the current run. Figure 7 shows an example of a complete run.

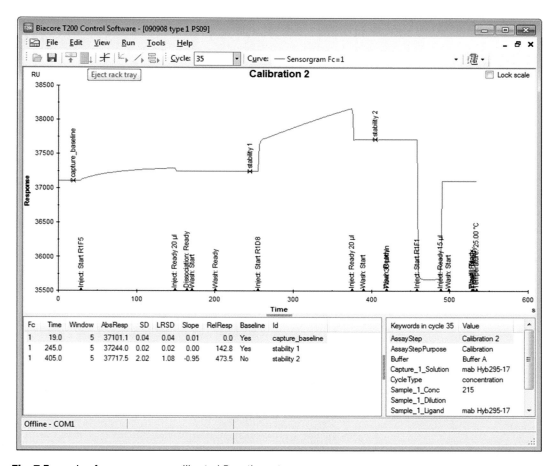

Fig. 7 Example of a sensorgram calibrated D-antigen assay

16. After the run is finished, the Biacore T200 Evaluation software starts automatically and shows an overlay of all generated sensorgrams (*see* Fig. 8) (*see* Subheading 3.4) (*see* **Note 17**).

3.3 Calibrated D-Antigen Assay for Two or More mabs and/or Serotypes in One Method

Because the same trivalent International Reference will be used in one method to quantify the D-antigen amount of two or three serotypes sequentially, it is necessary to build up a method in which the calibration curve can be allocated to a specific serotype. The concentration wizard for the calibrated D-antigen assay for one serotype can be used as base for the method builder.

1. Open the Biacore T200 control software.

2. Click **File**, then **Open/new Method**. Check **Open importable wizard templates** and **browse** to the calibrated D-antigen assay, e.g., "concentration assay type 1 .blm" (*see* **Note 18**).

3. Because the instrument settings are already imported by importing the concentration assay method, we can skip the general settings. Click **Assay steps** from the main menu **Method Builder**.

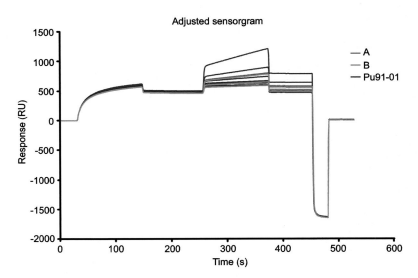

Fig. 8 Overlay of sensorgrams calibrated D-antigen assay with International Reference Pu9-01 and two vaccines A and B

4. Change the name **Calibration** in calibration 1 by clicking at the name in the flow diagram. The name appears in a window in the base settings where you can edit the name. Change also the name **Sample** in Sample A.

5. Copy in the flow diagram the two parameters **calibration 1**, and **sample A**. The flow diagram is now extended with two copies of the parameters. Change the names of these parameters in **calibration 2**, and **sample B** (*see* Fig. 9). Repeat instruction 5, if necessary and feasible (*see* **Note 12**), for the third serotype.

6. Click **Cycle type**. Cycle types define the series of injections and associated parameters that will be used in a cycle. Because we have imported the concentration assay method, the concentration cycle type is already available. Delete **Conditioning** (or any other cycle types) and add **Start up**. Define for Start **up** the series of injections: (1) On the **Commands** tab, select the command **Sample** from the drop-down list and click **Insert** to add the command to the cycle type definition. Next, specify the **Settings** for Sample 1 in the right-hand panel: First unclick in **Method variables** all possible variables. **Type** low sample consumption, **Sample solution** HBS-P, **Contact time** 120 (s), **Dissociation time** 0 (s), **Flow rate** (5 μl/min). (2) On the **Commands** tab, select the command **Regeneration** from the drop-down list and click **Insert** to add the command to the cycle type definition. Next, specify the **Settings** for Regeneration 1 in the right-hand panel: First unclick in **Method variables** all possible variables. **Type Regeneration solution** Glycine-HCl, **Contact time** 30 (s), **Flow rate** (30 μl/min).

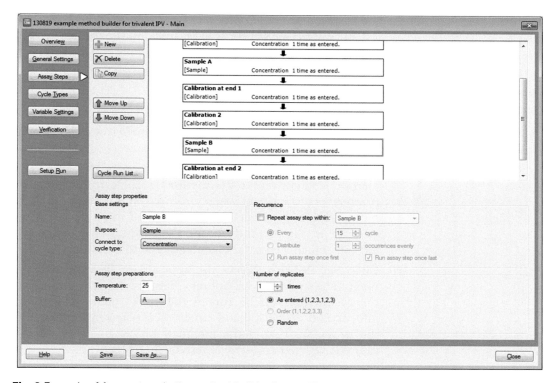

Fig. 9 Example of Assay steps in the method builder for a calibrated D-antigen assay with two or more mabs and/or serotypes

7. Check the **settings** of the **concentration cycle type**: **Capture 1: Capture solution:** variable (check **Method variables**), **Contact time** 120 (s), **Flow rate** (10 µl/min); **Sample 1: Type** low sample consumption, **Sample solution** variable (check **Method variables**), **Contact time** 120 (s), **Dissociation time** 15 (s), **Flow rate** (10 µl/min); **Regeneration 1: Type Regeneration solution** Glycine-HCl, **Contact time** 30 (s), **Flow rate** (30 µl/min).

8. Click **Variable settings**. Click in the left-hand panel assay step **Start up**, check in the right-hand panel, **define all values in method**. Repeat this for each assay step. Next, click in the left-hand panel assay step **Start up**. Enter in row 1, both columns **Capture solution** and **Solution** HBS-P. Click the next assay step **Calibration 1**. Enter in column **Capture solution** (in case of serotype 1) 8× Hyb295-17 (1 per row), column **Solution** 8× Pu91-01, and column **Concentration** 8 D-antigen concentrations of Pu91-01 (*see* Table 1) (*see* **Note 19**). Next, click assay step **Sample A**. Enter **Capture solution, Sample solution** and **Dilution** (of the sample) in the appropriate columns. Please note that the capture solution should always be the same solution as the capture solution for

the corresponding calibration curve, e.g., the mab against type 1. Control vaccine samples can be added to the Sample A table. Repeat the steps of calibration 1, and sample A, for calibration 2, and sample B. Please enter for all assay steps the same capture solution, e.g., the mab against type 2. Control vaccine samples can be added to the Sample B table. If there is enough room for a third series (*see* **step 5**), repeat the steps for calibration 3, etc. The capture solution for this series might be the mab against serotype 3 (*see* **Note 20**).

9. Click **Verification**. This workspace reports the results of method verification. If the method is correctly defined and parameters are entered where required, the message is: "The method has been verified and can be used to set up a run". If an error in the method occurs, the verification window will indicate which step is mistaken and how to solve this.

10. Click **Set up run**. Select the appropriate **flow path**. Click **Next**. The cycle run list appears. This workspace summarizes the cycles that will be run in the entire method. Use this information to check that the method is correctly defined, that the cycles are run in the intended order, and that variable values have been correctly entered. Click **Next**.

11. The window **System preparations** appears. From here, follow steps **10–16** of Subheading 3.2.

3.4 Biacore T200 Evaluation Concentration Assay

After the method run is finished, the evaluation software starts automatically (*see* **Note 17**). At first sight we see an overlay of all the sensorgrams depicting all cycle runs.

1. Check the overlay on irregularities, e.g., air spikes, drifting baselines. Identify which cycle causes the irregularity: Click **cycle <overlay>** and click cycles one by one (*see* **Note 21**). If an irregularity has occurred in one of the cycles, take that into consideration at the concentration analysis evaluation (e.g., skip the results of this specific cycle, *see* **Note 24**).

2. In the menu bar, click **Concentration analysis** and then **Using calibration**.

3. Choose the **Flow cell** (the immobilized flow cell), **report point** (stability 2; this is the response after binding of the sample to the capture antibody), (*see* also Fig. 10) and Response type (relative response) to be used for the calibration curves. All calibration curves in the evaluation item use the same settings.

4. Calibration curves are shown in the right-hand panel. Use the browse buttons or the selector button at the top of the panel to select which calibration curves to display. The available options depend on the way the calibration curves are used: for the Use average calibration curve option a single average calibration curve is displayed (*see* **Note 22**).

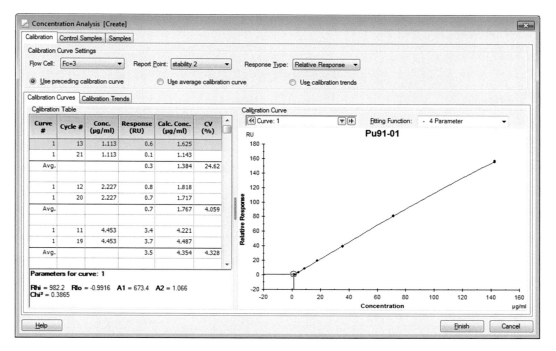

Fig. 10 Example of the BIAevaluation concentration analysis window

5. Choose the 4-parameter fitting as the **fitting function** for the calibration curve (*see* **Note 23**).

6. Click for sample and control results the relevant tabs. The **Sample tab** shows the results of concentration analysis for the unknown samples (*see* Fig. 11): Sample data is listed in the left-hand panel. The Calc.Conc. column in this table shows the measured concentration multiplied by the dilution factor (i.e., the concentration in the original sample before dilution). Concentrations are calculated according to the setting on the Calibration tab. The calibration curve used to calculate sample concentrations is listed in the Calib.Curve column. Samples that lie outside the range of the calibration curve are listed as above or below the limits of calibration as appropriate. The right-hand panel shows the calibration curve used for the currently selected sample. Calibration points are shown in red and samples in black (*see* **Notes 24** and **25**).

7. Click **Finish** to save the concentration analysis evaluation item.

3.5 Calibration-Free Concentration Analysis

This assay needs monoclonal antibodies that have a high affinity to polio epitopes (*see* **Note 26**). The mabs, mentioned in the materials, meet this criterion. The vaccine samples should be diluted to 0.5–1 µg/ml protein in HBS-P, mabs can be diluted 250× in HBS-P. For this method we also need the molecular weight of a polio particle (8250 kDa) and the diffusion coefficient (D) of poliovirus

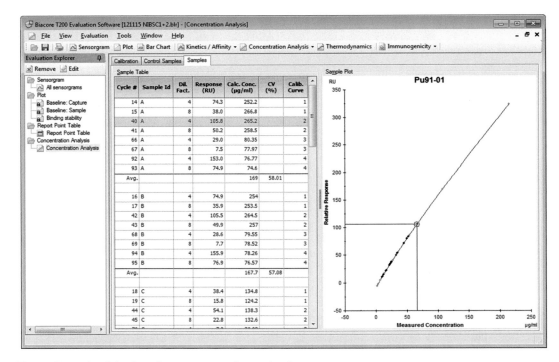

Fig. 11 Example of the D-antigen concentration evaluation

(1.44E–11 (m^2/s) at 20 °C) [11]. A CM3-sensorchip is used for the immobilization; however, besides the immobilization of Goat-anti mouse IgG Fc-antibodies, a blank immobilization of a second flow cell is needed. This flow cell serves as a reference flow cell. Regarding the immobilization procedure, *see* Subheading 3.1.

1. Open the Biacore T200 control software.

2. Click **File**, then **Open/new Method**. Click in directory **Biacore Methods** Calibration-free concentration (*see* **Note 27**). Click **Open**.

3. Click **General Settings**; choose **Concentration Unit** μg/ml.

4. Click **Assay Steps**; choose **Startup**, then enter **Number of replicates** 5.

5. Click **Blank**, then enter **Recurrence every** ten **cycle.**

6. Click **Cycle Types**, and then click cycle types **Sample**. On the **Commands** tab insert the command **Capture**. The sequence of commands for cycle type **sample** is: **Capture 1, Sample 1, Regeneration 1**. Next, specify the **Settings** for Capture 1 in the right-hand panel: First click in **Method variables** capture solution. **Contact time** 60 (s), Flow **rate** (10 μl/min), **Flow path** both.

7. Click **Variable settings**. Click **sample** in **Assay steps**. Delete "No" in the **Sample1 Blank** cell (the cell ought to be empty).

Table 2
Example of variable values for assay step "sample" in method builder
Calibration-free Concentration assay (*see* Subheading 3.5 step 9)

Capture solution	Sample solution	Flow rate (μl/min)	M_W (Da)	Diffusion coefficient (20 °C)	Dilution
mab 295-17	A	5	8,250,000	1.44E–11	10
mab 295-17	A	100	8,250,000	1.44E–11	10

8. Click **Verification.** This workspace reports the results of method verification. If the method is correctly defined and parameters are entered where required, the message is: "The method has been verified and can be used to set up a run". If an error in the method occurs, the verification window will indicate which step is mistaken and how to solve this.

9. Click **Set up run.** Select the appropriate **flow path** (either **2-1** or **4-3**, dependent of which flow cells are immobilized for this assay). Click **Next.** A **Sample and Position Import table pops up**. Click Cancel. Enter the variable values for **Assay step Sample** (Table 2). Note that the variables must be entered in duplicate because of the two different flow rates.

10. Enter the variable values for **Assay step Blank in** row 1 and 2: buffer. Click **Next.**

11. The cycle run list appears. This workspace summarizes the cycles that will be run in the entire method. Use this information to check that the method is correctly defined, that the cycles are run in the intended order, and that variable values have been correctly entered. Click **Next.**

12. The window **System preparations** appears. From here, follow **steps 10–15** of Subheading 3.2.

3.6 Biacore T200 Evaluation CFCA

After the method run is finished, the evaluation software starts automatically (*see* **Note 17**).

1. In the menu bar, click **Concentration analysis** and then **Calibration free**.

2. A window **Select samples** appears. Click on the checkmarks for **Expand all cycles, Show original sensorgrams**, and **Use reference subtracted data.**

3. Choose which samples to evaluate: The table in the top panel lists all samples that can be evaluated. Remove the checkmark from a sample row to exclude the sample from the evaluation, or from a cycle row to exclude the individual cycles. Use the

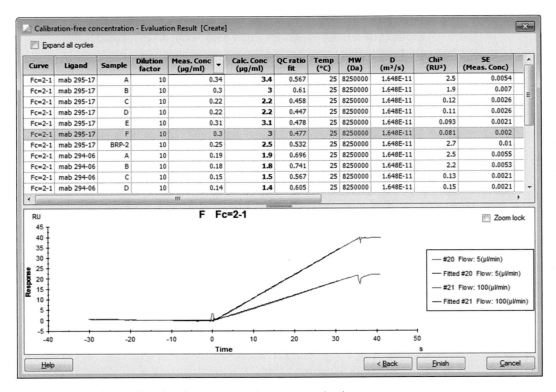

Fig. 12 Example of the calibration-free concentration assay evaluation

QC ratio (>0.20), Initial rate (<0.3 RU at Flow = 5) and the sensorgram appearance to help in deciding whether to exclude samples (*see* **Note 28**). Click **Next**.

4. This step shows the results of the evaluation. The results are presented as a table of samples with calculated concentration values (*see* Fig. 12). Use the **Expand all cycles** option to show or hide cycle details for each sample. Concentrations are reported as values measured in the injected sample (Meas. Conc) and values calculated for the sample before dilution (Calc. Conc = Meas. Conc × Dilution factor). Choose the unit for reported concentration in the column header for **Meas. Conc**.

4   Notes

1. A CM3 chip is recommended (instead of the regular CM5 chip) because an (inactivated) poliovirus particle is large (±30 nm) compared to individual protein molecule. The shorter dextran chains on the CM3 chip allow the interaction to take place closer to the surface which can improve sensitivity.

2. Because EDC is sensitive to hydrolysis, it is important to freeze the aliquots of EDC immediately.

3. Each laboratory will have an own panel of monoclonal antibodies. Especially for serotype 3, it is important to choose a mab that is site 1 specific because of a possible trypsin-induced degradation of site 1 which might be overseen if a site 2 or 3 specific mab would have been used.

4. The D-antigen concentration should be above the lower limit of quantification of 3-0.4-1.1 DU/ml for type 1, 2, and 3, respectively.

5. Although these tubes do not have the same size as micronic tubes, they do fit in a micronic rack, which makes it easier to handle a multichannel pipette.

6. The method is also suitable for Biacore T100 software.

7. To get rid of unimmobilized molecules and to stabilize the base line after immobilization faster, we adapted the amine coupling method by adding three injections of regeneration buffer: in the immobilization set-up window, press **Custom Methods**. A new window **Custom Methods** appears. Click from the methods **Amine**. Press **Copy**. In the methods window **Copy of amine** appears. Add to the command list (lower window) **Inject** by pressing the **inject button.** In the new window fill in **solution**: Glycine-HCl, **contact time** 30 (s), **flow rate 5** (μl/min). Press **Add**. Repeat this inject **command** twice. Change the **method name** in "Amine + regeneration 3×". Press **OK**. The new immobilization method is now added to the Method scroll menu in the immobilization set up window (Fig. 13).

8. You can change the reagent rack and microplate types in the drop-down lists above the respective illustrations. If you do so, all positions in the affected rack or plate will be cleared and must be reassigned either manually or automatically.

9. Solution volumes listed in the table are minimum volumes unless they are specified as **Exactly.**

10. The estimated run time and buffer consumption are minimum values. Estimated buffer consumption is calculated with a dead volume of 50 ml in the bottle and rounded up to the nearest 100 ml. We recommend diluting the 10× HBS-P buffer (running and dilution buffer for both immobilization and D-antigen c assays) to a volume of at least 1 l, to have enough for the concentration assays as well.

11. Immobilization levels to be expected are between 4000 and 6000 RU.

12. It is advisable to use Rack 1 in combination with a microplate. In this setting one serotype could be measured in one run with a maximum of 26 samples in three dilutions or in case of three serotypes (*see* also Subheading 3.3), 16 samples in three dilutions.

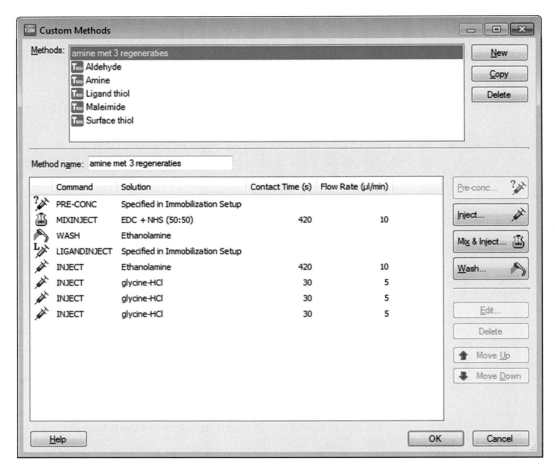

Fig. 13 Command for customized Amine method: Amine coupling with three regenerations

13. If you prefer to present control samples separately you may check **Run control samples**. Control samples are presented separately in evaluation of concentration analysis. Check **Repeat control sample(s) every ... sample cycles** and enter an interval to repeat the control samples at the specified interval. Controls will be run at the start and end of the assay and at the specified interval. If this option is not checked, controls will be run once at the start of the assay. Enter sample names and expected concentrations for the control samples. Measured concentrations for control samples are expressed relative to the expected concentrations in the evaluation. The unit for expected concentrations is the same as that chosen for the calibration curve. Concentrations for control samples should be within the range covered by the calibration curve, to allow the measured concentrations to be determined. Each row in the table represents one cycle. If you want to run replicate control samples, enter the same sample name on different rows.

14. Although it is optional, we recommend a twofold dilution series of three dilutions in HBS-P within the calibration curve range.

15. If using a microplate and a multichannel, it is more convenient to place the dilution series of calibration curves and samples in rows of descending order. Click the **Menu** button and the **Automatic Positioning**. Choose for the **Calibration** and **Samples** the row **Orientation, Anchor** bottom left and **First Sort** by content descending. Check if by choosing a larger **Vial Size** and **Pooling** yes, the content of two or more vials for one solution can be merged. Click **apply**.

16. In case of using other mabs than mentioned in the materials section, find a mab dilution that generates a signal above 100 Response Units (RU) after capturing the International Reference in the 1:1 dilution. This can be explored by the **Binding Analysis** wizard. Follow the instructions in Subheading 3.2, **steps 3–6**, but instead of checking the **Ligand Capture**, check **Sample** [2]. In the window **Injection Parameters**, the **Ligand Name** and **Sample** are replaced by **First Sample and Second Sample**. Enter in the sample table into the sample ID 1 column: mab A in a different dilutions, e.g., row 1 mab A 1:100; row 2 mab A 1:500; row 3 mab A 1:1000. Enter in column Sample Id 2 Pu91-01 (diluted 1:1) 3× in row 1, 2, and 3, respectively. Click **Next** and follow instructions according to Subheading 3.2, **steps 10–16**.

17. The evaluation software can also be started separately from a specific method.

18. It is easier to import a wizard that has been made for a calibrated D-antigen assay for one serotype and one mab.

19. Right-click on a row in the table for options for copying and pasting variable values and for inserting and removing rows.

20. Instead of capture mab solutions of serotype 1, 2, and 3, you may use two or three different site specific mabs for the same serotype. In this case, the purpose of the assay is not just D-antigen quantification but also epitope characterization. However, the capture mab solution should always be the same for both sample and calibration curve.

21. If you block the calibration curve cycles in the **cycle<overlay>**, the overlay will show just these cycles. By blocking certain cycles you will get a good impression of the quality of the cycle runs.

22. In the **calibration curve settings** different options for using calibration curves are available: **Use preceding calibration curve, Use average calibration curve**, and **Use calibration trends**. Normally the use of an average calibration curve is recommended, but in case of a single calibration curve of an increasing baseline it is better to choose the **Use preceding calibration curve** option.

23. If the calibration curve shows an obvious outlier, you may choose to exclude this point: right-click on a point in the plot or on a row in the calibration data table and choose Exclude cycle to exclude the point from the fitting. Excluded points are shown with open symbols. Right-click on a curve and choose Exclude curve to exclude the calibration curve. Excluding a curve automatically excludes all the calibration points in the curve. If you reinclude sufficient points to allow calibration curve fitting, the curve will still be excluded but will be shown as a broken line.

24. If a sample point shows an obvious outlier (*see* also Subheading 3.4 **step 1**) and you want to exclude this point: Right-click on a point in the plot or on a row in the sample data table and choose Exclude cycle to exclude the sample from the average and CV calculations. Excluded points are shown with open symbols.

25. In case of the calibrated D-antigen concentration assay with two or more mabs and/or serotypes, be aware that average and CV calculations in the sample data table are based on the results of one sample and could be derived from more than one calibration set, e.g., in case of a trivalent sample the average result could be the average D-antigen concentration of serotype 1, 2, and 3. In this case it is necessary to calculate average and CV per calibration curve.

26. The association and dissociation rates of the used mabs are unknown; however, these mabs show high QC ratios. Using a low affine mab translates into overlapping initial binding rates at the two different flow rates which results in a too low (<0.2) QC ratio.

27. The CFCA assay is included in the regular Biacore methods and contains a lot of default settings in the method builder. With regard to polio vaccines we have made a few adjustments to the method and only these adjustments will be dealt with.

28. The QC ratio provides an indication of the extent to which the initial binding rate is influenced by flow rate. The QC ratio in the results table is calculated from the fitted curves and not from the original sensorgram data. In this way, disturbances and excluded data in the original sensorgrams do not affect the reported QC ratio, except insofar as they affect the fitting. Exclude samples where the QC ratio is lower than about 0.2 and where the initial binding rate at the lower flow rate is too low (<0.3 RU) to be reliably measured. Check also the sensorgram appearance. Exclude cycles that are obviously disturbed. The sensorgram at the lowest flow rate should be approximately linear over the first 30 s of sample injection. Exclude samples where the sensorgram at the lowest flow rate

shows excessive curvature. Note that if you exclude individual cycle from a sample you have to exclude also the cycle with the second flow rate because analysis of each sample uses at least two different flow rates.

References

1. Wood DJ, Heath AB, Kersten GF et al (1997) A new WHO International Reference Reagent for use in potency assays of inactivated poliomyelitis vaccine. Biologicals 25:59–64
2. van Steenis G, van Wezel AL, Sekhuis VM (1981) Potency testing of killed polio vaccine in rats. Dev Biol Stand 47:119–128
3. Bevilacqua JM, Young L, Chiu SW et al (1996) Rat immunogenicity assay of inactivated poliovirus. Dev Biol Stand 86:121–127
4. Guideline on waiving the *in vivo* assay of poliomyelitis vaccine (inactivated) and its combinations. In: European Pharmacopoeia. 2720 7th ed: EDQM
5. Morgeaux S, Milne C, Daas A (2005) Feasibility study to develop a common in vitro D antigen assay for inactivated poliomyelitis vaccines. Pharmeuropa Bio 2005:19–26
6. Westdijk J, Brugmans D, Martin J et al (2011) Characterization and standardization of Sabin based inactivated polio vaccine: proposal for a new antigen unit for inactivated polio vaccines. Vaccine 29:3390–3397
7. Wood DJ, Heath AB, Sawyer LA (1995) A WHO collaborative study on assays of the antigenic content of inactivated poliovirus vaccines. Biologicals 23:83–94
8. Bakker WA, Thomassen YE, Van't Oever AG et al (2011) Inactivated polio vaccine development for technology transfer using attenuated Sabin poliovirus strains to shift from Salk-IPV to Sabin-IPV. Vaccine 29:7188–7196
9. Biacore concentration analysis handbook, BR1005-12 Edition AB/01-2007, GE Healthcare
10. van der Merwe PA (2001) Surface plasmon resonance in protein-ligand interactions. Oxford University Press, London
11. Koch F, Koch G (1985) The molecular biology of poliovirus. Springer-Verlag Wien, New York

Chapter 17

Identification and Analysis of Antiviral Compounds Against Poliovirus

Pieter Leyssen, David Franco, Aloys Tijsma, Céline Lacroix, Armando De Palma, and Johan Neyts

Abstract

The Global Polio Eradication Initiative, launched in 1988, had as its goal the eradication of polio worldwide by the year 2000 through large-scale vaccinations campaigns with the live attenuated oral PV vaccine (OPV) (Griffiths et al., Biologicals 34:73–74, 2006). Despite substantial progress, polio remains endemic in several countries and new imported cases are reported on a regular basis (http://www.polioeradication. org/casecount.asp).

It was recognized by the poliovirus research community that developing antivirals against poliovirus would be invaluable in the post-OPV era. Here, we describe three methods essential for the identification of selective inhibitors of poliovirus replication and for determining their mode of action by time-of-drug-addition studies as well as by the isolation of compound-resistant poliovirus variants.

Key words Drug resistance, Antiviral, Poliovirus, Phenotyping, Genotyping, Antiviral mechanism of action, Time-of-drug-addition studies

1 Introduction

In developing strategies to respond to and control poliovirus outbreaks during the final stages of eradication and in the post-eradication era, significant challenges arise. First, as vaccine-derived polioviruses (VDPVs) [1] continue to circulate, there will be a high probability of outbreaks of paralytic disease in the first several years after oral polio vaccine (OPV) cessation [2]. Second is the issue of how to deal with wild-type polio outbreaks. In this context, the role of antiviral drugs for poliovirus was considered at a workshop sponsored by the National Research Council (NRC) [3]. The workshop report, released in March 2006, concluded that the development of poliovirus antiviral drugs might be beneficial and probably essential to successfully eradicate polio. Within this context, it is of utmost importance to identify poliovirus inhibitors, which can be

Javier Martín (ed.), *Poliovirus: Methods and Protocols*, Methods in Molecular Biology, vol. 1387,
DOI 10.1007/978-1-4939-3292-4_17, © Springer Science+Business Media New York 2016

employed in the post-OPV era, and to elucidate their mode of action. Here, we provide a detailed description of methods that can be used to (A) identify selective antiviral compounds and (B) determine the mode of action of anti-poliovirus compounds by time-of-drug-addition studies as well by the isolation of compound-resistant poliovirus variants. For the identification of antiviral compounds, the MTS-based cytopathic effect (CPE) reduction assay is a reliable, easy-to-use and rapid method that is widely used for the identification of antivirals [4]. This cell-based assay relies on the ability of a compound to inhibit polio-virus-induced induction of CPE. When a compound has inhibitory activity on the replication of poliovirus, virus-induced CPE remains absent. As a consequence, the cells remain meta-bolic active and are able to convert a yellow tetrazolium salt substrate (MTS/PMS) to a brown-colored product (Formazan). When a compound does not show antiviral effects, the host cells are destroyed and—as a result of the lack of metabolic activity—the yellow substrate remains unmetabolized. Quantification of the colorimetric conversion allows calculation of the 50 % effective concentration (EC_{50}) values. A first approach to deter-mine the mode of action of antiviral compounds is the time-of-drug-addition study. Such experiments offer an easy and reliable tool to get an indication during which stage of the viral replica-tion cycle the anti-poliovirus drug exerts its antiviral activity. The method allows discriminating between early- and late-stage inhibitors. In principle, the method involves addition of the drug to cell cultures either before or after virus/host cell attach-ment/entry, and evaluates the antiviral effect in both instances. A drug that still elicits an antiviral effect after viral entry is likely to function as a non-capsid binder. In contrast, entry inhibi-tors/capsid binding drugs will exhibit antiviral activity only when added before the virus has entered the host cell and lose their activity when added at a later stage. Unraveling of the exact molecular mode of action of antiviral compounds however requires the isolation of resistant poliovirus variants. These studies are a very powerful tool in unraveling the mode of action of a particular polio antiviral compound. The method here described is tailored as such that one can start with a wild-type virus population and end up with an enriched and purified com-pound-resistant virus population. For this, five consecutive steps are needed. Three additional steps are added to characterize the phenotype and genotype of the variants. As outlined above, in the antiviral screening section of this chapter, a compound is only considered to be a selective antiviral when it completely inhibits virus-induced CPE following microscopic inspection (i.e., the treated infected condition resembles the untreated, uninfected condition). This is the prerequisite to be able to per-form the method described below.

2   Materials

2.1   For Identification of Poliovirus Antiviral Compounds with the MTS-Based CPE Reduction Assay

- Buffalo green monkey (BGM) cells.
- Growth medium (MEM supplemented with 10 % fetal calf serum, 1 % 200 mM L-glutamine, and 0.075 % sodium bicarbonate).
- Assay medium (MEM supplemented with 2 % fetal calf serum, 1 % 200 mM L-glutamine, and 0.075 % sodium bicarbonate).
- Phenol red- and glutamine-free MEM medium.
- 0.05 % Trypsin-EDTA.
- Dimethyl sulfoxide (DMSO).
- MTS solution containing 2 g/L 3-(4,5-dimethylthiazol-2-yl)-5-(3-carboxymethoxyphenyl)-2-(4-sulfophenyl)-2H-tetrazolium and 46 mg/L phenazine methosulfate in PBS (pH 6–6.5).
- Tissue culture flasks, 75 cm².
- 96-well microtiter plates.
- Microplate reader.

2.2   For Time-of-Drug-Addition Studies

- Buffalo green monkey (BGM) cells.
- Assay medium (MEM supplemented with 2 % fetal calf serum, 1 % 200 mM L-glutamine, and 0.075 % sodium bicarbonate).
- 24-well microtiter plates.
- Antiviral drug.
- qPCR machine.
- 2× One-Step RT-qPCR mixture.
- Poliovirus specific primers and probe.

2.3   For Selecting and Purifying Putative Compound-Resistant Poliovirus Variants

- Buffalo green monkey (BGM) cells.
- Assay medium (MEM supplemented with 2 % fetal calf serum, 1 % 200 mM L-glutamine, and 0.075 % sodium bicarbonate).
- 96-well microtiter plates.
- 25 cm² flask.
- Antiviral drug.

3   Methods

3.1   Identification of Poliovirus Antiviral Compounds with the MTS-Based CPE Reduction Assay

3.1.1   Subculture of BGM Cells

- Grow BGM cells to confluency in 75 cm² flasks.
- Aspirate the supernatant.
- Wash the cells with 0.05 % trypsin-EDTA.
- Trypsinize the cells with 2 mL trypsin at 37 °C.
- Resuspend the trypsinized cells in 10 mL growth medium.

- Pellet the cells during 6 min at 1100 rpm (200–250×g).

- Resuspend the pellet in 6 mL growth medium and transfer 1 mL to a 75 cm² flask with 20 mL growth medium.

3.1.2 Preparation
of Compound Solutions

- All compounds are dissolved in DMSO to a final concentration of 10 mg/mL and stored, depending on the stability and properties of the compound, at –20 or 4 °C until used.

3.1.3 Setting
Up the Antiviral Assay Plate

- Seed a microtiter plate 24 h in advance to allow the BGM cells to adhere to the plate (25,000 cells in 100 μL assay medium per well) (*see* **Note 1**).

- A maximum of six compounds, in eightfold serial dilutions can be assayed in parallel on a single microtiter plate (*see* Fig. 1). For every compound, a 1-to-2 serial dilution series will be performed. To this end, for each compound a start concentration has to be prepared, which should be 4× concentrated with regard to the final concentration.

- 100 μL of the compound solution is added to the first column (column 2) of the dilution series, mixed by pipetting and 100 μL of compound containing medium is transferred to the next well. Repeat this up to column 9. Column 10 will be used as virus control (VC; assay medium and virus), column 11 will

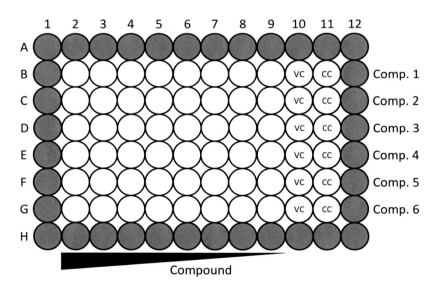

Fig. 1 96-well plate layout to allow quantification of the dose-response antiviral effect of six compounds in parallel on virus-induced cytopathic effects (to be quantified by methodologies that measure residual host cell metabolism or host cell death) or virus replication (to be quantified by real-time quantitative RT-PCR or titration for infectious virus content). The values obtained for the virus control (VC) and the cell control (CC) are to be used to verify the assay quality by calculation of the Z-factor and are necessary to convert the raw values into percentages of the controls. These percentages are subsequently used to calculate the EC_{50} values (concentration at which the compound elicits a 50 % antiviral effect) by logarithmic extrapolation from the dose-response curves

be used as cell control (CC; only assay medium). Wells at the edge of the microtiter plate will not be used to prevent edge effects that may influence the interpretation of the results, but will contain 100 μL medium.

- To each well of quadrant B2 to G10, add 100 μL of a 2× virus solution in assay medium (final concentration of one hundred 50 % cell culture infective dose $CCID_{50}$ of poliovirus per well).

- Add 100 μL of assay medium to column 11.

- Return the assay plate to an incubator (37 °C, 5 % CO_2, 95–99 % relative humidity) for 2–3 days, until full CPE is observed in the VC wells (*see* **Note 2**).

3.1.4 Evaluation and Quantification of Antiviral Activity

- At 2–3 days post infection, aspirate the assay medium.

- Add per well 100 μL of a 5 % MTS solution in phenol red and glutamine-free medium.

- Incubate for 1 h at 37 °C and 5 % CO_2 in incubator.

- Measure absorbance in each well at a wavelength of 498 nm using the microplate reader.

- At each concentration for a given compound, as well as for the virus and the cell control, calculate the percentage CPE as compared to the untreated cell control (*see* **Note 3**).

- Finally, for each compound, calculate the EC_{50} value, which is defined as the concentration of compound that induces 50 % protection against virus-induced CPE. This value is determined using the logarithmic interpolation from a dose-response curve (CPE versus compound concentration) (*see* **Note 4**).

3.2 Determining the Mode of Action of Polio Antiviral Compounds by Time-of-Drug-Addition Studies

From here on, the following items are defined as follows:

- Time point (TP) "0" ("zero") is arbitrarily defined as the moment where virus is added to the cell cultures.

- Concentration "*X*" refers to the final desired drug concentration.

3.2.1 Definitions

- The final poliovirus titer is designated as "*Y*" $CCID_{50}$.

3.2.2 Time-of-Drug-Addition Experiment

- One day prior to the time-of-drug-addition experiment, BGM cells are seeded to confluency in 24-well microtiter plates. Per compound to be evaluated, four wells should be prepared (*see* Fig. 2).

- Solutions to be prepared for each antiviral at the start of the time-of-drug-addition experiment are shown in Table 1.

- Optionally, the experiment could contain three control antivirals; a capsid binding agent such as pirodavir, a non-capsid agent such as rupintrivir (a protease inhibitor) and an antivirally inactive drug. The number and volume of assay solutions should be increased accordingly.

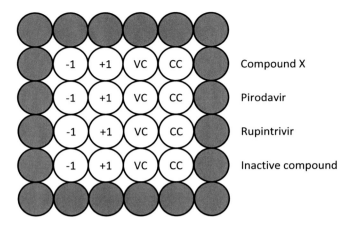

Fig. 2 24-well plate layout to perform a time-of-drug-addition experiment in which compound *X* is bench-marked against the two reference compounds pirodavir (an early-stage entry inhibitor) and rupintrivir (a late-stage replication inhibitor), as well as an inactive compound

Table 1
Solutions to be prepared for each antiviral at the start of the time-of-drug-addition experiment

	Poliovirus titer (10^6 CCID$_{50}$)	Drug concentration ($10\times$ EC$_{50}$)	Total volume required in assay medium (mL)
Solution A	–	*X*	3
Solution B	$2Y$	–	3
Solution C (starting from B)	$2Y$	*X*	1

- For each antiviral to be evaluated, mark the four wells in the 24-well plate with "TP −1h", "TP +1h", "VC" (virus control) and "CC" (cell control). VC refers to the condition where virus but no drug is added to the cell cultures, CC contains no virus and no antiviral drug.
- At time point −1 (1 h prior to addition of virus to the cell cultures):
  - Aspirate the medium from all wells in the 24-well plate.
  - To well "TP −1h", add 500 µL of solution A.
  - To the remaining three wells, add 500 µL of assay medium.
- At time point 0 (addition of virus):
  - To well "TP −1h", add 500 µL of solution C.
  - To wells "TP +1h" and "VC", add 500 µL of solution B.
  - To well "CC", add 500 µL of assay medium.
- At time point +1h (1 h post addition of virus):
  - Aspirate the medium from all wells.

- Wash all wells with 1 mL of assay medium.
- To wells "TP –1h" and "TP +1h", add 1000 µL of solution A.
- To wells "VC" and "CC", add 1000 µL of assay medium.

- At time point +7h (corresponding to one completed poliovirus replication cycle) (*see* **Note 5**):
  - Aspirate the medium from all wells.
  - Wash all wells with 1 mL of assay medium.
  - Lyse the cells and isolate total RNA.

3.2.3 Quantification of the Amount of Poliovirus RNA in Each Sample by Means of RT-qPCR

Per RNA sample, the following reaction has to be composed (assay performed with poliovirus type 1 strain Sabin): 12.5 µL of 2× reaction mixture, 0.125 µL RT enzyme, 900 nM forward primer (5′-CCTCCGGCCCCTGAATG-3′), 900 nM reverse primer (5′-ACCGGATGGCCAATCCAA-3′), 200 nM TaqMan probe (5′-FAM-CCGACTACTTTGGGTGTCCGTGTTTC-TAMRA-3′), 5 µL of viral RNA extract and RNase/DNase-free water to reach a final volume of 25 µL. The temperature cycling protocol consists of a 30′ reverse transcription step at 48 °C, 10′ of denaturation at 95 °C followed by 40 cycles of 15″ at 95 °C and 1′ at 60 °C.

Express the amount of poliovirus RNA from samples "TP –1 h" and "TP +1 h" relative to the amount of RNA in sample "VC". Sample "CC" should contain no poliovirus RNA. If both conditions "TP –1 h" and "TP +1 h" show a reduction in poliovirus RNA as compared to the virus control, the antiviral drug is likely to be a non-capsid binding agent, such as a protease inhibitor, a polymerase inhibitor or an inhibitor of morphogenesis (late-stage inhibitor). In contrast, if only condition "TP –1 h" but not "TP +1 h" show a reduction in poliovirus RNA, the compound is presumed to act as an inhibitor of attachment/entry and/or uncoating (early-stage inhibitor).

3.3 Determining the Mode of Action by Selection and Purification of Putative Compound-Resistant Poliovirus Variants

3.3.1 Step 1: Determination of the Most Optimal Compound Concentration and Virus Input

In this step, a combination matrix of compound concentrations and virus inputs is compiled (*see* Fig. 3).

- Seed a microtiter plate 24 h in advance to allow the BGM cells to adhere to the plate (25,000 cells in 100 µL/well).
- Set up six antiviral assays in a single 96-wells plate.
  - Prepare in row B to G, six parallel 8-step 1-to-2 compound dilution series (starting at column 2 and ending with 9). Choose an as high as possible (nontoxic) compound concentration so that the known EC_{50} is included in the envisaged concentration range. Column 10 will be used as virus control (VC; medium and virus), column 11 as cell control (CC; only medium). Wells at the edge of the microtiter plate will not be used, but will contain 100 µL medium.

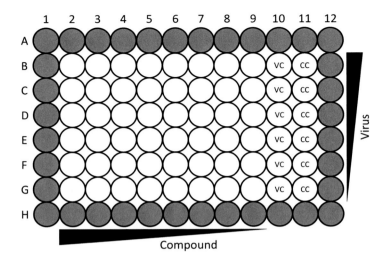

Fig. 3 96-well plate layout to determine the optimal assay conditions (i.e., amount of virus and concentration of compound) that have to be use to obtain compound-resistant virus variants. For this, the antiviral activity of a compound is determined in the presence of different virus inputs (MOI = multiplicity of infection). The assay condition with the highest amount of virus and lowest concentration of compound for which still full protection against virus-induced cytopathic effects can be observed after microscopic inspection will be used for the next step of the selection process

- – Prepare six virus dilutions (e.g., 1/100, 1/300, 1/1000, 1/3000, 1/10,000, 1/30,000) and add 100 μL of each to one of rows B to G (starting at well 2, ending with 10). Add 100 μL assay medium to column 11 so that all wells will contain 200 μL in total.

- – Incubate the assay plate at 37 °C and 5 % CO_2.

- When the virus controls show 100 % CPE, collect quantitative data using MTS readout (*see* CPE reduction assay discussed earlier in this chapter). In addition, score each assay well by microscope to discriminate between virus-induced cell death or, for instance, compound-induced toxicity. Calculate the EC_{50} values from each dose-response curve: this provides information on the multiplicity of infection (MOI)-dependency of the antiviral effect of the compound.

The most optimal assay condition for the next step of the protocol is the one with the lowest concentration of compound (XμM) and the highest virus input ($1/Y$) for which still 100 % inhibition of virus-induced cell death is observed.

3.3.2 Step 2: Selection of Compound-Resistant Virus Variants

In this step, dozens of identical assays are set up in parallel with the optimal compound concentration (XμM) and virus dilution ($1/Y$), as determined in **step 1**. Even though one would not expect to observe virus-induced cell death in any of these assay conditions at all, because the poliovirus population that is used in this assay is an infectious virus quasi-species that contains random mutations throughout the genome because a virus input is used that is as high as possible and because the compound pressure is kept to the absolute minimum, any virus variant that carries the prime mutation to be able to replicate in the presence of the compound should produce clearly visible CPE.

Set up at least three virus selection plates.

- Seed three microtiter plates 24 h in advance to allow the BGM cells to adhere to the plate (25,000 cells in 100 µL/well).
- Set up the virus selection plates.
    - Prepare a 4× concentrated compound solution and add 50 µL to each of the wells in quadrant B2 to G9 as well as to wells E11 to G11 (the latter will serve as treated, uninfected controls to be able to check for toxicity-induced changes in cell and monolayer morphology).
    - Prepare a 4× virus inoculum and add 50 µL to each of the wells in quadrant B2 to G9 as well as to the wells B10 to G10 (the latter will serve as untreated infected controls = virus controls).
    - Normalize the rest of the volumes with assay medium to a total of 200 µL.
    - Incubate the microtiter plates at 37 °C and 5 % CO_2.
- Once per day, check each assay well for the emergence of virus-induced CPE. Incubate the assay plate for as long as possible to allow putative compound-resistant virus variants to replicate for as long as possible (*see* **Note 6**).
- Collect the supernatant from the assay wells in which the development of CPE is observed most quickly. Do not mix supernatant from different assay wells as they count for independent-selected virus variants. As a control sample for the passaging, collect the supernatants of an untreated, infected control.

3.3.3 Step 3: Semi-purification of the Compound-Resistant Virus Variants

The harvested supernatant samples contain both virus from the original virus inoculum, one or even more putative compound-resistant virus variants, and compound-sensitive virus that could replicate because of trans-complementation effects. The next step is designed to purify and enrich the putative compound-resistant virus variants. To this end, the samples harvested from **step 2** are titrated in the presence and absence of compound.

For the putative-resistant virus variants:

- Seed as many microtiter plates as you have collected supernatant samples 24 h in advance to allow the cells to adhere to the plate (25,000 cells in 100 µL/well).

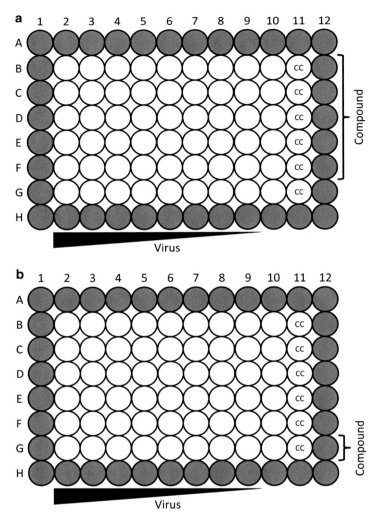

Fig. 4 (a) Layout of the 96-well plate that is used to enrich the putative compound-resistant virus variant that is present in the supernatant of the samples that have been obtained in the previous step. This is accomplished by performing of a virus titration in five-fold in the presence of compound at the same concentration as in the previous step. In row B–G, the virus sample is diluted in the presence of the compound with the purpose of collecting, from each row, the supernatant of the wells with the lowest virus input for which still clear virus-induced cytopathic effects can be observed. This supernatant is expected to contain an enriched population of the putative compound-resistant virus variant. In total, for each original sample, five enriched samples are collected. In row G, the putative compound-resistant virus isolate is diluted in the absence of compound to obtain an indication of the amount of infectious virus that is present in the sample. **(b)** 96-well plate layout in which the untreated virus is titrated in fivefold in the absence of compound and once in the presence of the compound. This allows to obtain a first insight in the resistance phenotype of the putative compound-resistant virus variants compared to that of the original virus

- Set up five virus titrations series in the *presence* of compound for each supernatant sample (*see* Fig. 4a).

– Prepare a 1-to-5 virus dilution series in the medium, row B to F, column 2–10.

– Prepare a 2× compound dilution and add 100 µL to each of the wells in quadrant B2 to F10.

- Set up one virus titration in the *absence* of compound for each supernatant sample.

  – Prepare a 1-to-5 virus dilution series in the medium, row G, column 2–10.

  – Add 100 µL assay medium to each well from G2 to G10.

- Normalize the rest of the volumes with assay medium by adding 100 µL assay medium to column 11; the untreated and uninfected cell controls.

- Once per day, check each assay well for the emergence of virus-induced CPE. Incubate the assay plates for as long as possible to allow putative compound-resistant virus variants to replicate for as long as possible (*see* **Note 7**). Collect the supernatant of the assay wells with the highest virus dilution/lowest virus input for which still clear signs of CPE are visible in the presence of compound (*see* **Note 8**).

  For the untreated, infected virus control:

- Except for one alteration, the protocol for the untreated, infected virus control is identical to that of the putative compound-resistant virus variants. Instead of performing five titrations in the presence and one in the absence of compound, now five titrations in the absence and one in the presence of compound have to be performed (*see* Fig. 4b).

- Incubate the assay plate for the same period as the putative resistant virus plates. Collect the supernatant of the assay wells with the highest virus dilution/lowest virus input for which still clear signs of CPE are visible in the absence of compound (referred to in the next step as original virus control) (*see* **Note 9**).

3.3.4 Step 4: Larger-Scale Culture of the Putative Compound-Resistant Virus Variants

The supernatant collected in **step 3** contains a purified or enriched putative compound-resistant virus variant. In the next step, this sample is used to culture a larger virus stock for resistance phenotyping and, later on, for genotyping.

- Seed, per condition (see below), one 25 cm² cell culture flask one day in advance, 2×10^6 BGM cells per flask in 5 mL.

- Prepare the following conditions

  – An untreated, uninfected control condition (cell control).

  – An untreated, infected control condition (original virus control).

  – A treated control condition infected with the original virus control sample.

  – For each of the samples obtained in **step 3**, a treated condition infected with the putative compound-resistant virus variant.

  – For each of the samples obtained in **step 3**, an untreated condition infected with the putative compound-resistant virus variant.

  – Incubate the cultures and check once or twice per day for the development of CPE. Register your observations.

• Aliquot and freeze the supernatant of the putative compound-resistant virus variants cultured in the presence of the compound.

3.3.5 Step 5:
Determination of the Virus
Titer of the Putative
Compound-Resistant
Virus Stocks

Perform a standard virus titration to acquire information on the infectious virus content of each of the virus stocks that have been produced and to determine the optimal virus dilution for the phenotypic resistance experiment.

• Seed as many microtiter plates as you have collected supernatant samples 24 h in advance to allow the cells to adhere to the plate (25,000 cells in 100 μL cells/well).

• Set up, in parallel on the same plate, six virus titrations for each supernatant sample as well as for the original virus sample.

  – Prepare a 1-to-10 virus dilution series in the assay medium.

  – Normalize the volumes to 200 μL with assay medium.

  – Incubate the assay plate until full CPE has developed in as many assay wells as possible or for as long as the cell control conditions can tolerate.

  – Perform an MTS readout and microscopic scoring of the CPE.

• These results allow quantifying (by the method of Reed and Muench) differences in virus titer between the original virus and the different putative compound-resistant virus variants. Based on these results, the virus inputs can be normalized for the phenotypic resistance experiment.

3.3.6 Step 6:
Phenotyping of the Putative
Compound-Resistant Virus
Variants

Essentially, repeat **step 1** but now with the putative compound-resistant virus variants in parallel with the original virus. This experiment will allow to confirm resistance to the compound, will allow to calculate the fold resistance, as well as to measure the effect of different virus MOI's on the residual activity of the compound, if there is any.

3.3.7   Step 7: Genotyping of the Compound-Resistant Virus Variants

Perform Sanger sequencing to identify putative compound-induced resistance mutations. The observation of mutations in the viral genome of the compound-resistant virus variants will allow to identify the viral protein(s) that is (are) involved in the mechanism of action of the compound. Most frequently, the compound-resistant virus variants possess one (or more) mutation(s) the original virus variant does not have. However, some compounds may induce different sets of mutations in the same or even different proteins. Perform a BLAST search with the mutations that have been identified to verify whether or not they are unique for poliovirus and if not, with which frequency and within which context they have been reported (sometimes, other wild-type viruses naturally contain a particular mutation, …). Mutations due to serial passaging can be excluded by comparing the sequence of original virus samples that underwent the process in the absence of the compound (virus control) with the sequence of the original input virus.

3.3.8   Step 8: Reverse-Engineering of the Mutations into a Wild-Type Infectious Clone

To confirm the link between a particular mutant genotype and phenotypic resistance to the antiviral effect of a compound, reverse-engineer the mutations that have been identified in the above-outlined procedure into an infectious clone of poliovirus by site-directed mutagenesis. In case multiple mutations are detected, the single mutants as well as the combinations thereof have to be engineered. This way, insight can be obtained in which mutation(s) is (are) essential for phenotypic resistance and which mutation(s) is (are) secondary and may have for example been acquired by the virus to correct for a replication fitness decrease because of the primary mutation.

Because the resistance selection procedure outlined above requires only very few passages as compared to a serial passaging approach which may take over 20 passages, the mutations that are identified, are most likely the predominant mutations that the virus has to acquire to adapt its replication to become less (or in-)sensitive to the inhibitory effect of the compound (*see* **Note 10**).

4   Notes

1. Depending on the cell type that is used, the number of cells and the time point of seeding of the cells may have to be adjusted. It is important that the cell cultures reach ~100 % confluency at the time the antiviral assays is initiated.

2. Depending on the virus/cell type that is being used, this incubation time may have to be prolonged/shortened. The endpoint is where the virus control exhibits full CPE.

3. CPE values are calculated as follows: $\%CPE = 100 \times (OD_{CC} - OD_{PV+Compound})/(OD_{CC} - OD_{VC})$. In this formula, OD_{CC} corresponds to the optical density (OD) of the uninfected and untreated cell

cultures, OD_{VC} represents the OD of the virus-infected and untreated cell cultures, and $OD_{PV+Compound}$ represents the OD of the virus-infected cell cultures that have been treated with a given concentration of compound.

4. Since the MTS readout does not discriminate between CPE and compound-related toxicity, a microscopic inspection of each well with scoring of the CPE and/or toxicity should be performed in parallel to allow the identification of selective antiviral compounds.

5. For viruses with a different replication cycle length, the endpoint should be adapted accordingly.

6. It may be required to incubate the assay plate for a longer time then it takes for the virus control to cause full CPE. Additionally, it is very important to keep track of the order in which the cells show first signs of CPE as this information may later give a clue about the importance of the discovered mutations.

7. It may be required to keep the assay plate for a longer time then the virus control to allow progression to full CPE.

8. For each assay plate (=each original supernatant sample), select the most promising sample for the next step and store several others as backup.

9. Select the most promising sample for the next step and store several others as backup.

10. In case only partial resistance is observed, it may be advisable to use these virus isolates as starting point for a serial passaging effort to verify whether or not the virus can acquire additional mutations to further increase its resistance to the compound.

References

1. Minor P (2005) Vaccine-derived poliovirus (VDPV): impact on poliomyelitis eradication. Vaccine 20:2649–2652
2. Ken O, Sutter RW, de Gourville EM, Dowdl WR, Pallansch MA (2005) Vaccine-derived polioviruses and the endgame strategy for global polio eradication. Annu Rev Microbiol 59: 587–635
3. Couzin J (2006) Report concludes polio drugs are needed, after disease is eradicated. Science 311:1539
4. http://www.promega.es/~/media/files/resources/promega%20notes/75/a%20high throughput%20system%20for%20screening%20potential%20antiviral%20compounds.pdf?la=es-ES

INDEX

Javier Martín (ed.), *Poliovirus: Methods and Protocols*, Methods in Molecular Biology, vol. 1387,
DOI 10.1007/978-1-4939-3292-4, © Springer Science+Business Media New York 2016

Printed in the United States
By Bookmasters